Classification of Knowledge in Islam

Classification of Knowledge in Islam

A Study in Islamic Philosophies of Science

OSMAN BAKAR

Foreword by
SEYYED HOSSEIN NASR

ISLAMIC TEXTS SOCIETY

First edition published 1992 by
Institute for Policy Research, Kuala Lumpur, Malaysia

This edition published 1998 by
The Islamic Texts Society
Miller's House
Kings Mill Lane
Great Shelford
Cambridge CB22 5EN,
United Kingdom
www.its.org.uk

British Library Cataloguing-in-Publication Data.
A catalogue record for this book is available from the British Library.

ISBN 978 0946621 71 2 paper

Printed and bound in the UK by TJ Clays, Padstow, PL28 8RW.

Cover illustration: Ottoman book cover, late 15th century
© V&A Picture Library CT12689, L.531-1983.

Cover design: Imtiaze Ahmed Manjra

Dedicated in humble gratitude to my Parents

Haji Bakar bin Yusof and Hajah Besah bint Taib

who although both illiterate made great sacrifices

for the sake of their children's education

CONTENTS

	Page
Foreword by Seyyed Hossein Nasr	xi
Acknowledgments	xvi
List of Transliterations	xvii
Introduction	1
General Domain of Study	1
Classifications Chosen for Study	1
The Nature and Scope of the Present Study	3
Contemporary Scholarship on the Subject	5

PART I. AL-FĀRĀBĪ

Chapter

	Page
1. The life, Works and Significance of al-Fārābī	9
1.1. Introduction	9
1.2. Al-Fārābī's Educational Background and Scholarly Life	10
1.3. Al-Fārābī's Works and Significance	21
2. Al-Fārābī's Psychology in Its Relation to the Hierarchy of the Sciences	43
2.1 The Idea of the Unity and Hierarchy of the Sciences	43
2.2 The Bases of the Hierarchy of the Sciences	46
2.3 The Hierarchy of the Faculties of the Human Soul	48
2.3.1 The Sensitive Faculty	50
2.3.2 The Imaginative Faculty	51
2.3.3 The Rational Faculty	54
3. The Methodological Basis of the Hierarchy of the Sciences	69
3.1 Revelation, Intellect, and Reason	69
3.2 Religion, Philosophy and the Sciences	79
3.3 Al-Fārābī's Theory of Knowledge	83

4. The Ontological and the Ethical Bases of the Hierarchy of the Sciences 95
4.1 The Ontological Basis 95
4.1.1 The Subject-Matter of Metaphysics 97
4.1.2 The Subject-Matter of Natural Science 99
4.1.3 The Subject-Matter of Mathematics 100
4.1.4 The Subject- Matter of Political Science 103

4.2 The Ethical Basis 106
4.2.1 Al-Fārābī's Theory of Virtue 107

5. Classification and Description of Linguistic Science and Logic 121
5.1 Classification and Enumeration of the Sciences 121
5.2 Characteristics of Al-Fārābī's Classification 124
5.3 Division of Linguistic Science and Logic 126
5.3.1 The Science of Language 127
5.3.2 Logic 128

6. Classification and Description of the Philosophical Sciences 137
6.1 The Mathematical Sciences 137
6.2 Natural Science 139
6.3 Metaphysics 142
6.4 Political Science 143
6.5 Jurisprudence and Dialectical Theology 145

PART II. AL-GHAZZĀLĪ

Chapter

7. The Life, Works and Significance of al- Ghazzālī 155
7.1 Religious and Political Background of al-Ghazzālī's Period 155
7.2 Al-Ghazzālī's Early Education and Intellectual Interest 157

7.3 Al-Ghazzālī's Intellectual Crisis 158
7.4 Post-Crisis Intellectual Life and Works 159
7.5 Al-Ghazzālī's Spiritual Crisis 162
7.6 Spiritual Retreat and Scholarly Output 163
7.7 The Authenticity of Some Works Attributed to al-Ghazzālī 165
7.8 Significance of al-Ghazzālī's *Iḥyā* and His Sufism 171

8. Al-Ghazzālī's Classification of Seekers After Knowledge 181
8.1 Basis of Classification 181
8.2 Al-Ghazzālī's Views Concerning the Four Classes 182
8.2.1 The Mutakallimūn 182
8.2.2 The Philosophers 185
8.2.3 The Ta'līmites 189
8.2.4 The Sufis 194
8.3 Significance of the Classification 196

9. Al-Ghazzālī's Classification of the Sciences 203
9.1 Basis of the Division into Theoretical and Practical Parts 204
9.2 Basis of the Division into "Presential" and Attained Knowledge 204
9.3 Basis of the Division into Religious and Intellectual Sciences 205
9.4 Basis of the Division into "farḍ 'ayn" and "fard kifayah" Sciences 206
9.5 Classification of the Religious and Intellectual Sciences 207
9.5.1 Religious Sciences 207
9.5.2 Intellectual Sciences 208
9.6 Nature and Characteristics of the Religious Sciences 210
9.7 The Ethico-Legal Status of the Intellectual Sciences 213
9.8 Nature of the Theoretical-Practical Division 217

9.9 The Worldly and Other-Worldly Rational Sciences 217
9.10 Significance of the Division into "Presential" and "Attained" knowledge 218
9.11 Conclusion 219

PART III. QUṬB AL-DĪN AL-SHĪRĀZĪ

Chapter

10. The Life, Works and Significance of Quṭb al-Dīn al-Shīrāzī 229
10.1 Quṭb al-Dīn's Education and Intellectual Life 229
10.2 Quṭb al-Dīn's Works 236
10.2.1 Philosophical Works 237
10.2.2 Religious Works 242
10.3 General Significance of His Works 243

11. Quṭb al-Dīn's Classification of the Sciences 249
11.1 "Ḥikmat" as the Basis of Classification 250
11.2 Divisions of "Ḥikmat" 251
11.2.1 Theoretical Philosophy and Its Divisions
11.2.2 Practical Philosophy and Its Divisions 256
11.3 Non-Philosophical or Religious Sciences 257
11.3.1 Sciences of Fundamental Principles of Religion 258
11.3.2 Sciences of Branches of Religion 258

Conclusion 263
The Philosophical Bases of the Three Classifications: Similarities and Differences 263

A Selected Bibliography 271

FOREWORD

The disorder which rules over the modern educational curriculum in most Islamic countries today is to a large extent due to the loss of the hierarchic vision of knowledge as one finds in the traditional Islamic education system. In the Islamic intellectual tradition, there existed a hierarchy and inter-relation between various disciplines which made possible the realization of unity in multiplicity not only in the domain of religious faith and experience but also in the realm of knowledge. The discovery of order and the appropriate relationship between various disciplines was the goal of the leading Islamic intellectual figures, from theologians to philosophers, from Sufis to historians, many of whom devoted much of their intellectual energy to the subject of the classification of the sciences. This subject is, therefore, a key for the understanding of a major dimension of the Islamic intellectual tradition and the Islamic education system not to speak of its role in enabling contemporary Muslim educators to cast an objective eye upon the confusion and chaos that reigns in current education curricula with their blind emulation of Western models blended often in *ad hoc* fashion with what has survived of the *madrasah* system.

Considering the significance of the subject of the classification of the sciences in Islam, it is rather remarkable that the present book is the first which deals with the subject in a scholarly fashion and at the same time in a contemporary language. Moreover, the author remains faithful to the Islamic intellectual tradition itself with its emphasis upon the unity and hierarchy of forms of knowledge which necessitates the classification of the sciences, and he is able to bring out the metaphysical and philosophical background which underlies the various classifications considered in this study.

In the traditional Islamic universe, both the subject and object of knowledge are considered to be hierarchic. Object reality is not only the spatio-temporal world available to the senses. There is first of all the Absolute Reality, Allah, who alone *Is* in the absolute sense of the word. Then there are the angelic orders, the intermediate imaginal world (the *'ālam al-khayāl*), then the world of the *jinn* and the men and finally the natural world. The Qur'ān constantly refers to these realities as well as the heavens and the earth (*al-samāwāt wa'l-arḍ*), the heavens, it needs to be emphasized, being in the plural. There is, therefore, within the more general distinction between the Creator and creation, the whole hierarchy of beings standing below the Divine Throne "from the Plieades (*al-thurayyā*) to the lowly dust". Obviously from the Islamic points of view the science which deals with God is not on an equal footing with the science of the human soul nor that science with the science of minerals in contrast to the modern educational system where theology, psychology and geology are placed horizontally alongside each other like so many drawers in a cabinet each containing a certain amount of information.

Furthermore, in the Islamic perspective there exists a hierarchy within the subject who knows. Man is not simply the Cartesian subject of the *cogito* who "knows" on the single level of what is called the mind. Man can know through the senses, through the imaginal faculty, through reason with its own several levels of activity, through the heart-intellect so often mentioned in the Qur'ān and finally through revelation which is the objective counterpart of intellection with the eye of the heart (*'ayn al-qalb*). As the final revelation of the Word of God, the Qur'ān contains all of knowledge in principle precisely because it stands at the apex of the hierarchy of modes and sources of knowing followed again in a hierarchic manner by other modes of knowing.

The Islamic intellectual authorities were fully aware of both hierarchies of the object and the subject of knowledge and in the light of these realities sought to classify the sciences derived not only from the Qur'ān and *Ḥadith*, but also developed by Muslim scientists and scholars on the basis of what they had inherited according to the destiny of Islam from earlier civilizations such as those of the Greeks, Persians and Indians. They developed these schemes of classification according to the intellectual perspective to which they themselves belonged and not according to individual whim and fancy, for in the Islamic tradition what is said (*mā qāl*) always precedes he who said it (*man qāl*). It is the major intellectual traditions which are important in determining the views of Islamic thought concerning a particular subject.

Dr. Osman Bakar was faced with a formidable choice in

selecting from numerous authorities belonging to major intellectual perspectives, figures who would best present the meaning of the classification of the sciences and who wielded the greatest influence upon later generations of Muslims scholars and scientists. He was also confronted with the problem of choosing only a number of the pertinent Islamic schools of thought seeing that the subject is so vast that it was impossible to deal in one study with all the intellectual perspectives even if he did limit himself to a single figure from each school.

The choice of the author has been judicious. Among numerous Peripatetic (*mashshā'ī*) philosopher-scientists who have dealt explicitly with the classification of the sciences including al-Kindī, Abu'l-Hasan al-'Āmirī, Ibn Sīnā, Ibn Rushd and Naṣīr al-Dīn Ṭūsī, he selected al-Fārābī whose classification was one of the most influential in the earlier period of Islamic history and who also exercised much influence upon the medieval West. Then he chose al-Ghazzālī, at once Ash'arite theologian and Sufi, who was among the most influential of all Islamic thinkers and whose classification again exercised much influence from the Seljuq period onward. Finally, Dr. Bakar selected Quṭb al-Dīn Shīrāzī, at once a commentator of Suhrawardī and the doctrines of the School of Illumination (*al-ishrāq*), a philosopher immersed in Ibn Sīnā's philosophy and one of the leading scientists in the history of Islam. In analyzing for the first time in English the section of Shīrāzī's *Durrat al-tāj* on the classification of the sciences, the author has revealed how a later leading intellectual figure who was himself both a philosopher and a scientist looked upon the problem of classification.

The pages which follow continue many enlightening analyses which are very pertinent to Islamic intellectual history as well as Islamic education and this book must be considered as an important contribution to a better understanding of certain significant aspects of classical Islamic thought. The work is also an important contribution to the Islamic philosophy of science, for this subject cannot but be concerned even in its modern context with modes and methods of knowing as far as they concern the sciences.

Besides its philosophical and historical value, moreover, the work of Dr. Bakar is an important contribution to the present discussion on the Islamization of knowledge being carried out throughout much of the Islamic world today. How can one Islamicize knowledge without being concerned with the traditional Islamic classification of the sciences? How can an Islamic education system accept a situation in which there is no hierarchy between the knowledge of the angels and of molluscs or between the method of knowledge based upon reason wed to the external senses and knowledge which derives from the certitude (*yaqīn*) derived from heart-knowledge? The views of classical Islamic thinkers ably analyzed by Dr. Bakar here speak very directly to the current debate on the Islamization of knowledge and in fact provide an absolutely necessary dimension without which talk of this subject cannot proceed much beyond mere chatter.

Dr. Bakar's method in the present work is to go back to the original texts and let the primary sources speak for themselves. Then on the basis of the words of the masters, Dr. Bakar provides an analysis which is based upon both scholarship and an a genuinely Islamic philosophical perspective. Such a combination is indeed rare today. Many younger scholars who have developed a careful scholarly method of dealing with texts of an intellectual nature lack the philosophical perspective. And when they do have a philosophical perspective it is usally a Western one totally alien to the Islamic world-view with results which are often outright distortions. Dr. Bakar brings to this task not only a solid knowledge of the Western philosophy of science and general problems connected with the relation between the sciences in the West, but also a thorough grounding in the Islamic intellectual tradition. He belongs to a rising group of Muslim scholars who possess a traditional Islamic perspective while being well acquainted with modern Western thought. The significance of this book lies precisely in this fact as a result of which the author is able to make accessible to the contemporary reader an important facet of Islamic intellectual life in a language comprehensible to those who cannot benefit directly from the classical sources and in a manner which is authentic and itself belongs to the Islamic intellectual tradition.

Osman Bakar is to be congratulated for having achieved such a work. May Allah provide him with the knowledge and energy to make available to the contemporary world other important facets of traditional Islamic thought. And may this work find many readers in quest of an authentic knowledge of the Islamic intellectual tradition and in search of means to answer the intellectual challenges of the modern world which threaten the very citadel of Islamic life and thought.

Seyyed Hossein Nasr
Bethesda, Maryland
U. S. A.
November 1990

Acknowledgments

This book was presented to the Department of Religion, Temple University, Philadelphia in June 1988 as a doctoral thesis. It is now presented in its original form and, with the exception of few minor changes, no alterations have been made in the text of the book.

I would like to thank the members of my committee, Professors G. S. Sloyan, N. Samuelson, J.W.Morris and M. Ayoub, the careful attention they gave this dissertation and for their invaluable suggestions. I wish to express my special gratitude to my committee chairman, Professor Seyyed Hossein Nasr for having given generously of his time, guidance and moral support. I also owe thanks to my friends – students, professors, and administrative staff – at the Department of Religion, Temple University, who supported me in various ways during my stay in Philadelphia.

I am also happy to express my thanks to my employer, The University of Malaya, Kuala Lumpur, Malaysia, which granted me a four-year study leave with generous financial support. Finally, and most of all, I am grateful to my wife Badariah Ahmad for the extraordinary patience and understanding she had shown toward my work, burdened as she was with the heavy responsibilility of caring after our five children. The children were the "delights of the eye," and a source of inspiration.

As far as the publication of this book is concerned, it affords me great pleasure to record my sincere thanks to the Institute of Policy Research, Kuala Lumpur. Without its support and encouragement, the publication of this book would not have materialized.

Transliteration

Arabic Letter	Transliteration
ء	'
ب	b
ت	t
ث	th
ج	j
ح	h
خ	kh
د	d
ذ	dh
ر	r
ز	z
س	s
ش	sh
ص	ṣ
ض	ḍ
ط	ṭ
ظ	ẓ
ع	'
غ	gh
ف	f
ق	q
ك	k
ل	l
م	m
ن	n
ه	h
و	w
ي	y
ة	t

Short	Vowels
ـَـ	a
ـُـ	u
ـِـ	i

Long	Vowels
اَ	ā
وُ	ū
يِ	ī

Diphthongs	
ـَوْ	aw
ـَيْ	ay
ـِيَّ	īy
ـُوَّ	uww

INTRODUCTION

1. General Domain of Study

One of the recurring themes in traditional Islamic scholarship is the classification and description of the sciences (*al-'ulūm*). From al-Kindī in the third/ninth century to Shāh Waliallāh of Delhi in the twelfth/eighteenth century, successive generations of Muslim scholars have devoted a considerable deal of their intellectual talents and genius to expositions of this theme.

Some of the classifications were as influential as they were original. Others were mere repetitions of earlier ones and were simply forgotten. The authors of these classifications were scholars of diverse religious and philosophical persuasions, representing almost the whole spectrum of the Islamic intellectual tradition. Philosopher-scientists, theologian-jurists and Sufis, Sunnis and Shī'ites, were all represented in this enterprise of classifying the sciences. The primary motive behind this whole intellectual enterprise appears to be the concern with the means of preserving the hierarchy of the sciences and with the delineation of the scope and position of each science within the total scheme of knowledge.[1] The general conviction, which was shared by many medieval Jewish and Christian thinkers, was that the above goal could best be achieved through the classifications of the sciences.

My first encounter with some of the classifications suggested to my mind the possibility of a "concrete philosophy" underlying each of them. Upon further reflection on this question I became more convinced that it would be a worthwhile and interesting investigation to discover the philosophical basis of each classification and to see in what way this basis is related to the intellectual perspective of its author. This philosophical interest led me to the present study of three major Muslim classifications of the sciences.

2. Classifications Chosen for Study

For the purpose of my study I have chosen the classifications written by the following thinkers: al-Fārābī (258/870 – 339/950), al-Ghazzālī (450/1058 – 505/1111), and Quṭb al-Dīn al-Shīrāzī

(634/1236 – 710/1311). The choice of these figures is based upon several considerations. First, each thinker was either a founder or an eminent representative of a major intellectual school in Islam. The philosophical ideas which dominate his thinking constitute a particular intellectual perspective that is shared by many thinkers. Nasr has shown that in Islamic intellectual history the identification of an individual school with a particular figure is legitimate.[2] The choice of the above figures therefore enables my study to embrace the major Islamic intellectual perspectives. Each individual classification may then be seen as the embodiment not only of the philosophical perspective of its author but also of the intellectual school he represents.

Al-Fārābī was generally regarded as the founder and one of the most prominent representatives of a major school of Islamic philosophy, namely the *mashshā'ī* (Peripatetic) school of philosopher-scientists. Al-Ghazzālī was a famous theologian or representative of *kalām*, jurist and Sufi. As for Quṭb al-Dīn al-Shīrāzī, he represented the *ishrāqī* (illuminationist) school of philosophy. He was also a gifted scientist.

My second consideration in choosing the above figures concerns the significance of the periods of Islamic history in which they flourished. Al-Fārābī represented that important period which witnessed the beginning of intense activity in the study of the philosophical sciences, including mathematics and the natural sciences. Al-Ghazzālī lived two centuries later in a period characterized by intellectual tension between *falsafah* and *kalām*, political and religious tension between Sunnism and Shi'ism, and spiritual tension between the esoteric Sufis and the exoteric jurists. He played an important role in resolving some of these tensions.

Quṭb al-Dīn appeared on the intellectual scene of Islam two centuries after al-Ghazzālī. He represented one of the most challenging periods of Islamic history. He was witness to the fall of Baghdad and the destruction of many intellectual and religious centers in the eastern lands of Islam at the hands of the Mongols. Not long after this tragic event, there was a new flowering of the philosophical sciences. Quṭb al-Dīn and his teacher, Naṣīr al-Dīn al-Ṭūsī, were at the forefront of the intellectual movement which helped to revive these sciences.

Between al-Fārābī and Quṭb al-Dīn there was a historical span of four centuries. During this whole period remarkable and significant developments and progress occurred in the realm of the sciences, philosophical as well as religious. The pre-Islamic sciences were Islamized and developed and new sciences were created. It would be an interesting study to investigate whether or not these developments and changes had affected or influenced in any significant manner the fundamental basis and structure of the classifications under study, written as they were under different philosophical and religious climates. As far as I know, no one has ever formulated this question before, let alone answered it. My present study is partly an attempt at answering the question.

In the case of al-Fārābī there is another reason for choosing his classification. His was the first influential classification in Islam, for which he was honoured with the title "The Second Teacher" (*al-mu'allim al-thānī*). That classification became the model for all later authors. For this reason, the greater portion of my study is devoted to his classification.

3. The Nature and Scope of the Present Study

In the following eleven chapters I have undertaken a detailed philosophical study of the three classifications I have chosen. The primary aim of this study is to formulate the underlying philosophical basis of each classification and to relate this basis to certain principles contained in the Islamic revelation. I have also given an analytical treatment of the following questions: (1) the major distinguishing features of each classification, (2) the attitude of each thinker to the philosophical and religious sciences and how he envisaged the distinction between them.

Descriptions of the different sciences as given by each author are only discussed insofar as they help to illustrate the main thesis presented in this study. The thesis is that the classifications are based at once upon philosophical ideas which are common to all intellectual schools in Islam and ideas which are specific to the intellectual and religious world-view of its author and of the school he represents.

For each thinker there is a chapter (chapters one, seven, and ten) devoted to his life, works and significance. This account provides knowledge of the education and training of each thinker in the different sciences, his philosophical interest and scholarly output, his intellectual circles, and the intellectual and religious climate of his time. Such knowledge serves as a necessary background for a proper analysis of the classification each had composed.

Since al-Fārābī's classification is presented as the "model" classification and is the first to be analyzed in this study, I have devoted more chapters (chapters two, three, and four) to establishing and explaining its philosophical basis than I have given to the other two classifications. The most important philosophical idea which al-Fārābī applied to his classification is the hierarchy of the sciences. Chapters two, three, and four deal with the three criteria listed by al-Fārābī, by means of which the hierarchy of the sciences might be established. To clarify the three different bases of this hierarchy, it is necessary to refer to al-Fārābī's psychology, logic, ontology, and ethics even if this means having to incorporate certain materials which are well-known to scholars of medieval philosophy. What is of central importance is that in this discussion an attempt has been made for the first time to establish a conceptual relationship between certain ideas in al-Fārābī's psychology, logic, ontology, and ethics and his classification of the sciences.

In chapters five and six I seek to summarize al-Fārābī's classification and description of the sciences and point out the significance of the position he had accorded to each science in his classification.

My study of al-Ghazzālī's classifications of the sciences includes a chapter (chapter eight) on his classification of knowers or seekers after knowledge. He divides knowers into four main groups: theologians (*mutakallimūn*), Ta'līmites or Ismā'īlīs, philosophers (*falāsifah*) and Sufis. This chapter seeks to establish al-Ghazzālī's epistemology through an analysis of his critique of the epistemological and methodological claims of the four groups. This analysis shows that in al-Ghazzālī the problem of the relationship between reason and supra-rational experience and between religion and philosophy is envisaged from a point of view significantly different from the one encountered in al-Fārābī. This chapter provides necessary material for the understanding of al-Ghazzālī's classifications of the sciences analyzed in chapter nine.

In the final chapter (chapter eleven) I seek to establish the philosophical basis of Quṭb al-Dīn's classification. Although the fundamental idea underlying this classification is again the distinction between philosophical and religious sciences, Qutb al-Dīn's points of emphasis on the distinction are different from those of al-Fārābī and al-Ghazzālī.

My study of the above classifications of the sciences is essentially a study in Islamic philosophies of science. The term "science" (*'ilm*) is used in this study in the comprehensive sense of an organized body of knowledge that constitutes a discipline with its distinctive goals, basic premises, and objects and methods of inquiry. I am therefore referring to a philosophy of science which embraces a far wider meaning and domain of study than does the modern discipline of the same name.

4. Contemporary Scholarship on the Subject

To the best of my knowledge, no attempt has yet been made at a philosophical study of Muslim classifications of the sciences in the manner and on the scale I have just described. Quṭb al-Dīn's classification, which was composed in Persian and which has not yet been translated into any European language, is almost untouched by scholars. As for al-Ghazzālī's classifications, no systematic study specifically devoted to it has yet been attempted. The few discussions we have of it[3] have been done in a somewhat summary fashion to serve as a background for the study of some other aspects of al-Ghazzālī's thought.

It is al-Fārābī's classification which has received the most attention from scholars. But even in al-Fārābī's case the studies so far undertaken[4] have dealt only briefly either with the historical aspects of the classification or with the problem of the relationships between certain sciences discussed by al-Fārābī. The fundamental philosophical basis of that classification has not been explored in these studies.

From a more general point of view, Nasr and Rosenthal[5] have made some general conclusions concerning the meaning and significance of the Islamic classifications and their implications for the cultivation of the sciences in Islamic history. Their general observations on this matter have inspired me to undertake the present study. But neither of them offered a comparative philosophical study of specific classifications.

Endnotes

Introduction

[1]S.H.Nasr, *Science and Civilization in Islam*, Harvard University, Cambridge, 1968, p.59
[2]S.H.Nasr, *An Introduction to Islamic Cosmological Doctrines*, Shambhala Publications, Boulder, 1978, rev. edn., p.22. Hereafter this work is cited as *IICD*.
[3]See M. Umaruddin, *The Ethical Philosophy of al-Ghazzali* Lahore, 1970; and M.A.Sherif, *Ghazzali's Theory of Virtues*, SUNY Press, Albany, 1975.
[4]I have particularly in mind M. 'Abd al-Rāziq, *Failasūf al-'arab wa'l-mu'allim al-thānī*, Cairo 1945; 'U. Amīn in his introduction to his edition of *Iḥṣā' al-'ulūm*, Dār al-fikr al-'arabī, Cairo, 1949; and M. Mahdi, "Science, Philosophy and Religion in al-Farabi's *Enumeration of the Sciences*." in *The Cultural Context of Medieval Learning*, ed. J.E.Murdoch and E.D.Sylla, Dordrecht, 1975, pp. 113-47.
[5]F.Rosenthal, *The Classical Heritage in Islam*, trans. from German by E.Marmorstein and J.Marmorstein, Routledge and Kegan Paul Ltd., London, 1975; S.H.Nasr, *Science and Civilization in Islam*.

Verily with God is full knowledge
and He is acquainted with all things.
The Qur'ān: Luqman, Verse 34

PART I

Al-Fārābī's Classification

CHAPTER 1

THE LIFE, WORKS AND SIGNIFICANCE OF AL-FĀRĀBĪ

1.1 Introduction

The aim of this opening chapter is to present and highlight those aspects of al-Fārābī's life and works which together provide an indispensable background for a clearer understanding of the major issues that lie at the heart of my study of his classification of the sciences. With the help of the brief biography which follows – and which does contain several new clarifications of episodes in his life that are obscure and problematic – we will be able to see that, in a sense, al-Fārābī's enumeration and exposition of the various sciences were a kind of commentary upon his own educational and intellectual experience. For these sciences were either actually studied by him at different stages of his life or were originally founded in Islam by him.

In marked contrast to the life of several other great philosophers of Islam who enjoyed similar fame and influence in both the Islamic world and the Latin West – among them Ibn Sīnā (370-428/980-1037) – very little is known with certainty about al-Fārābī's family background or early life, training and education. Even as it concerns his later life, there are too many episodes which are at present not known in a definitive manner. Although al-Fārābī had a few immediate disciples,[1] he did not undertake to dictate his autobiography to any one of them in the way Ibn Sīnā did to his favourite disciple, al-Juzjānī. Neither, unlike Ibn Khaldūn, did al-Fārābī undertake to write an autobiography. In fact, his lost work *Kitāb fī ẓuhūr al-falsafah* (*Book on the Appearance of Philosophy*),[2] seems to be a semi-autobiography in much the same way as al-Ghazzālī's *al-Munqidh min al-ḍalāl* (*Deliverance from Error*) may be regarded as one. The few fragments that have survived[3] provide important data about certain phases of his educational and intellectual pursuits. We also know of some of his immediate teachers in logic and philosophy who formed a link in the long chain of transmission of philosophical teachings from the Athens of Aristotle to the Baghdad of his own day.

For the present account of al-Fārābī's life and works, I have had to rely primarily on the traditional biographies[4] that were written by Muslim scholars and historians centuries after him. To supplement these sources there are a number of accounts of his life which came from the pens of modern scholars.[5] The latter, in their efforts to gain a better understanding of the various facets of his life, ideas and teachings, have sought since the beginning of the second half of the nineteenth century to improve upon these traditional biographies by resolving certain contradictions among them. They have also supplemented them with new data and sources of information.

1.2 Al-Fārābī's Educational Background and Scholarly Life

Abū Naṣr Muḥammad ibn Muḥammad ibn Ṭarkhān ibn Awzalagh[6] al-Fārābī, better known in traditional Islamic sources as simply Abū Naṣr, the second outstanding representative of the Muslim Peripatetic (*mashshā'ī*) school of philosopher-scientists[7] after al-Kindī (185-260/801-873), was born in Wasīj, a small village in the district of the city of Fārāb[8] in the province of Transoxiana, Turkestan around the year 257/890.[9] Al-Fārābī's birthplace did not become part of *dār al-islām* until only about three decades before his birth when the district of Isbījāb in which it was situated was conquered and Islamized by Nūḥ ibn Asad, a member of the Sāmānid family, in 225/839-840. This means that, in all likelihood, al-Fārābī's grandfather was a convert to Islam.

Although his father is mentioned in certain sources as being of noble Persian descent, the family has generally been considered as Turkish. Not only did they speak *Sogdian* or a Turkic dialect, but their manners and general cultural habits were Turkish.[10] That al-Fārābī must have come from at least a respectable family, if not also a rich one, is asserted by D.M.Dunlop.[11] His view is based upon his consideration of al-Fārābī's grandfather's name, Ṭarkhān, which in Turkish not only signifies a military officer but is also associated with certain feudal privileges and exceptions. If this was indeed so, then a family tradition of distinguished military career appeared to have just been established when the intellectually-gifted al-Fārābī departed from that tradition and opted for a scholarly life, for his father also was a military officer (*qā'id jaysh*) as described by Ibn Abī Usaibi'ah.[12]

Most probably al-Fārābī's father served in the army of the Sāmānid rulers who were then governing much of Transoxiana, and from 260/874 the whole of it, as an autonomous province within the 'Abbāsid Caliphate. There is no compelling reason for us to be inclined to the suggestion by R.Walzer that his father may have belonged to the Turkish bodyguard of the Caliph. This would have brought al-Fārābī to Baghdad early in life.[13] On the contrary Walzer's suggestion raises more questions than it solves about al-Fārābī's early life and education. The conflicting reports by the traditional biographers about al-Fārābī's early life and education can be resolved by examining which of them is most consistent with his overall life.

In Ibn Khallikān's account, al-Fārābī passed his youth in Fārāb[14] while Ibn Abī Usaibi'ah mentions Damascus as the place where this philosopher grew up (on the basis of a report by a certain Abu'l-Ḥasan al-Āmidī).[15] In my opinion the former seems to be the more likely if we take their whole chronological accounts of al-Fārābī into consideration.[16] In the absence of any compelling account to the contrary we may accept Ibn Khallikān's account as authentic. It was thus in Fārāb, whose residents were mostly followers of the Shāfi'ite school of law, that al-Fārābī received his elementary education. His father, as we have seen, was certainly in a position to provide the best education possible for the young Abū Naṣr. He is described as having "possessed from the beginning a keen intelligence and a great gift for mastering nearly every learned subject."[17]

What al-Fārābī learned at the primary level of education, either under the private tutoring of teachers at their homes or in formal sessions at the mosque, could not have been much different from the traditional curriculum imparted to every Muslim child of his age and time.[18] Its basis, of course, would have been the Qur'ān, for in the words of Ibn Khaldūn:

> It should be known that instruction of children in the Qur'ān is a symbol of Islam. Muslims have, and practice, such instruction in

all their cities, because it imbues hearts with a firm belief [in Islam] and its articles of faith, which are derived from the verses of the Qur'ān and certain Prophetic traditions. The Qur'ān has become the basis of instruction, the foundation for all habits that may be acquired later on. The reason for this is that the things one is taught in one's youth take root more deeply [than anything else]. They are the basis of all later knowledge. The first impression the heart receives is, in a way, the foundation of [all scholarly] habits.[19]

These potentially powerful and pervasive effects of the Qur'ān clearly manifested themselves in the intellectual and spiritual life of al-Fārābī, although the forms these manifestations took in his case undoubtedly differed from those manifested in other thinkers of Islam. The particular intellectual and spiritual perspectives he expounded may be identified historically with sources of non-Islamic origin, but from the point of view of their reality they ultimately correspond to a particular dimension of the Islamic revelation. If their Islamicity has been questioned by various authorities, both traditional (including al-Ghazzālī) and modern, that is because it has been judged in the light of the criteria dictated by another dimension of that revelation.[20] The question of the Islamicity of al-Fārābī's ideas and perspectives will be treated more fully in the following sections. I have raised the question at this juncture only to stress the point that those profound Qur'ānic effects of which Ibn Khaldūn spoke, made possible by the central role accorded to the Book in a Muslim's education, were no less evident or no less real in his life and thought than in that of any other thinker identified with Islamic orthodoxy. The fact that he was later to become a *faylasuf par excellence* does nothing to alter this.[21]

Apart from receiving instruction in the Qur'ān, al-Fārābī must have also learned grammar, literature, the religious sciences [especially jurisprudence (*fiqh*), exegesis (*tafsīr*) and science of the traditions (*'ilm al-ḥadīth*)], and elementary arithmetic.[22] This view has its basis in the following considerations: First, Ibn Khaldūn, in his survey on the different methods of instruction of children employed in the Muslim cities, described the curriculum in the Eastern lands of Islam as mixed. He spoke of the curriculum as primarily concerned with teaching the Qur'ān and "the works and basic norms of (religious) scholarship once (the children) are grown up."[23] Second, according to Ibn Abī Usaibi'ah, al-Fārābī worked as a judge (*qāḍī*) for some time.[24] This could only mean that the basis of his early intellectual training was in the religious sciences, mainly *fiqh*. Third, later historians recorded the claim that he was supposed to have made toward the end of his life that he knew nearly every language.[25] I interpret this to mean that he must have known many languages. He certainly knew Turkish, Arabic and Persian and very likely a number of the Central Asiatic dialects and local languages, as well as Syriac and Greek.[26] His early education in several of these languages must have included their grammar and their popular literature.

The above curriculum served as the basis of al-Fārābī's more advanced level of education. We do not know whether his native Fārāb could still afford him the opportunity to study at this level.[27] More likely, as suggested by M. Mahdi,[28] he moved to Bukhārā to pursue advanced study of *fiqh* and other religious sciences. Bukhārā at the time was the capital city and also the intellectual and religious center of the Sāmānid dynasty. This, Ibn Khallikān described as "one of the best ever to have ruled."[29] It has generally been associated in Islamic history with learning. Al-Fārābī grew up under the reign of Naṣr I ibn Aḥmad (260-279/874-892). The rise of the Sāmānid dynasty, which considered itself Persian and claimed descent from the Sassanids, marks the active beginning of the Persian literary and cultural Renaissance in Islam.[30] In it the new Persian language[31] flowered side by side with literary productivity in the Arabic language. Al-Fārābī's knowledge of New Persian was most likely acquired during this second phase of his education in Bukhārā. It was also here that, according to Mahdi, he first began to study music.[32] This was a field in which he was later to establish himself as an undisputed authority.

Upon completing his study of the religious sciences, al-Fārābī became a *qāḍī* (judge). Quoting earlier scholarly reports, Ibn Abī Usaibi'ah claimed that al-Fārābī abandoned that job when he came to know of the availability of instruction in the philosophical sciences and soon thereafter began to be wholly immersed in their study.[33] In my view, this only goes to show that he had already developed a certain amount of interest in philosophy and in particular in logic even prior to his becoming a judge. If it was only later that he turned in earnest to these sciences – though not as late as was believed by many scholars – it was partly because of the problem of the availability of teachers in those subjects. About a century later, in the very same city of Bukhārā, there was apparently only one teacher with whom Ibn Sīnā could study arithmetic. It was only when the famous mathematician, Abū 'Abdallāh al-Nātilī, happened to come to that city that Ibn Sīnā had the opportunity to master the *Almagest*, the *Elements* of Euclid and some logic.[34]

It appears that al-Fārābī's interest in logic and other philosophical sciences, including metaphysics, was first kindled in the course of his advanced study of the linguistic and religious sciences. Prior to his time, both *kalām* (dialectical theology)[35] and *uṣūl al-fiqh* (principles of jurisprudence) had developed to the point of employing systems of logic which appear to share many common elements and features with Stoic logic.[36] This jurisprudential-theological logic, known under the name of *ādāb al-kalām* or *ādāb al-jadal* has been shown by Makdisi to be an integral part of the religious sciences curriculum before and during al-Fārābī's time.[37] Moreover, it is evident from several of his works on logic[38] that al-Fārābī knew this *ādāb al-jadal* well and that he was dissatisfied with its nature and scope. Both his acquaintance and dissatisfaction with that logic dated back to the period of his study of the religious sciences. What he knew of Aristotelian logic (*manṭiq*) by this time was mainly restricted to those elements which had been either appropriated or criticized by the disciplines of *kalām* and *uṣūl al-fiqh*.[39] I strongly believe that it was mainly this intellectual dissatisfaction which led him to seek authoritative instruction in Aristotelian logic,[40] even while he was discharging his duties as a *qāḍī*.

Similarly, his study of the linguistic sciences must have posed to his philosophical mind a number of fundamental questions which could not be resolved within those disciplines. One such question concerns the nature of the relation between logic and grammar. This problem was soon to become a source of much contention between the logicians (*al-manṭiqīyūn*) and the grammarians (*al-naḥwīyūn*),[41] a problem to which al-Fārābī later gave detailed treatment.[42]

The issues cited above, I believe, constitute the substance of his intellectual concerns prior to his initiation into the study of *manṭiq*.

When the opportunity of access to books and teachers in that discipline came, without hesitation he left his work and his native province (apparently never to return) to embark on the next phase of his intellectual life. This move marks the beginning of a lifelong journey completely dedicated to the pursuit of knowledge and the scholarly life. It took him to different cities of the Islamic world and even, according to certain reports, to the land of the Greeks. This moving out was to be of decisive importance for the subsequent intellectual history of Islam.

The question of where al-Fārābī was first initiated into the study of Aristotelian logic and philosophy is a subject of contention among modern scholars. There is no definitive statement on the matter in the traditional sources. A few scholars like Walzer[43] and Mahdi[44] have mentioned the city of Merv (Marw) in Khurāsān as a possible place. A greater number of scholars, including M.Fakhry,[45] F.E.Peters,[46] De Boer,[47] and F.W.Zimmermann,[48] are of the opinion that al-Fārābī first began his study of logic and philosophy in Baghdad. I am inclined to accept the first view as correct. In my opinion the traditional accounts, once their problem of chronological disorder and inconsistency is resolved, furnish ample evidence in its support. The most precious piece of evidence comes from a surviving fragment of al-Fārābī's previously mentioned "autobiography," as quoted by Ibn Abī Usaibi'ah. In this part of the treatise, al-Fārābī provides information about his teacher in logic and philosophy, as well as about a few others who had helped to keep alive the tradition of logico-philosophical learning. It should be noted that he was initiated into that tradition after it had become almost extinct. Ibn Abī Usaibi'ah relates:

> Then [i.e., after the rise of Islam] the instruction was moved from Alexandria to Antioch and remained there for a long time until at last but one teacher remained. With him there studied two men. They moved away, taking the books with them. One of them was of the people of Harrān, the other of the people of Marw. As to the one of the people of Marw, there studied with him two men, one of whom was Ibrāhīm al-Marwazī and the other Yūḥannā ibn Ḥailān. With the man of Harrān studied the bishop Isrā'īl and Quwairī, both of whom went to Baghdad. Now Ibrāhīm [*sic*, in error for Isrā'īl] occupied himself with religion, but Quwairī took up instruction. As for Yūḥannā ibn

> Ḥailān, he also occupied himself with his [i.e., Christian] religion. Ibrāhīm al-Marwazī went down to Baghdad and settled there. With al-Marwazī studied Mattā ibn Yūnān [i.e., Abū Bishr Mattā ibn Yūnus. That which was taught [in logic] at that time was up to the end of the assertoric figures [of the syllogism]. But Abū Nasr al-Fārābī says about himself that he studied with Yūhannā Ibn Hailān up to the end of *Analytica Posteriora* (*Kitāb al-burhān*).[49]

Al-Fārābī's above account of the last phase of the migration of the Greek philosophical school of Alexandria to Baghdad has generally been accepted. It is also generally agreed that this migration took place about 276/900 during the Caliphate of al-Mu'taḍid (892-902). This account, confirmed by other sources, clearly establishes Ibn Ḥailān as al-Fārābī's principal teacher. It is also clear that Mattā Ibn Yūnus, his only other teacher in logic known to us, was at this time only a fellow student associated with the same philosophical school. Al-Fārābī could not have commenced his study of logic in Baghdad with either one of them. Neither could he have studied with any other teacher of the newly born school, since around this time he was reported to have taught logic and music to the famous Baghdad scholar of language, grammar and poetry, Ibn al-Sarrāj (d.316/929). This he did in exchange for lessons in advanced Arabic grammar.[50] Al-Fārābī, therefore, arrived at Baghdad already equipped with a sound knowledge of logic which he had acquired under the instruction of Ibn Ḥailān at the logico-philosophical school of Merv.

We are informed in the above account that Ibn Ḥailān, unlike Quwairī and al-Marwazī, did not take up instruction in Baghdad but devoted himself instead to religious duties. Yet al-Qifṭī and Ibn Abī Usaibi'ah reported that Ibn Ḥailān taught al-Fārābī logic in Baghdad during the Caliphate of al-Muqtadir (908-932).[51] In my view, both accounts are true. These accounts lend further support to my contention that al-Fārābī could not have commenced his study of logic in Baghdad. The reason al-Fārābī was able to study with Ibn Ḥailān in Baghdad and also in Harrān, according to Ibn Khallikān – despite the latter's devotion to religious duties after leaving Merv – is that he had previously studied with Ibn Ḥailān in Merv. In Baghdad and Harrān, al-Fārābī was merely continuing the previous study in order to master "some particular applications of the art of logic."[52]

Concerning the philosophical school of Merv itself, it was probably intimately associated with the great Nestorian monastery at Masergasan north of the quarter known under Islamic rule as Sulṭān-Qal'a.[53] It is a well-known fact that the Nestorians had played an instrumental role in the spreading of Greek learning, chiefly in Syriac translation, to lands as far east as Persia. Merv, in fact, had long been an outpost of Greek learning. During the Caliphates of Hārūn al-Rashīd (786-809) and al-Ma'mūn (813-833), who were both educated in that city, it produced some of the earliest translators of Greek scientific works into Arabic.[54] It was in this city that Ibn Ḥailān, himself a Nestorian, studied logic and philosophy. Al-Fārābī, under his guidance, read the canonical texts of Aristotelian logic, including the *Analytica Posteriora* which no Muslim before him had studied, under a special master. This raises the question of what language was being used by Ibn Ḥailān in his teachings and commentaries on the *Organon* in view of the fact that the *Analytica Posteriora* was translated into Arabic by Mattā Ibn Yūnus only after the foundation of the Baghdad philosophical school.[55]

Mahdi gave several reasons to support his view that al-Fārābī must have studied with Ibn Ḥailān in Syriac or Greek or both.[56] Madkour, Zimmermann, and Gonzales Palencia,[57] among others, reject the possibility that al-Fārābī knew the Greek language based on his presumed faulty interpretation of the word *safsaṭah*. This is not the place to go into a detailed discussion of this issue. Based on the present state of our knowledge of al-Fārābī, suffice it to say that while the counter-argument of Madkour, Zimmermann and Palencia is insufficient and inconclusive, there are other reasons, in addition to those cited by Mahdi, to support the view that al-Fārābī very likely knew Greek. In several of his works, particularly *Kitāb al-ḥurūf* which deals among other things with the question of the origin and development of language in a particular nation, al-Fārābī frequently indulges in comparative linguistics. These exercises involve mainly the Greek and Arabic languages.[58] Even Zimmermann's study of al-Fārābī's Commentary and short treatise on Aristotle's *De Interpretatione* – which for him provides justification for the above view – reveals a number of points which can be used as arguments in support of the other view.[59] More important, however, is a report by al-Khaṭṭābī (931-998), a younger contemporary of al-Fārābī, that the latter studied philosophy in Constantinople. The report, most likely based on al-Fārābī's lost autobiographical treatise, says:

> After this [i.e., after completing the study of Aristotle's *Posterior Analytics* with Ibn Ḥailān] he traveled to the land of the Greeks and stayed in their land for eight years until he completed [the study of the] science [s] and learned the entire philosophic syllabus.[60]

A study of the philosophic syllabus at the University of Constantinople during his time would be a further strong indication that al-Fārābī possessed a fair knowledge of the Greek language.

Al-Fārābī's journey to Constantinople took place after a number of years of study in Baghdad. During this time he sought to master the Arabic language and devoted his mind to the philosophical sciences. As we have seen, he had arrived in that city from Merv with Ibn Ḥailān around 287/900. Very likely, it was Ibn Ḥailān himself who played an instrumental role in influencing al-Fārābī to visit and study in Constantinople at the philosophical school that was intimately linked to that of Alexandria.[61] Al-Fārābī left Baghdad during the Caliphate of al-Muqtadir (908-932), going first to Harrān in the company of his teacher. According to Ibn Khallikān, al-Fārābī also studied in Harrān with Ibn Ḥailān. It is possible, therefore, that he spent some time there before continuing his journey to Constantinople.

Al-Fārābī returned to Baghdad some time between 297/910 and 307/920. His second stay there lasted more than two decades, during which time he devoted himself to philosophical learning, teaching and writing. Upon his arrival in Baghdad, he found that Mattā Ibn Yūnus (d.328/940), the Nestorian philosopher who was once a student of al-Marwazī and Quwairī and was then far advanced in age, had gained the highest reputation in the field of logic. He still attracted large crowds of pupils to his public lectures on Aristotelian logic.[62] Al-Fārābī became his student. According to some authorities, "the abilities which Abū Naṣr al-Fārābī displayed in rendering the most abstract ideas intelligible and expressing them in the simplest terms could only be attributed to the tuition of Abū Bishr (Mattā)."[63] His principal teacher Ibn Ḥailān, with whom he continued to study after his return to Baghdad, died in this city sometime before 320/932. Al-Fārābī's teaching and writing during this period soon established his reputation as the leading Muslim philosopher, indeed the philosophic authority after Aristotle. Both al-Qifṭī and Ibn Abī Usaibi'ah point out that al-Fārābī, despite his relative youth, soon outstripped his teacher Mattā in the field of logic.[64]

Al-Fārābī apparently kept himself aloof from the political turmoils and religious and sectarian conflicts which beset Baghdad during this period. His only contact with the society of the court was with the viziers (*wazīr*) who patronised the philosophic sciences, such as Ibn al-Furāt, 'Alī ibn 'Īsā, and Ibn Muqlah.[65] His major work on music, *Kitāb al-mūsīqā al-kabīr* was written at the request of Abū Ja'far Muḥammad ibn al-Qāsim al-Karkhī who became *wazīr* to Caliph al-Rāḍī in 324/936.[66] His intense devotion to teaching the philosophical sciences during this period of his stay in Baghdad is best indicated by the traditional report that he read Aristotle's *Physics* forty times and his *De anima* two hundred times.[67] The modern philosopher Hegel reacted to this report with the statement that al-Fārābī "must have had a strong stomach" but as pointed out by Rescher "this report does not mean that al-Fārābī read these works so frequently for his private edification, as Hegel understood it to say, but that he gave regular courses of explanatory lectures on them."[68] Because of the deteriorating political situation in Baghdad, which forced Caliph al-Mutaqqī and his *wazīrs* and bodyguard to flee the city in 330/942, al-Fārābī moved at the end of the same year to the more peaceful Damascus which was then ruled by the Ikhshīdid dynasty. In the *Fuṣul al-madanī* the last of al-Fārābī's extant political works and perhaps also the last he ever wrote (as reasonably argued by D.M. Dunlop) he seemed to allude to the political disasters and corruption which he personally witnessed in Baghdad. In the light of those circumstances he appeared to regard his departure from that city as his personal *hijrah* (emigration). He writes:

> Therefore it is wrong for the virtuous man to remain in the corrupt polities, and he must emigrate to the ideal cities, if such exist in fact in his time. If they do not exist, then the virtuous man is a stranger in the present world and wretched in life, and to die is preferable for him than to live.[69]

At the beginning of his stay in Damascus, al-Fārābī worked, according to some reports,[70] as a garden keeper during the day and devoted himself to reading and writing philosophical books at night. Later, he used to spend the greater part of his time "near the borders of some rivulet or in a shady garden where he composed his works and received the visits of his pupils."[71] He stayed in Damascus for about two years before departing for Egypt[72] in the midst of another political conflict, this time between the Hamdānids of Mosul and the Ikhshīdids of Syria. This conflict resulted in the occupation of Aleppo and Damascus by the former in 333/945 and 334/946 respectively. Nothing is known about his activity in Egypt except that there is a reference by Ibn Abī Usaibi'ah to a political work which he completed there in 337/948-49 and a similar reference by Ibn Khallikān.[73] Al-Fārābī returned to Damascus in 338/949.

It was during this second stay in Damascus that the philosopher was welcomed to the court of the new ruler of Syria, the Hamdānid prince Sayf al-Dawlah (d.967), at Aleppo. Not long after his residence in Aleppo and Damascus, that is by 335/947, Sayf al-Dawlah began to surround himself at court with a circle of learned men, which later was to include such famous poets as al-Mutanabbī (d.965), Abū Firās (d.968), Abu'l-Faraj (d.968), and the grammarian Ibn Khālawayh.[74] In his first encounter with Sayf al-Dawlah, al-Fārābī impressed him with his command of several languages, his mastery of the philosophical sciences, and his musical talent.[75] He came to be highly respected by his patron, at whose court he spent the rest of his days as a scholar. Living an ascetic Sufi life, he did not avail himself of all the lucrative advantages that he could reap from his influence at court except for a daily pension of four dirhams out of the public treasury. He refused to comply with the regulations for dress when at court. At times, he would dress in Sufi garb or in his Central Asiatic attire with a large fur hat. Yet at other times, he would appear dressed better than anyone else. Al-Fārābī's practice of alternating between these types of dress is probably not unconnected with a certain Sufi attitude alluded to by Nasr:

> The practice of alternating between simple and highly ornate dress was carried on also by a number of well-known Sufis like Abu'l-Hasan al-Shādhilī, the founder of the famous Shādhilīyah Order, perhaps to show their independence not

only of the world but also of renunciation of the world, or what Rūmī calls "renunciation of renunciation" (*tark-i tark*).[76]

In Rejab 339/December 950, al-Fārābī died in Damascus at the age of eighty.[77] He was buried in the cemetery outside the southern or minor gate (*al-bāb al-ṣaghīr*) of the city.[78] It was Sayf al-Dawlah himself who led a number of his courtiers[79] in the funeral prayers for al-Fārābī, the scholar who was one of the earliest and also the most famous member of the "circle of Sayf al-Dawlah".

1.3 Al-Fārābī's Works and Significance

Al-Fārābī wrote numerous works[80] on almost every branch of science known to the medieval world, with the notable exception of medicine.[81] Traditional bibliographers have attributed to him more than one hundred works of varying length, most of which have survived.[82] A number of these are extant only in their Hebrew or Latin translations. Although, of late, Farabian studies have increased considerably, the greater portion of these extant works remain only in manuscript form. Of those which have been edited and published only a few have been seriously studied. Consequently, it is not possible at present to provide a comprehensive account of the various facets of his works and thought.

Apparently al-Fārābī wrote all his works in Arabic. Scholars, both traditional and modern, have generally praised his simple and clear Arabic philosophic prose, in contrast, say, to the difficult style of his predecessor al-Kindī. It is generally agreed that most of his works were composed in Baghdad and Damascus. There is no evidence to indicate that he wrote any work while he was in Khurāsān. Since the man's writings deal with such a vast range of subjects, many modern scholars have sought to identify the general contours and patterns, likewise the central focus of his intellectual concerns, by classifying these works into a number of groups.[83] The resulting divisions vary from one scholar to another because many of al-Fārābī's works may fall into more than one category. In my view, a more fruitful division of his works (especially in the context of the present study), would be one which corresponds to his division of the sciences presented in *Kitāb ihsā' al-'ulūm* (*The Book of the Enumeration of the Sciences*). If an account of his education has given an indication of the extent to which he studied the various sciences enumerated in the above work, so a general survey of his works classified according to the above division helps to reveal his intellectual attitude and contributions to each of the sciences.

In the *Iḥṣā' al-'ulūm*, al-Fārābī classifies the various branches of knowledge under eight headings: the linguistic, the logical, the mathematical (propaedeutic), the physical, the metaphysical, the political, the juridical. The juridical happens to receive the shortest treatment in his descriptions of the sciences. It is true, however, that he dealt with certain philosophical questions pertaining to the *Sharī'ah* (Islamic Law), particularly in his political works. It was in the field of logic, however, more than any other branch of philosophy or science, that his intellectual efforts were mainly spent. He wrote commentaries on the entire Aristotelian *Organon*, namely the Categories (*al-maqūlāt*), Hermeneutics (*al-'ibārah*), Prior Analytics (*al-qiyās*), Posterior Analytics (*al-burhān*), Topics (*al-jadal*). Sophistics (*al-mughāliṭah* or *al-safsaṭah*), Rhetoric (*al-khiṭābah*) and Poetics (*al-shi'r*). He also commented on the Isagoge (*al-īsāghūjī*) of Porphyry which, in the Syriac logical tradition to which al-Fārābī became heir, was placed at the head of the *Organon* as an introduction.[85]

The commentaries were written in the triplicate manner of the Alexandrian school, consisting of short (*ṣaghīr*), medium (*awsaṭ*) and long (*kabīr*) commentaries. Besides these commentaries, al-Fārābī wrote a number of short treatises devoted to special aspects of logic. These include the *Risālat ṣudira bihā al-kitāb* (*Treatise with Which the Book Begins*), *Risālat fī jawāb masā'il su'ilā 'anhā* (*Treatise on Answers to Questions Put to Him*), and *Risālat fī qawānīn ṣinā'at al-shi'r* (*Treatise on the Canons of the Art of Poetry*). An interesting and significant work which may be placed under the first category of his writings – although it does not deal with pure logic as such – was his discourse on those prophetic *ḥadīths* which he had collected with the express aim of demonstrating that the art of Aristotelian logic (*ṣinā'at al-manṭiq*) is in fact recommended by them.[86] This attempt at providing a religious foundation to logic, which is consistent with his scheme of integrating all knowledge into an organic unity within the world-view of Islam, was later followed even more intensely by al-Ghazzālī.

The Turkish scholar Ahmet Ateš, in his bibliographical survey of al-Fārābī's writings,[87] puts the number of his logical works at over forty. Many of these survive. Rescher[88] has rightly called al-Fārābī "the first specialist in logical studies among the Arabic-speaking people." The great significance of his logical works lies in the fact that they express Aristotelian logic, for the first time, in an "appropriate and exact Arabic terminology which henceforth became a heritage of nearly all branches of Islamic learning."[89] In this project he used examples that were familiar to and current among the people of his time.[90] His eminent position in the field of logic was described by a fifth/eleventh century Andalusian jurist and historian of ideas, Ṣa'id ibn Aḥmad al-Andalusī (d.462/1070), in these terms:

> He then excelled all the people of Islamism and surpassed them by his real acquirements in that science (i.e., logic); he explained its obscurities, revealed its mysteries, facilitated its comprehension and furnished every requisite for its intelligence, in works remarkable for precision of style and subtlety of elucidation, noticing in them what al-Kindī and others neglected such as the art of analysis and the proper modes of conveying instruction. In these treatises he elucidated in plain terms the five main principles of logic, indicating the manner of employing them with advantage and the application of the syllogistic forms to each of them. His writings on this subject are therefore highly satisfactory and possess the utmost merit.[91]

Al-Fārābī's mastery of the art of logic won praise not only from later generations of Muslim thinkers but also from the great figure of medieval Jewry, Maimonides (d.1204). In a famous letter to Ibn Tibbon (d.1230), the translator of his *Guide for the Perplexed*, Maimonides wrote:

> Do not busy yourself with books on the art of logic except for what was composed by the wise man Abū Naṣr al-Fārābī. For, in general, everything that he composed – and particularly his book on the *Pinciples of Beings* – is all finer than fine flour. His arguments enable one to understand and comprehend, for he was very great in wisdom.[92]

The second major division of al-Fārābī's writings, which number a dozen or so, are those dealing with physics (*tabī'īyat*), understood in its traditional and particularly peripatetic sense, or natural philosophy. In this branch, we may distinguish three main categories of al-Fārābī's writings. First we have the commentaries on a number of works of Aristotle and of some of his Greek commentators. The most important of these is *Sharḥ kitāb al-samā' al-ṭabī'ī li-arisṭūṭālīs* (*Commentary on Aristotle's Physics*). Among the others, we may particularly mention *Sharḥ kitāb al-samā' wa'l-'ālam li-arisṭūṭālīs* (*Commentary on Aristotle's Book of the Heavens and the Universe*) and *Sharḥ maqālat al-iskandar al-afrūdīsī fi'l-nafs* (*Commentary on Alexander of Aphrodisias' Treatise on the Soul*). The second category consists of a number of independent scientific treatises of an expository nature on such subjects as psychology, zoology, meteorology, the nature of space and time, and the vacuum. These include the *Risālat fi'l-khala'* (*Treatise on the Vacuum*), *Kalām fī a'ḍā' al-ḥayawān* (*Discourse on Animal Organs*), and *Kalām fi'l-ḥaiz wa'l-miqdār* (*Discourse on Space and Measure*). We may also include in this category of works the highly influential *Maqālat fī ma'ānī al-'aql*[93] (*Treatise on the Meanings of the Intellect*). Although the latter is essentially a metaphysical work, it is of importance for physics. During the medieval period it was translated into Latin under the name of *De Intellectu et Intellecto*, and also into Hebrew. In our own time it has been rendered into several languages.[94]

In the third and last category we have a few works composed as refutations of the views of certain philosophers and theologians on particular aspects of natural philosophy. The titles given by Ibn Abī Usaibi'ah are: *Kitāb al-radd 'alā jālīnūs fī mā ta'awwalahu min kalām arisṭū* (*Book of Refutation of Galen's Interpretations of Aristotle's Discourse*),[95] *al-radd 'alā ibn al-rāwandī fī ādāb al-jadal* (*Refutation of Ibn al-Rāwandī's Account of Dialectic*),[96] *al-radd 'alā yaḥyā al-naḥwī fī mā raddahu 'alā arisṭū* (*Refutation of John the Grammarian's Criticism of Aristotle*),[97] and *al-radd 'alā al-rāzī fi'l-'ilm al-ilahī* (*Refutation of al-Rāzī's Metaphysics*).[98]

Of the above writings of al-Fārābī on natural philosophy, his psychological treatises are particularly significant from the point of view of the present study. They constitute important sources, among others, for our knowledge of his epistemology, a subject which will be fully treated in the next chapter. Al-Fārābī's discussion of psychology is, however, by no means restricted to the writings which I have classified under the works dealing with the physical sciences. As we shall see in the next chapter, it is also to be found in a number of works which, in my adopted scheme of classification, are characterized as either political or metaphysical. One important branch of knowledge related to natural philosophy – but which is not considered by al-Fārābī as a branch of it – is medicine. He apparently displayed little overall interest in medicine, to judge by his writings, and whatever interest he had in it was of a theoretical nature.[99] Even then he was mainly concerned with medicine's methods and principles with a view to demonstrating its place in the hierarchy of the sciences. As a practicing Sufi, he was more of a physician of the soul than of the body. In fact he wrote an important treatise on the therapeutic effects of music on the soul.

If al-Fārābī's major works in the logical and physical sciences have earned him the title of the first great Muslim commentator on Aristotle, much of the rest of his works will in turn demonstrate that he was not just a follower or commentator on the Stagirite. Although he wrote treatises on all branches of the Pythagorean *Trivium* and *Quadrivium*, besides composing commentaries on Ptolemy's *Almagest* and the difficulties of the introductory matter to Books I and V of Euclid's *Elements*,[100] his best known mathematical works were in the field of music. Among his several musicological works,[101] the already-mentioned *Kitāb al-mūsīqā al-kabīr*, regarded by many as the greatest piece of work on music theory in the Middle Ages, constitutes a significant advance on the musical theory of the Greeks.[102] In the East this treatise became an indispensable reference for almost every writer on music from Ibn Sīnā in the fifth/eleventh century to Ṭanṭāwī in the fourteenth/twentieth century.[103] It was used, according to Ibn 'Aqnin (d. 1226), as a textbook even in the Jewish schools.[104]

As for the influence of al-Fārābī's teaching on music in the West, the research of H.G. Farmer, the well-known historian of music, has shown that his definitions of music were known as far away as England by the end of the twelfth century through Daniel of Morlay, a pupil of Gerard of Cremona (d.1187). Al-Fārābī's definitions continued to be quoted by musicologists until the sixteenth century.[105]

In the field of music, al-Fārābī was more than an expert theoretician. He was an excellent composer and performer. Thus, his legacy in this field has survived not only in the form of treatises but also the spiritual concert or *samāʻ* of some of the Sufi orders. For example, the Mawlawī of Anatolia continue to perform his compositions until the present time.[106]

In addition to the musicological work, we should mention his treatise on judicial astrology (*aḥkām al-nujūm*), called *Fī mā yaṣiḥḥ wa mā lā yaṣiḥḥ min aḥkām al-nujūm* (*On the True and the Untrue in Judicial Astrology*).[107] This treatise, written at the request of the astrologer Abū Isḥāq Ibrāhīm ibn ʻAbdullāh al-Baghdādī, has been cited by a number of scholars (including Mahdi) as indicative of al-Fārābī's disavowal of astrology as a science. The question of his attitude toward astrology will be examined in chapter four when I discuss his enumeration and exposition of the various sciences. Suffice it to say here that, in my view, al-Fārābī did not reject astrology as such. His main criticism was leveled against its contemporary practitioners in whom he had little confidence. He writes:

> The most illustrious among the astrologers are the least prone to manage their own affairs in the light of their own astrological findings; consequently, we must assume that their prognosis is inspired by the quest for profit or is merely the result of ingrained habit.[108]

The next category of his writings, dealing with the various metaphysical sciences and numbering more than fifteen works, must be counted as the crown of his intellectual output. They reveal not only the foundation of his intellectual edifice and the source of unity of his rich philosophical ideas but also his real worth as a metaphysician. In this category, also, the writings are in the form of commentaries, refutations and independent treatises. Of the commentaries, the most important are the two closely related works, the *Maqālat fī aghrāḍ mā baʻd al-ṭabīʻah* (*Treatise on the Aims of Aristotle's Metaphysics*)[109] and the previously cited *Kitāb al-ḥurūf* (*Book of Letters*).[110] A significant work of challenge in this domain is the previously mentioned *Refutation of al-Rāzī's Metaphysics.* It is counted as a work on the physical sciences because it deals with al-Rāzī's anti-Aristotelian views of matter, time, space and atoms. However, this work also contains al-Fārābī's metaphysical defense of prophecy.

The remaining metaphysical treatises of al-Fārābī include the well-known *Fuṣuṣ al-ḥikam* (*Bezels of Wisdom*), *Kitāb fi'l-wāḥid wa'l-waḥdah* (*The Book on the One and the Unity*), and a set of works aimed at harmonizing and unifying the wisdom of Plato and Aristotle such as *Falsafat arisṭūṭālīs* (*The Philosophy of Aristotle*), *Kitāb falsafat aflāṭūn wa ajzā'hā* (*The Philosophy of Plato and Its Parts*), and *Kitāb al-jam' baina ra'yai al-ḥakīmain aflāṭūn al-ilāhī wa arisṭūṭālīs* (*The Book of Harmony Between the Ideas of the Two Sages, the Divine Plato and Aristotle*). I have also designated as metaphysical al-Fārābī's few works on theory of knowledge and the first principles of particular sciences, e.g., the *Kitāb fī uṣūl 'ilm al-ṭabī'ah* (*The Book on the Principles of Physics*) and the *Iḥṣā' al-'ulūm*, since his definition of metaphysics includes the inquiry into the principles of the observational and particular sciences.[111] Finally there are those works which deal with ethics and happiness. They are at once metaphysical and political, but they will be classified under the latter category.

As regards the *Fuṣūṣ al-ḥikam*, its authenticity has been questioned by a number of modern scholars, including Leo Strauss, S. Pines, Khalil Georr, and F. Rahman.[112] These have influenced a host of other scholars, including Walzer and Rescher, to view it as a probable work of Ibn Sīnā. But, as pointed out by Nasr, the reasons advanced by them, including the so-called internal evidence of a striking inconsistency between this work and other works of al-Fārābī, do not seem sufficient to prove their point.[113] In the Islamic world, however, this treatise has generally been regarded through the centuries as an authentic work of al-Fārābī. As a work on ontology, it contains the first explicit reference by a Muslim thinker to the Aristotelian distinction between essence or quiddity (*māhīyah*) and existence (*wujūd*). Al-Fārābī formulated the relation between essence and existence and discussed it in the light of the distinction between Creator and creature.[114]

The *Fuṣūṣ al-ḥikam* is the most continuously influential work of al-Fārābī in the East in the field of metaphysics. This is attested to by the fact that it has been taught and read in the *madrasahs* (colleges) to the present day. To the already long list of commentaries written on this work over the centuries, a leading contemporary *ḥakīm* (sage) and gnostic of Persia, Ilāhī Qumsha'ī, added an important one of his own, viewing the original work as a summary of the doctrines of gnosis (*'irfān*).[115]

In addition to the last set of works cited above, we may mention *Ta'līqāt fī'l-ḥikmah* (*Explanatory Remarks on Wisdom*) and *Kitāb fī ẓuhūr al-falsafah* (*On the Appearance of Philosophy*). These works deal with the metaphysical origin and historical manifestations of *sophia* or *ḥikmah,* that is ultimately of Divine origin. These became the model for that category of works by Islamic philosophers, which, says Schuon, "had the merit of integrating these great Greeks in one and the same synthesis, for what interested them (i.e., Islamic philosophers) was not systems, but truth in itself."[116] It is in the light of this remark that al-Fārābī's synthesis of Plato, Aristotle, Plotinus and other Greeks, as contained in many of his works, should be understood. This should render invalid the accusation leveled against al-Fārābī and others after him that they had missed the meaning of these great Greek thinkers.

Although al-Fārābī wrote a few works which touch on special aspects of the linguistic and theological sciences, the only major category of his writings left to be considered concerns political philosophy or the science of society (*al-'ilm al-madanī*). He is considered to be its true founder in Islam. Al-Fārābī's works on ethics belong to this category. The central theme in his political philosophy is happiness, which for him is "earthly happiness in this life and supreme happiness in the life beyond." His works in this domain include the *Kitāb ārā' ahl al-madīnat al-fāḍilah* (*The Book of Opinions of the People of the Ideal City*), *Kitāb al-siyāsat al-madanīyah*[117] (*The Book on the Government of the City State*), *Kitāb al-millat al-fāḍilah* (*The Book of The Excellent Community*), the already mentioned *Fuṣūl al-madanī* (*Aphorisms of the Statesman*), *Talkhīṣ nawāmīs aflāṭūn* (*Epitome of Plato's Laws*), *Risālat fī'l-siyāsah* (*Epitome on Politics*), and *Kitāb taḥṣīl al-sa'ādah* (*On Attaining Happiness*). In all these writings al-Fārābī achieves a remarkable synthesis of the views of the ancient Greek sages, principally Plato, and Islamic doctrines embodied in the Qur'ān and the Tradition (*Sunnah*) of the Prophet.

Most of the modern studies of these works have sought to demonstrate their overwhelming Platonic or Aristotelian inspirations. This is especially true of the sections on metaphysics which usually precede al-Fārābī's actual discussion of human society. In my view, the fundamental ideas which he discusses in his political works (e.g., the perfect society and the perfect ruler, the necessity of a revealed law (*sharī'ah*) and the attainment of happiness as the ultimate aim of political science, were mainly inspired by his own vision of the unfolding in history of the Islamic Revelation through the earthly career of the Prophet. That career culminated in the birth of *Madīnat al-nabī* (*The City of the Prophet*) which, for Muslims, constitutes the epitome of the socio-political order of Islam.

Al-Fārābī's political works, which were apparently the last to be written and against the background of a steep decline in the quality of political life, constitute the first attempt at a comprehensive portrayal in the scientific and philosophical mold. He explicated his views of what would constitute the perfect society and government, something Muslims generally associate with the Prophet's community and the rule at *al-Madīnah*. His knowledge of Plato's *Republic* and *Laws* provides him with the necessary intellectual materials for his philosophical formulations and analysis to the extent that these Greek classics agreed with Islamic political and religious doctrines. The figure of the prophet-king of Plato became identified with the prophet and law-giver of the Abrahamic tradition. For al-Fārābī as well as for Ibn Sīnā after him, Plato's investigation of Greek divine law could serve as a guide to the study and understanding of all divine laws, including the *Sharī'ah* of Islam.[118]

The political writings of al-Fārābī exercised a considerable influence upon many Muslim and Jewish thinkers, especially from the thirteenth century onward.[119] Ibn Rushd (520/1126-595/1198), in whom that influence is most visible among the Muslim philosophers, defended al-Fārābī's theory of prophecy against al-Ghazzālī's criticism. In his *Paraphrase of Plato's Republic* he mentions al-Fārābī's *Fusūl al-madanī* as one of his important sources.[120] Maimonides, as we have seen, greatly appreciated the *al-Siyāsat al-madanīyah,* which was translated into Hebrew in the middle of the thirteenth century. There was also a Hebrew paraphrase of the *Tahsīl al-sa'ādah,* which was included by Shemtob ben Falaquera (1225-1290) in his *Introduction to Science*. This in turn was translated into Latin. Other political treatises of al-Fārābī translated into Latin include *al-Tanbīh 'alā sabīl al-sa'ādah* (*Reminder of the Way of*

Happiness) and his commentary on Plato' [s] *Laws*.[121]

Also worthy of mention are some of al-Fārābī's treatises on esoteric sciences like alchemy and the interpretation of dreams. He wrote an alchemical treatise called *Fī wujūb ṣanā'at al-kīmiyā'* (*On the Necessity of the Art of Alchemy*)[122] which was translated into German early in this century by E.Wiedemann. As far as al-Fārābī's treatment of the phenomena of dreams is concerned, we do not know of any independent treatise written by him, but he is known to have dealt with the subject in a number of his works.[123]

It is clear why al-Fārābī is considered to be one of Islam's greatest philosophers. Among later Muslim scholars he was to gain the distinguished title of *al-Mu'allim al-thānī* (The Second Teacher).[124] Modern scholars have suggested different reasons why the above honorific was conferred on him.[125] In my view, the soundest explanation is that given by Nasr:

> The term "teacher" or *mu'allim* as used in this context does not mean one who teaches or is a master of the sciences. Rather, it means one who defines, for the first time, the boundaries and limits of each branch of knowledge and formulates each science in a systematic fashion. That is why Aristotle, who was the first in Greece to have classified, defined, and formulated the various sciences, is called "The First Teacher," and Mīr Dāmād, who performed the same task on a smaller scale within the consolidated Twelve-Imām Shi'ah world of the Safavids is referred to by many in Persia as the Third Teacher.
>
> As for al-Fārābī, it was because his *Iḥṣā' al-'ulūm*, the Latin *De Scientiis* was the first classification widely known to the Muslims – the effort of al-Kindī in this direction not being generally recognized by later generations – and because he really molded and formulated the various branches of knowledge in a complete and permanent form within Islamic civilization that he gained the title of "The Second Teacher."[126]

ENDNOTES

Chapter 1

[1]Several of the traditional sources speak of al-Fārābī having many students. See, for example, Ibn Khallikān, *Kitāb wafayāt al-a'yān*. Eng. tr. by W. MacGuckin de Slane under the title *Ibn Khallikān's Biographical Dictionary*. vol. III, Paris: Oriental Translation Fund, p. 309; repr. New York & London: Johnson reprint Corporation,

1961; Ibn Abī Usaibi'ah, *'Uyūn al-anbā' fī ṭabaqāt al-aṭibbā'*, Beirut, 1963, p. 603. However, only two of these students are known to us by name. They are the famous Jacobite Christian theologian and philosopher Abū Zakariyā' Yaḥyā ibn 'Adī (893-974) and his brother Ibrāhīm who was still with al-Fārābī in Aleppo shortly before the latter's death.

[2]This was the title given by Ibn Abī Usaibi'ah, who appears to be the first scholar to have mentioned it. See his *'Uyūn*, p. 604. I believe that this is the same treatise which is known to Muslim bibliographers as *Kitāb fī ism al-falsafah wa sabab ẓuhūrihā (On the Name of Philosophy and the Cause of Its Appearance)*.

[3]I believe that the account of the chain of transmission of philosophical teachings from Aristotle to al-Fārābī given by the latter's contemporary, Abu'l-Ḥasan al-Mas'ūdī (d.345/956), in his *al-tanbīh wa'l-ishrāf* (Cairo, 1938), was based on this "autobiography" of al-Fārābī, although he did not mention this by name. For an English translation of the above account, see S.M.Stern, "al-Mas'ūdī and the philosopher al-Fārābī," *al-Mas'ūdī Millenary Commemoration Volume*, ed. S.Maqbul and A.Rahman, Aligarh, 1980, pp. 28-41.

The other source, which contains a much longer quotation, is Ibn Abī Usaibi'ah's *'Uyūn*, pp. 604-5. Mahdi refers to yet another surviving fragment of this lost treatise, which is distinct from that preserved by either al-Mas'ūdī or Ibn Abī Usaibi'ah. This is in an Arabic manuscript attributed to a younger contemporary of al-Fārābī by the name of al-Khaṭṭābī (931-98) which is currently preserved at the Kabul (Afghanistan) Library of the Ministry of Information. See M. Mahdi, "al-Fārābī," *Dictionary of Scientific Biography*, ed. C.G.Gillispie, New York, IV (1970), 526. This larger work is hereafter cited as *DSB*.

[4]All the known traditional biographies of al-Fārābī are now available in a single book through the efforts of H.A. Maḥfūẓ, Professor of Oriental Studies at the University of Baghdad. See his *al-Fārābī fi'l-marāji' al-'arabīyah*, Baghdad, 1975.

Of these traditional accounts of al-Fārābī's life and works, see in particular Ṣā'id Ibn Aḥmad al-Andalusī, *Ṭabaqāt al-umam*. Najaf, 1967, pp. 70-72 (cf. Maḥfūẓ, pp. 58-59); Ibn Khallikān, *op. cit.*, pp. 307-11; Ibn Abī Usaibi'ah, *op. cit.*, pp. 603-09 (cf. Maḥfūẓ, pp. 98-111); al-Bayhaqī, *Tārīkh ḥukamā' al-Islām*, Damascus, 1946, pp. 30-35 (cf. Maḥfūẓ, pp. 68-70) and also his *Tatimmat suwān al-ḥikmah*. Lahore (1935), pp. 16-20 (cf. Maḥfūẓ, pp. 71-75); and al-Qifṭī, *Tārīkh al-ḥukamā'*, Leipzig, 1903, pp. 277-80 (cf. Maḥfūẓ, pp. 90-96).

[5]For modern accounts of his life, see M.Steinschneider, *al-Fārābī, Des Arabischen Philosophen Leben und Schriffen*, Saint Petersburg, 1869; I.Madkour, "al-Fārābī," *A History of Muslim Philosophy*, ed. M.M.Sharif, Wiesbaden, 1963, I, 450-68; R.Walzer, "al-Fārābī", in *Encyclopaedia of Islam*, 2nd ed., Leiden-London (1960-), II, 778-81; M.Mahdi, *op. cit.*, pp. 523-26; A.'Abd al-Wāhid Wāfī, "al-Fārābī," *Turāth al-insānīyah*, Egypt, II, 569-82 (cf. Maḥfūẓ, pp. 342-53).

[6]The traditional sources do not agree about al-Fārābī's line of ancestors. I have adopted the description given by Ibn Khallikān, which has generally been accepted by modern scholars as authentic. However, both names, *Tarkhān* and *Awzalagh*, appear in most of these sources whenever the name of al-Fārābī is mentioned.

Furthermore, modern scholars do not agree about the correct pronounciation of the name of al-Fārābī's greatgrandfather. Various pronunciations have been suggested, viz., *Uzalāj* (Rescher), *Uzlagh* (Langhade, De Boer), *Uzluk* (M. Turker Kuyel), and *Awzalagh* (de Slane, Walzer, Mahdi). In my judgment *Awzalagh* is the best rendering into Arabic of the corresponding Turkish name.

[7]The term "philosopher-scientist" was first coined by S.H. Nasr. Justifying his use of this term as the most fitting description of this school of Islamic philosophy, he writes:

> In this school, science was combined with philosophy and in fact was considered as a branch of it just as in another sense philosophy began with the classification of the sciences. The great figures of this school, like al-Kindī himself, were philosophers as well as scientists, although in some cases, like that of Abū Sulaimān al-Sijistānī, philosophy dominated over science, and in others, like that of al-Bīrūnī, science prevailed over philosophy.

S.H.Nasr, *Three Muslim Sages*, Harvard University Press, 1969, pp. 9-10.

In the case of al-Kindī and al-Fārābī, for example, it is their possession of the common traits associated with the philosopher-scientists, such as having a universal interest in all the sciences, which justifies their inclusion in the same intellectual school, despite the fact that their philosophical views exhibit major distinguishing features.

[8]Most authorities, both traditional and modern, have traced al-Fārābī's place of origin to Fārāb, the modern Otrar, in Transoxiana. One notable exception is Ibn al-Nadīm, the author of *Kitāb al-fihrist* (*The Book of Index*), who lists al-Fārābī's birthplace as al-Fāriyāb in Khurāsān. See Ibn al-Nadīm, *The Fihrist of al-Nadīm*, ed. and trans. B.Bodge, Columbia University Press, New York & London, 1970, II, 629.

The earliest reference to Fārāb as al-Fārābī's birthplace was made by the fourth/tenth century geographer, Ibn Ḥawqal (d. 367/978). See his *Ṣūrat al-arḍ*, Beirut, p. 418. On the history and significance of Fārāb in Islam, see G.Le Strange, *The Lands of the Eastern Caliphate*, New York, 1966, pp. 484-85; also W.Barthold, "Fārāb," *Encyclopaedia of Islam*, 2nd ed., II, 778.

[9]Al-Fārābī's date of birth cannot be established with certainty. The given date has generally been accepted by modern scholars based on Ibn Khallikān's report (*op. cit.*, p. 310) that al-Fārābī died at Damascus in 339/950 aged upwards of eighty years.

[10]M.Mahdi, *op. cit.*, p. 523.

[11]D.M.Dunlop, *Arab Civilization up to 1500 A.D.*. Beirut, London, 1971, p. 184.

[12]Ibn Abī Usaibi'ah, *'Uyūn*, p. 603 (cf. Maḥfūẓ, p. 98).

[13]R.Walzer, "al-Fārābī," *Encyclopaedia of Islam*, 2nd ed., p. 778.

[14]Ibn Khallikān, *op. cit.*, p. 307.

[15]Ibn Abī Usaibi'ah, *op. cit.*, p. 603.

[16]I believe that al-Amidī's report of al-Fārābī working as a gardener in Damascus while being diligently engaged in reading and writing philosophical books at night by means of the watchman's candle is probably true. Chronologically, this activity must have taken place during his first stay in that city after having left Baghdad, and not in his younger days. Thus, Ibn Abī Usaibi'ah's account, on the particular issue in question, could be reconciled with that of Ibn Khallikān.

[17]S.H.Nasr, *Three Muslim Sages*. p. 14. See also Ibn Abī Usaibi'ah, *op. cit.*, p. 603.

[18]Concerning the Muslim traditional curriculum during al-Fārābī's time, see G.Makdisi, *The Rise of Colleges: Institutions of Learning in Islam and the West*, Edinburgh University Press, 1981, pp. 75-81; S.H.Nasr, *Science and Civilization in Islam*, pp. 65-79.

[19]Ibn Khaldūn, *The Muqaddimah: An Introduction to History*, translated from the Arabic by F.Rosenthal, Bollingen Foundation, New York, 1958, 3, 300.

[20]That such a judgment is indeed false from the point of view of the Islamic revelation understood in its totality has been forcefully stated by Nasr:

> The Islamic revelation possesses within itself several dimensions and has been manifested to mankind on the basic levels of *al-islām, al-īmān* and *al-iḥsān* and from another perspective as *Sharī'ah, Ṭarīqah* and

> *Haqīqah*.... We must avoid the mistake made only too often by many Orientalists during the past century of identifying Islam with only the *Sharī'ah* and *kalām* and then studying the relationship of "philosophy" or metaphysics with that particular dimension of Islam. Rather, in order to understand the real role of "philosophy" in Islam we must consider Islam in all its amplitude and depth, including especially the dimension of *al-Haqīqah*, where precisely one will find the point of intersection between "traditional philosophy" and metaphysics and that aspect of the Islamic perspective into which *sapientia* in all its forms has been integrated throughout Islamic history.

See S.H. Nasr, "The Meaning and Role of *Philosophy* in Islam." *Studia Islamica*, 36 (1973), 58-9.

[21]In the words of Ibn Abī Usaibi'ah, "he (may God's mercy be upon him) was a true philosopher and an eminent authority in the philosophical sciences." See his *'Uyūn*, p. 603.

[22]This has been asserted by a number of scholars. See I. Madkour, *op. cit.*, p. 451; A.A.Wāfī, *op. cit.*, p. 344; and S.H.Nasr, *Science and Civilization in Islam*, p. 47.

[23]Ibn Khaldūn, *The Muqaddimah*, III, 302.

[24]Ibn Abī Usaibi'ah, *'Uyūn*. p. 604.

[25]Al-Fārābi was reported to have claimed to have known up to seventy languages when he first encountered his patron, the Hamdānid prince Saif al-Dawlah, at the latter's Court in Aleppo. See Ibn Khallikān, *op. cit.*, p. 309.

[26]M.Mahdi, *op. cit.*, p. 523; A.A.Wāfī, *op. cit.*, (Mahfūz, p. 344); S.H.Nasr, *op. cit.*, p. 47; T.J.De Boer, *The History of Philosophy in Islam*, London, 1965, p. 108. The question of the possibility of al-Fārābī knowing either Syriac or Greek or both will be specifically discussed later.

[27]A.A.Wāfī (*op. cit.*, p. 344) is of the opinion that the whole of al-Fārābī's pre-Baghdad education and learning was in his native city. I consider this as very unlikely in view of the fact that Islam in Fārāb was only a few decades old when al-Fārābī had reached this stage of his education.

[28]M.Mahdi, *op. cit.*, p. 523.

[29]Ibn Khallikān, *op. cit.*, p. 313.

[30]E.G.Brown, *Literary History of Persia*, Cambridge University Press, Cambridge, 1951, I, 339-76.

[31]Besides the above work of Brown, see Sa'id Naficy, "Persian Literature," *A History of Muslim Philosophy*, ed. M.M.Sharif, II, 1043-57.

[32]M.Mahdi, *op. cit.*, p. 523.

[33]Ibn Abī Usaibi'ah, *'Uyūn*, p. 604.

[34]S.H.Nasr, *IICD*, p. 178.

[35]The term *kalām* is translated by many scholars as dialectical theology. I have adopted this translation because al-Fārābī himself identified *kalām* with the dialectical method.

[36]For traditional accounts of logic used in early *kalām*, see *Kitāb naqd al-nathr*, ed., Tāhā Husain and 'Abd al-Hamīd al-'Abbādī, Cairo, 1938. The edited text, which forms part of a larger work written by a contemporary of al-Fārābī, Ibn Wahb al-Katīb, but earlier wrongly attributed to Qudāmah Ibn Ja'far (d. 337/948), and whose real title was *al-Burhān fī wujūh al-bayān*, was later published in *Majallat al-majma' al-'ilmī al-'arabī*, Damascus (1949), no. 24; see also al-Maqdisi, *Kitāb al-bad' wa'l-ta'rikh*, ed., C.Huart, Paris, 1899, Vol.I, Chap.I; al-Qirqisānī, *Kitāb al-anwār*, trans., G.Vajda, published in *Revue des Etudes Juives* (hereafter cited as *REJ*), no. 122 (1963); and Ibn

'Aqīl, *Kitāb al-jadal*, ed., G.Makdisi, published as "Le Livre de la Dialectique d'Ibn 'Aqīl" in *Bulletin d'Etudes Orientales*, Damascus, no. 20 (1967).

For a traditional account of logic used in early *uṣūl al-fiqh* see A.S.al-Nashshār, *Manāhij al-bahth 'inda mufakkirī al-islām wa naqd al-muslimīn li'l-manṭiq al-aristūṭālisī*, 2nd ed., Cairo (1967). This is a modern study of al-Zarkashī's *al-Bahr al-muhīṭ fī uṣūl al-fiqh* cited by C. Brockelmann in his *Geschichte der Arabischen Litteratur* 2nd ed., Leiden, 1943, II, 112, and Supplement II, 108.

For a detailed modern treatment of the above subjects, see the various articles in G.E. von Grunebaum (ed.), *Logic in Classical Islamic Culture*, Wiesbaden, 1970, especially that of Josef van Ess entitled "The Logical Structure of Islamic Theology," pp. 21-50.

[37]G.Makdisi, *op. cit.*, p. 81.

[38]These include (1) *Kitāb al-ṣaghīr fi'l-manṭiq 'alā ṭarīqat al-mutakallimīn* (*Short Compendium on Logic in the Manner of the 'Mutakallimūn'*). According to Rescher, this is the same work known under the title *Kitāb al-qiyās al-ṣaghīr*, ed., M.Turker, "Farabi'nin bazi mantik eserleri," *Ankara Universitesi Dil ve Tarih-Cografya Fakultesi Dergisi*, vol. 16 (1958), pp. 165-286; trans., N.Rescher, *al-Fārābī's Short Commentary on Aristotle's 'Prior Analytics'*, Pittsburgh, 1963. Hereafter, this translation will be cited as *Short Pr. Anal.* See section 8, pp. 93-111, where al-Fārābī discusses the method of "transfer" and the *ṭard wa 'aks* method ("method of coextensiveness and coexclusiveness") used by the *mutakallimūn* and the *fuqahā'* of his times. Al-Fārābī, however, calls the latter method *ṭarīq al-wujūd wa'l-irtifā'* ("method of finding and raising").

(2) *Kitāb al-radd 'alā ibn al-rāwandī fī ādāb al-jadal* (*Book of Refutation of Ibn al-Rāwandī's Account of Dialectic*). This work was directed against Ibn al-Rāwandī when the latter was still a Mu'tazilite theologian and had not yet expounded his unorthodox views of prophecy and revelation;

(3) *Kitāb al-khaṭābah* (*Middle Commentary of Aristotle's Topics*, trans., G.Vajda, "Autour de la theorie de la connaissance chez Saadia" in *REJ*, Vol. 126 (1967).

[39]See I.Madkour, "La logique d'Aristotle chez les Mutakallimūn," in *Islamic Philosophical Theology*, ed., P.Morewedge, SUNY Press, Albany, 1979, pp. 58-68; Josef van Ess, *op. cit.*, p. 32; also R.M. Frank, "*Kalām* and Philosophy, A Perspective form One Problem." in P. Morewedge *op. cit.*, p. 74.

[40]The reason cited in Ibn Abi Usaibi'ah's *'Uyūn* (cf.Maḥfūẓ, p. 99) that al-Fārābī decides to read philosophy following an encounter with a man who has in his possession a number of Aristotle's books does not invalidate my claim here but rather lends support to it. For his readiness to immerse himself in that study presupposes some measure of interest in philosophy on his part.

[41]In al-Fārābī's own lifetime, this tension between the grammarians and the proponents of the newly established Aristotelian logic (*manṭiq*) was to result in the celebrated debate in Baghdad in 320/932 between his teacher, Mattā ibn Yunus (d. 328/940) and Abū Sa'id al-Sirāfi (280-368/893979). The former defended logic (*manṭiq*) as a universal art which is superior to and independent of grammar, while the latter, a famous philologist and religious scholar in that city, defended grammar as a comprehensive discipline which encompasses the logic of the *manṭiqiyūn*. On this debate, see M. Mahdi, "Language and logic in Classical Islam," in Grunebaum (ed.). *op. cit.*, 51-83; D.S.Margoliouth, "The Discussion between Abū Bishr Matta and Abū Sa'id al-Sīrāfī on the Merits of Logic and Grammar," in *Journal of the Royal Asiatic Society* (hereafter cited as *JRAS*), 1905, pp. 79-129.

[42]I shall deal with al-Fārābī's views on this question in chapter 5.

[43]R.Walzer, "al-Fārābī," *Encyclopaedia of Islam*, 2nd ed., p. 779.

[44]M.Mahdi, "al-Fārābī," *DSB*, p. 523.
[45]M.Fakhry, *A History of Islamic Philosophy*, New York & London, 1983, 2nd ed., p. 108.
[46]F.E.Peters, *Aristotle and the Arabs: The Aristotelian Tradition in Islam*, New York – London, 1968, p. 161.
[47]T.J.De Boer, *op. cit.*, p. 107.
[48]F.W.Zimmermann, *al-Fārābī's Commentary and Short Treatise on Aristotle's De Interpretatione*, Oxford University Press, 1981, p. cvi, n. 1. Hereafter, this work is cited as *Commentary on De Interpretatione*.
[49]N.Rescher, "al-Fārābī on Logical Tradition," in *The Journal of the History of Ideas*, vol. 24 (1963), p. 129.
[50]Ibn Abī Usaibi'ah, *'Uyūn* (cf. Maḥfūẓ, p. 103); for a detailed discussion of al-Fārābī's association with Ibn al-Sarrāj, see M. Mahdi's introduction to his edition of al-Fārābī's *Kitāb al-ḥurūf*, Dar el-Mashreq Publishers, Beirut (1970), pp. 44-7.
[51]al-Qifṭī, *Tārīkh al-ḥukamā'*, p. 277 (cf. Maḥfūẓ, p. 90); Ibn Abī Usaibi'ah, *'Uyūn* (Maḥfūẓ, p. 101).
[52]Ibn Khallikān, *op. cit.*, p. 307.
[53]DeLacy O'Leary, *How Greek Science Passed to the Arabs*, London, 1949, p. 117.
[54]On these first translators, see *ibid.* pp. 155-64.
[55]N.Rescher, *op. cit.*, p. 132. The question of whether there was another Arabic translation of the *Posterior Analytics* before and besides that of Mattā ibn Yūnus has been raised by R.Walzer. Walzer refers to a recent study of the Hebrew and Latin translations of the *Posterior Analytics*, based on Ibn Rushd's three treatments of this Aristotelian work, which shows that Ibn Rushd and also Gerard of Cremona (d. 1187) knew and used another translation besides that of Mattā ibn Yūnus. Its translator is not known. But Walzer has conjectured that he may well be the translator Marāyā mentioned in the well-known Paris MS. of al-Ḥasan Ibn Ṣuwār's edition of the *Categories* of Aristotle, and that al-Fārābī may have read the *Posterior Analytics* with Yuhannā ibn Haylan using Marāyā's Arabic translation. See R. Walzer, *Greek into Arabic: Essays on Islamic Philosophy*, Cambridge, 1962, p. 99.

Walzer himself admits that his above suggestion is purely conjectural. But even if it turns out to be true, that is to say Marāyā's Arabic version of the *Posterior Analytics* in fact existed in al-Fārābī's youth, the question I have posed regarding the possible language of instruction remains, for as pointed out by Mahdi, we do not know whether Yuhannā Ibn Ḥaylān knew Arabic and whether he had access to Marāyā's translation if it indeed existed at the time he taught al-Fārābī. See M. Mahdi, "al-Fārābī and the Foundation of Islamic Philosophy," *Essays on Fārābī*, ed. I.Afshar, Tehran (1976).
[56]"The complete silence of Arabic sources about Ibn Ḥaylān in any connection except as the teacher of al-Fārābī, Ibn Ḥaylān's isolation from the intellectual life of Baghdad where Arabic was the main language of instruction in philosophy, and the report that al-Fārābī arrived at Baghdad knowing Turkish and a number of other languages but not Arabic (that is, he did not know Arabic well enough to study philosophy in that language) all indicate that he must have studied with Ibn Ḥaylān in Syriac or Greek or both. It is unlikely that the language of instruction (which included elaborate commentaries on Aristotle's Organon) could have been in any of the Turkic dialects, in Sogdian, or even in New Persian." M.Mahdi, "al-Fārābī," *DSB*, p. 523.
[57]I.Madkour, *La Sophistique (Logique d'al-Shifā')*, Cairo (1958), Preface, p.v; also his "al-Fārābī," *A History of Muslim Philosophy*, p. 451; and F.W.Zimmermann, *op. cit.*, p.xlvii; and al-Fārābī, *Catalogo de las Ciencias*, edicion y traduccion castellana por A. Gonzales Palencia, Madrid, 1932, p. 27. At several places in a number of his works, al-Fārābī has given an explanation of the origin of the name of the sophistic art (*safsatah*).

See for example, his *Iḥṣā' al-'ulūm*, ed. 'Uthmān Amīn, Cairo, 1949, p. 65; and *Risālat ṣudira bihā al-kitāb*, ed. and trans., D.M. Dunlop "al-Fārābī's Introductory Sections on Logic" in *The Islamic Quarterly*, 3,4 (Jan. 1957), pp. 226 (text) and 231 (trans.). Those who maintain the view that al-Fārābī could not have known Greek have argued that his explanation is etymologically incorrect and that he made the mistake of confusing a *nomen agentis* with a *nomen actionis* for, in the words of Zimmermann, "sophist could never mean sophistry." In my view it is insufficient to conclude that al-Fārābī did not know Greek on the basis of this "etymological error" alone.

[58]Al-Fārābī, *Kitāb al-ḥurūf* (*The Book of Letters: Commentary on Aristotle's Metaphysics*, ed. with introduction and notes by M. Mahdi, Beirut, 1969. For Greek terms used in this work, see p. 252; see also Zimmermann, *op. cit.*, p. xlvii.

[59]For example, Zimmermann admits that al-Fārābī did possess some knowledge of Greek grammar (*op. cit.*, p. cxxxvi) but posits that the knowledge was possibly gained from an Arabic adaptation of the Syriac version of Dionysius Thrax's *Ars Grammatica.* As to whether there was such an Arabic adaptation during al-Fārābī's time, nothing is known. That the Syriac translation was the direct source of his knowledge of Greek grammar is very likely but could not have been the only source, for H. Gatje ("Die Gliederung der sprachlichen Zeichen nach al-Fārābī" in *Der Islam*, vol. 47 (1971), pp. 124) has shown that there are material agreements as well as disagreements between al-Fārābī and Dionysius Thrax.

[60]M.Mahdi, "al-Fārābī" in *DBS*, pp. 523-24.

[61]In 618 A.D. Stephanus of Alexandria, who belonged to the school of Olympiodorus and was the last to occupy the Alexandrian chair of philosophy, was appointed an Imperial Professor at Constantinople by Heraclius (reign: 610-641). This appointment of Stephanus established an important link between the Alexandrian school and that of Constantinople. See K. Gyekye, *Arabic Logic: Ibn al-Tayyib's Commentary on Porphyry's 'Eisagoge'*, SUNY Press, Albany, 1979, p. 14.

[62]Ibn Khallikān, *op. cit.* p. 307

[63]*Ibid*, p. 307

[64]al-Qifṭī, *op. cit.*, p. 278 (Maḥfūẓ, p. 91); Ibn Abī Usaibi'ah, *Uyūn*, (Maḥfūẓ, p. 101)

[65]On these figures associated with the last days of 'Abbāsid political power, see Montgomery Watt, *The Majesty that was Islam*, London, 1976, pp. 156-8.

[66]Ibn Abī Usaibi'ah, *'Uyūn* (Maḥfūẓ, p. 108).

[67]Ibn Khallikān, *op. cit.*, pp. 307-308. According to Hegel, however, it was Aristotle's *Rhetoric* and not his *De anima* that al-Fārābī read two hundred times. F.Rosenthal remarked that Hegel's direct source could not be traced. See F. Rosenthal, "The Technique and Approach of Muslim Scholarship," in *Analecta Orientalia*, 24 (1947), 4.

[68]N. Rescher, "al-Fārābī on Logical Tradition," *The Journal of the History of Ideas*, 24 (1963), 131, n. 17.

[69]D.M.Dunlop, ed. and trans. with introduction and notes, *al-Fārābī: Fuṣūl al-madanī* (*Aphorisms of the Statesman*), Cambridge University Press, 1961, p. 72. Hereafter cited as *Fuṣūl al-madanī*.

[70]According to Ibn Abī Usaibi'ah, the report was that of al-Āmidī. See n. 16 above. There is also the report that al-Fārābī practiced medicine during his stay in Aleppo and Damascus. See S.H. Nasr, *Islamic Science: An Illustrated Study*, World of Islam Festival Publishing Company, London, 1976, p. 177. (Hereafter cited as *Islamic Science*). We do not know whether this medical practice took place during al-Fārābī's first or second stay in Damascus (see below), if the report is indeed true.

[71]Ibn Khallikān, *op. cit.*, p. 309.

[72]M.Mahdi, *op. cit.*, p. 524.

[73]According to Ibn Khallikān, *op. cit.*, p. 308, the political work in question is *Kitāb al-*

siyāsat al-madanīyah whose composition al-Fārābī first began while he was still in Baghdad. In Ibn Abī Usaibi'ah's account, however, the completed work refers to the six *fusūl* (sections) which al-Fārābī had been asked to give to his work previously completed at Damascus, namely *Kitāb al-madīnat al-fāḍilah*, as a summary of it. See the *'Uyūn* (Mahfūz, 107). That al-Fārābī has indeed written a summary of the *al-Madīnat al-fāḍilah* in six chapters is now established beyond any doubt by the editon of the Arabic text by M. Mahdi. See al-Fārābī, *Kitāb al-millat wa nuṣūṣ ukhrā* (*Book of Religion and Related Texts*), ed. with introduction and notes by M. Mahdi, Beirut (1968), pp. 77-86. This disproves the thesis held earlier by D.M. Dunlop that the six *fusūl* is none other than the *al-siyāsat al-madanīyah*. On Dunlop's arguments that led him to this view, see *Fusul al-madanī*, pp. 11-3. It seems to us that Dunlop has always understood both Ibn Khallikān and Ibn Abī Usaibi'ah to refer to one and the same political work that was completed by al-Fārābī in Egypt. But it becomes apparent now, assuming Ibn Khallikān's report to be true, that each biographer had a different work of al-Fārābī in mind. Its implication is that al-Fārābī had two unfinished works, *al-Siyāsat al-madanīyah* and *al-Madīnat al-fāḍilah*, at the time he left Baghdad for Syria.

[74]On these various literary figures associated with the famous Sayf al-Dawlah circle of learned men, see E.G. Brown, *op. cit.*, pp. 370-71. On the circle itself, see Ibn Khallikān, *op. cit.*, II, 334.

[75]Ibn Khallikān, *op. cit.*, III. 309.

[76]S.H.Nasr, *Three Muslim Sages*, p. 136, n. 23.

[77]Most traditional biographers had not mentioned the exact circumstances of al-Fārābī's death. But al-Bayhaqī gave the following account: "al-Fārābī was journeying from Damascus to Ascalon, and was met by a company of the thieves called "the Lads" (*fityān*). Al-Fārābī said to them, "Take what I have of riding animals, arms and clothing, and let me go." But they refused and determined to kill them. Seeing that there was no escape, Abū Nasr (al-Fārābī) dismounted and fought till he was slain with his friends. This greatly displeased the rulers of Syria (i.e. the Hamdānids) who buried Abū Nasr, pursued the thieves, and crucified them on tree-trunks closeby his grave." Al-Bayhaqi, *Tārīkh ḥukamā' al-islām*, pp. 33-4 (c.f. Mahfūz, pp. 69-70), trans. D.M. Dunlop, *op. cit.*, pp. 14-5. I. Madkour has rejected the above report by al-Bayhaqī as incredible (see his "al-Fārābī," *A History of Muslim Philosophy*, p. 452) while Dunlop finds no reason to doubt the account (see his *Arab Civilization to A.D.1500*, p. 185).

[78]I was told by Dr. Nasr that al-Fārābī's tomb was discovered recently in Damascus by a Persian scholar, Muhammad Javad Mashkar, who in his search made use of the descriptions of its location given in the historical documents as his guide. The tomb was found outside the southern wall of the city just as the historical documents have described it.

[79]The number mentioned by Ibn Khallikān (*op. cit.*, III, 310) is four while Ibn Abī Usaibi'ah (Mahfūz, p. 99) puts it as fifteen.

[80]The most up-to-date list of works attributed to al-Fārābī is that of the Turkish scholar, Mujgān Cunbur. See his *Fārābī Bibliografyasi*, Ankara, 1973. This work is highly useful since it provides in a single volume important information about the locations of extant manuscripts of works attributed to al-Fārābī, what manuscripts have been edited and what have been translated into other languages and studied. It also contains a fairly complete list of books and articles on al-Fārābī in various languages up to the time of its writing. Its author consulted all of the major modern bibliographies before it. These include: A.Ateš, "Fārābī'nin Eserlerinin Bibliografyasi," *Turk Tarih Kurumu Belleten*, Ankara, XV: 57 (1951), 175-92; N.Rescher, *al-Fārābī: An Annotated Bibliography*, Pittsburgh University Press, 1962; and K. Georr, *Bibliographie Critique d'al-Fārābī*, Paris, 1964. A more recent

bibliography, organized on more or less the same pattern as that of M. Cunbur, is H.A. Maḥfūẓ and J.A. Yasīn's *Mu'allafāt al-fārābī*, Baghdad, 1975. See also the still indispensable account of al-Fārābī's works by M. Steinschneider, *op. cit.*, pp. 11-135, especially for medieval Hebrew versions.

[81]See n. 99 below.

[82]On the traditional bibliographies, see the works mentioned in n. 4 above. For a discussion of those works traditionally attributed to al-Fārābī but whose authenticity has been questioned by a number of modern scholars, see A Sayili, *Belleten*, XV (1951), 60-4.

[83]Rescher, for example, classifies the writings of al-Fārābī that are in print under seven different headings: logic, rhetoric and poetics, theory of knowledge, metaphysics and general philosophy, physics and natural science, music, and ethics and political philosophy. See his *al-Fārābī: An Annotated Bibliography*, pp. 42-7. See also Walzer's classification in "al-Fārābī", *Encyclopaedia of Islam*, pp. 780-81.

[84]'Uthmān Amīn (ed.), *Iḥṣā' al-'ulūm*, p. 43.

[85]On the place of the *Isagōgē* in the Syriac curriculum of Aristotelian logic inherited by al-Fārābī, see *Short Pr. Anal.*, pp. 21-2.

[86]Ibn Abī Usaibi'ah, *'Uyūn*, p. 609.

[87]A.Ateš, *op. cit.*

[88]"He deserves to be classified as the first specialist in logical studies among the Arabic-speaking peoples, with the possible exception of his teacher, Abū Bishr Mattā Ibn Yūnus, who, however, was rooted in the Syriac milieu and was primarily rather a translator of logical texts than a student of logic." *Short Pr. Anal.*, p. 12.

[89]S.H.Nasr., *Three Muslim Sages*, p. 14.

[90]Says al-Fārābī: "We shall see to it that the canons which we shall lay down here are exactly those which Aristotle contributed to the art of logic. (However), we shall strive to express these matters, as much as possible by means of words familiar to people who use the Arabic language. We shall use for the explanation of these matters examples familar to people of our day. For Aristotle, when he laid down these matters in his books, expressed them by means of works customary among the people of his language, and used examples that were familiar to and current among the people of his day." *Short Pr. Anal.*, p. 49.

[91]Ibn Khallikān, *op. cit.*, III, 308. This passage was quoted by Ibn Khallikān from Ṣa'id Ibn Aḥmad al-Andalusī, *op. cit.*, pp. 70-71 (cf. Maḥfūẓ, p. 58).

[92]Quoted by L.Strauss, "Quelques remarques sur la science politique de Maimonide et de Fārābī," in *REJ*, 100 (1936), 5. I have cited the English translation by F.Najjar, ed. with introduction and notes, *Kitāb al-siyāsat al-madanīyah (al-Fārābī's 'Political Regime')*, Beyrouth, 1964, Preface, p. 9.

[93]This appears to be the better known title of al-Fārābī's treatise on the intellect. (See M. Cunbur, *op. cit.*, p. 3). However, the title mentioned in the bio-bibliographies of al-Qifṭī and Ibn Abī Usaibi'ah, the only primary Muslim sources to have referred to this work, is *Kitāb fi'l-'aql*. The latter two, moreover, state that al-Fārābī wrote two versions of this work, one long, the other short. There is yet another title by which the work was possibly known, as indicated by medieval Latin and Hebrew translations. The Latin title *De Intellectu et Intellecto* is a rendering of the Arabic *al-'aql wa'l-ma'qūl* (*The Intellect and the Intelligible*). In the Hebrew version, we have the plural *ma'qūlāt* instead of its singular form. On the various Latin translations, its merits and demerits, see E. Gilson, "Les Sources Greco-Arabes de L'Augustinisme Avicennisant," in *Archives d'histoire doctrinale et litteraire du moyen age*, vol.IV (1929), Appendice I (Le Text Latin Medieval du *De Intellectu* d'alfarabi), pp. 108-12. Gilson's French translation of his own critical edition of *De Intellectu* appears on pp. 126-41. As for the

various Hebrew versions of the above work of al-Fārābī, see L. Massignon, "Sur le Texte Original Arabe du "De Intellectu" d'al-Fārābī," in the same issue of the above journal, pp. 151-2.

[94]In our own times, al-Fārābī's above treatise has been translated into German, French, Italian, Russian and Turkish. On these translations, see al-Fārābī, *Epistola Sull'Intelletto*, trans. Francesca Lucchetta, Padova, 1974, pp. 14-5; also, M. Cunbur, *op. cit.*, p. 4.

[95]This work has been edited by the Egyptian scholar, 'Abd al-Raḥmān Badawī, and published with other philosophical treatises under the title *Rasā'il falsafiyat li'l-kindī wa'l-fārābi wa ibn bājjah wa ibn 'adi* (*Philosophical Treatises of al-Kindī, al-Fārābī, Ibn Bājjah and Ibn 'Adī*), University of Libya Press, Benghazi, 1973. For a discussion of the aims and nature of this work, see 'A.Badawī, "al-Fārābī, defenseur d'Aristote contre Galien," *Essays on Fārābī*, pp. 25-34.

[96]See n. 38 above, pt. (2). No extant manuscript of this work has been reported.

[97]This treatise has been edited and translated into English by M. Mahdi. See his "al-Fārābī Against Philoponus," in *The Journal of Near Eastern Studies*, 26 (1967), 223-60. It deals mainly with John Philoponus' criticism of Aristotle's views on the eternity of the world and motion.

[98]No manuscript of this work has yet been located.

[99]On al-Fārābī's medical interests, see M. Plessner, "al-Fārābī uber Medizin, eine ubersehene und seine neuentdeckte Quell," *XXI Congress Internazionale di Storia Medicina*, 1970, pp. 1533-9.

[100]On al-Fārābī's commentary upon Books I and V of Euclid's *Elements*, see H. Suter, *Die Mathematiker und Astronomen der Araber*, 1900, p. 55. This commentary was translated into Hebrew probably by Moses Ibn Tibbon. See M. Steinschneider, *op. cit.*, p. 73. See also Euclid, *The Thirteen Books of the Elements*, trans. with introduction and commentary by Sir Thomas L. Heath, vol. 1 (Books I and II), p. 88; and A. Kubesov and B.A. Rosenfeld, "On the Geometrical Treatise of al-Fārābī," in *Archives Internationales d'Histoire des Sciences*, 21 (1969), 50.

[101]On the various musicological works of al-Fārābī, see H.G. Farmer, *Al-Fārābī's Arabic-Latin Writings on Music*, Glasgow, 1934, pp. 4-6; also his *A History of Arabian Music to the Thirteenth Century*, London, 1929, pp. XV, 176, 264. Al-Fārābī's greatest work in music, *Kitāb al-mūsiqā al-kabīr*, was translated into French by Baron R. d'Erlanger, *La Musique Arabe*, Paris, 1930-1935, vols. I and II.

[102]See H.G.Farmer, *al-Fārābī's Arabic-Latin Writings on Music*, p. 4; and O. Wright, "al-Fārābī: Music," in *Encyclopaedia of Islam*, 2nd edn., p. 526.

[103]H.G.Farmer, *op. cit.*, p. 6.

[104]M.Steinschneider, *op. cit.*, p. 81. See also *Hebrew Union College Jubilee Volume*, Cincinnati, 1925, pp. 263-315; and H.G. Farmer, *op. cit.*, p. 6.

[105]For a discussion of al-Fārābī's influence on musical theorists of medieval Europe, see H.G. Farmer, "Clues for the Arabian Influence on European Musical Theory," in *JRAS* (1925), pp. 61-80 ; also his "The Influence of al-Fārābī's *Iḥṣā' al-'ulūm* (*De Scientiis*) on the writers on Music in Western Europe," in *JRAS* (1932), pp. 561-92.

[106]S.H.Nasr, *Three Muslim Sages*, p. 16.

[107]This treatise has seen several editions and has been translated into Turkish, German and Russian. See N. Rescher, *al-Fārābī: An Annotated Bibliography*, p. 46; and Maḥfūẓ and al-Yasīn, *op. cit.*, p. 470. It is also known by the name *Risālah fī faḍīlat al-'ulūm wa'l-ṣinā'āt* (*On the Excellence of the Sciences and the Arts*), derived apparently from the same expression used in *fasl* (section) I of the treatise. In this section, al-Fārābī maintains that the excellence of one science over another is by virtue of at least one of three things: the nobility of its subject matter, the demonstrative nature of its mode of

inquiry, or the great benefits which accrue from it whether these are anticipated ones or they are already present. As for astronomy (*'ilm al-nujūm*), says al-Fārābī, its excellence lies in the nobility of its subject matter. See lines 9-15 of the Arabic text in F. Dieterici, *al-Fārābī's Philosophische Abhandlungen*, Leiden (1896 edn.), p. 105.

[108]Majid Fakhry, *A History of Islamic Philosophy*, p. 114.

[109]This is also known as *Maqālat fī aghrāḍ arisṭūṭālīs fī kul maqālat min kitāb al-mauṣūm bi'l-ḥurūf* (*Treatise on the Aims of Aristotle in Each Chapter of the Book Designated by Letters*), for each of the twelve chapters of Aristotle's *Metaphysics* is known by a Greek letter. See Ibn Abī Usaibi'ah, *Uyūn*, (Maḥfūẓ, p. 110): and F. Dieterici, *op. cit.*, p. 34. According to Mahdi, al-Fārābi's above treatise also bears in some manuscript copies the title of *Risālat al-ḥurūf*, the same title mentioned at the end of the other commentary, *Kitāb al-ḥurūf*, which clearly indicates the intimate link between these two studies of the same work of Aristotle. See M. Mahdi, *Kitāb al-ḥurūf*, p. 36.

[110]This appears to be the same work as that listed in the bibliographies of Ibn Abī Usaibi'ah and al-Sufdī under the title of *Kitāb al-alfāẓ wa'l-ḥurūf*. In earlier sources, such as in several writings of Ibn Rushd and Maimonides, it is simply known as *Kitāb al-ḥurūf*. On the possible reasons for the later addition of the word *al-alfāẓ* to the title, see M Mahdi, *op. cit.*, pp. 34-35.

On the importance of this work, Mahdi writes: "Students of the history of the Arabic language will immediately recognize the importance of this work for a better grasp of the history and meanings of scientific terms in that language. Its date and volume, the position of its author in the development of Arabic and Islamic philosophy, and the paucity of other sources on this subject, suffice to recommend it to the student of the origin and development of the language of science in medieval Islam. It is necessary to point out, however, that the work is equally important for the student of premodern linguistic theory, and theories of the origin and development of religion, science, and philosophy." *Ibid.*, Preface, p. xi.

[111]'Uthman Amīn (ed.), *Iḥsa' al-'ulūm*, p. 99.

[112]See S.Pines, "Ibn Sīnā et l'auteur de la *Risālat al-fuṣūṣ fī'l-ḥikma*," *Revue des Etudes Islamiques* (hereafter cited as *REI*), vol. 19 (1951), pp. 121-6: F.Rahman, *Prophecy in Islam. Philosophy and Orthodoxy*, University of Chicago Press, 1958 & rep. 1979, pp. 21-2; K.Georr, "Fārābī est-il l'auteur de Fucūc-al-hikam?," *REI*, vol. 15 (1941-46), pp. 31-9. In the case of Pines, he argues for its Avicennian origin mostly because in certain manuscripts it has been attributed to Avicenna. Rahman and Georr, on the other hand, argue against the authenticity of the treatise mainly by citing the inconsistency between al-Fārābī's terminological usage and doctrinal expressions in this work, especially pertaining to psychology, and those to be found in his other works.

[113]See S.H.Nasr, *Three Muslim Sages*, p. 136, n. 25. Here, Nasr argues, in reply to Pines, that there are many other works, especially of the nature of the *Fuṣūṣ al-ḥikam*, by various authors falsely attributed during later periods to Ibn Sīnā. And to Georr's argument (n. 110 above) Nasr replies that it is not unusual to find al-Fārābī expressing ideas in one way in one work and another in the next. (cf. my discussion of al-Fārābī's psychology in chap. 2). Insofar as the ideas expounded in the *Fuṣūṣ al-ḥikam* are concerned, they may, without doubt, easily fit into al-Fārābī's realm of thought as it has been understood and studied in the East over the centuries. I would further add that the totality of al-Fārābī's works displays a richness of intellectual perspectives which could not be ignored in any attempt to judge the authenticity of any of his works.

[114]See T.Izutsu, *The Structure of Sabzawarian Metaphysics*, pp. 56-61.

[115]Nasr, *op. cit.*, p. 137, n. 26.

[116]F.Schuon, *Sufism. Veil and Quintessence*, World Wisdom Books, 1981, p. 119.

[117]This work which has been edited by F. Najjar (see n. 90 above) is also known as *Mabādi' al-mawjūdāt* (*Principles of Beings*), the very title cited by Maimonides in his letter to Ibn Tibbon. The medieval Hebrew translation of this work, attributed to Moses (d. 1283), the son of (Samuel) Ibn Tibbon, was published in 1850 by Professor Philoppowski in his *Sepher ha-Asiph*. See S. Munk, *Mélanges de Philosophie Juive et Arabe*, Paris, 1859, pp. 344-5.

Although this work is eclipsed in contemporary scholarship by *al-Madīnat al-fāḍilah* (see F. Najjar, *op. cit.*, p. 10, Preface), it has often been cited by leading Muslim scholars and historians of the medieval period as one of al-Fārābī's greatest works. See Sa'id al-Andalusī, *Tabaqāt al-umam*, p. 72.

[118]See al-Fārābī, *Plato's Laws*, trans. M. Mahdi, in *Medieval Political Philosophy: A Source-book*, ed. R. Lerner and M. Mahdi, The Free Press of Glencoe, Canada, 1963, pp. 83-94; (the latter work is hereafter cited as *Medieval Political Philosophy*). See also L. Strauss, "How Fārābī read Plato's Laws," in *Mélanges Louis Massignon*, Damascus, 1957, vol. III, pp. 319-44. On Ibn Sīnā's views on this question, see for example M. Mahdi, "Avicenna: On the Divisions of the Rational Sciences," in Lerner and Mahdi (eds.), *op. cit.*, p. 97.

[119]This was especially true of the Spanish school of philosophy, among both Muslims and Jews. See D.M. Dunlop, *Fuṣūl al-madanī*, pp. 7, 18-19. On the influence of al-Fārābī's political philosophy upon Maimonides, see L. Strauss, "Quelques Remarques sur la Science Politique de Maimonide et de Farabi," *op. cit.* On the general reception of al-Fārābī's political ideas in the Islamic world, see T.W. Arnold, *The Caliphate*, Oxford, 1924.

[120]See E.I.J. Rosenthal, *Averroes' Commentary on Plato's Republic*, Cambridge, 1956, pp. 208, 283; also D.M. Dunlop, *op. cit.*, p. 19.

[121]The medieval Latin version of *al-Tanbīh 'alā sabīl al-sa'ādah* is known by the name *Liber exercitationis ad vian felicitatis* whose edition by D.H. Salman appeared in 1940. Al-Fārābī's commentary on Plato's *Laws* was translated as *Alfarabius Compendium Legum Platonis* whose text was edited and translated into modern Latin by F. Gabrieli, and published in London in 1952.

[122]See E. Wiedemann, "Zur Alchemie bei den Arabern" in *Journal fur praktische chemie*, N.F., 76 (August 1907), 115-22. The Arabic text was published for the first time by A. Sayili, together with its Turkish translation. See *Belleten*, XV: 57 (1951), 69-79.

[123]See, for example, *Kitāb arā' ahl al-madīnat al-fāḍilah*, ed., A.N.Nadir, Beirut (1968); hereafter cited as *al-madīnat al-fāḍilah*. Chapter 24 of this work, entitled *On the Cause of Dreams*, is wholly devoted to the theory of dreams. In his *The Philosophy of Aristotle*, al-Fārābī makes a distinction between the study of dream-interpretation and the investigation into the nature and causes of dreams. See *al-Fārābī's Philosophy of Plato and Aristotle*, trans. M. Mahdi, Cornell University Press, Ithaca (1969), sec. 87, p. 121. (This work is hereafter cited as *The Philosophy of Plato and Aristotle*). For a modern discussion of al-Fārābī's theory of dreams, see M. Wali-ur-Rahman, "al-Fārābī and His Theory of Dreams," *Islamic Culture*, 10 (1936), 137-51.

[124]On the question of the transmission of al-Fārābī's writings to the medieval Jews and readers of Latin, and the influence of al-Fārābī upon the Scholastics and other European philosophers, see the various works cited by N. Rescher, *al-Fārābī: An Annotated Bibliography*, pp. 50-1. On the Latin translations, see in particular D.H. Salman, "The Medieval Latin Translations of al-Fārābī's works," in *The New Scholasticism*, 13 (1939), 245-61.

[125]For a discussion of these different explanations, see Nasr, "Chirā Fārābīrā mu'alim-i thānī khāndihand?" in *Essays on Fārābī*, First Part, pp. 14-9.

[126]Nasr, *Three Muslim Sages*, p. 134, n. 13.

CHAPTER 2

AL-FĀRĀBĪ'S PSYCHOLOGY IN ITS RELATION TO THE HIERARCHY OF THE SCIENCES

2.1 The Idea of the Unity and Hierarchy of the Sciences

No proper, in-depth philosophical study of al-Fārābī's classification of the sciences, indeed of any worthy classification, is possible without a prior investigation into certain aspects of epistemology. This investigation must provide the philosophical basis of classification. It is only in the light of such an epistemological paradigm that we can unveil the true significance of why, for example, al-Fārābī chose the scheme of classifying the sciences that he did, why a number of sciences in his day, such as alchemy and the interpretation of dreams, were excluded from his enumeration (although he wrote treatises on them), and why the religious sciences of jurisprudence (*'ilm al-fiqh*) and *kalām* (dialectical theology) do not seem to occupy a central position in his classification. N. Rescher, in fact, treated the *Iḥṣā' al-'ulūm*, the treatise in which the above classification is given, as one of al-Fārābī's important works on epistemology.[1]

The most fundamental idea related to traditional epistemology to which al-Fārābī fully subscribed is that of the unity and the hierarchy of the sciences. The profound relationship between this idea and traditional epistemology may be expressed by saying that in one sense the idea is the fruit of traditional inquiry into epistemology, while in another sense it is a basis of that inquiry. The former is true, because this idea results from the application of the doctrine of *Tawḥīd* (Unity of the divine principle)[2] to the whole domain of human intelligence and its activities of thinking and knowing. In inquiring into the problem of how a person knows – that is, the methodology of knowledge (*al-'ilm*) in its most comprehensive sense – one cannot but be confronted by the hierarchic nature and reality of the subjective and objective poles of knowledge. We are, in other words, confronted with the hierarchy of the faculties and powers of knowing within the human knowing subject and the world of beings that are knowable and known.

This hierarchy in both the microcosmic and macrocosmic orders of reality represents many manifestations of the divine principle. The idea of the hierarchy of the sciences, al-Fārābī would say, is rooted in the nature of things. The sciences constitute a unity because, as will be explained in chapter three, their ultimate source is one, namely the divine intellect. This is true regardless of the intermediary agencies through which people may have acquired these sciences.

The idea of the unity and hierarchy of the sciences may also be regarded as a basis for traditional epistemology. This is true in a human society closely bound to revelation, like the society in which al-Fārābī lived and thought.[3] There, the idea of a hierarchy of reality is very much alive. Thanks to the teaching of revelation, that idea is accepted as an axiomatic philosophical truth. It is evident that the idea of hierarchy is rooted in Islamic revelation from the following teachings of the Qur'ān and *ḥadīths*. First, the Qur'anic verses themselves are of various grades in respect of value although all of them are believed to be of divine origin. This is because they deal with different levels of reality.

The celebrated Verse of the Throne (*āyat al-kursī*)[4] was described by the Prophet as the chief (*sayyidah*) of the Qur'anic verses.[5] As explained by al-Ghazzālī, the reason for this is that the verse is exclusively "concerned with the divine essence, attributes and works" and that "it contains nothing other than these."[6] Moreover, according to another prophetic *ḥadīth*, the greatest divine name (*al-ism al-a'ẓam*) lies in the Verse of The throne.[7] The Prophet also said that the Chapter of Sincerity or Purity (*Sūrat al-ikhlāṣ*), which is made up of four short verses, equals one third of the Qur'ān. The high position occupied by this chapter is due to the fact that it concerns the knowledge of the *Ḥaqīqah* or the divine reality, which is the most excellent of the three fundamental forms or levels of knowledge contained in the Qur'ān. The other two divisions of the Qur'anic verses deal respectively with the *Ṭarīqah* and the *Sharī'ah*, both of which reflect the *Ḥaqīqah* at their own levels. The *Ṭarīqah*, the esoteric spiritual path to God, is the qualitative and vertical extension of the *Sharī'ah,* the divine law which is the general path to God.

The scriptural evidence cited above indicates that the hierarchic structure of the Qur'ān reflects the structure of objective reality. There are, however, numerous verses which refer directly to the hierarchy of creation. These speak of the tripartite division of the universe into the heavens, the earth, and the intermediate world.[8] There is also the lower heaven[9] and the higher angelic world which is nearer to God.[10] Angels, according to both the Qur'ān and *ḥadīths,* have been created by God with different ranks.[11]. Revelation also teaches that both Paradise and Hell are characterized by degrees.[12]

One finds numerous references in the Qur'ān and *ḥadīths* to the idea of degrees of intellectual and spiritual realization or the subjective experience of reality. We have, for example, a hierarchy of believers and knowers, as testified by the following verse: "God raises in degrees those of you who believe and those to whom knowledge is given."[13] According to Ibn 'Abbās (d.68/687-688), a companion of the Prophet, the learned rank seven hundred grades above ordinary believers.[14] There is, further, the hierarchy of witnesses of divine unity. Says the Qur'ān: "God bears witness that there is no god but Him, and so do His angels and those endowed with knowledge standing firm on justice."[15] We may also mention here the three main categories or divisions of mankind after the Day of Judgment. In Qur'anic terminology, these are (1) those nearest to God (*al-muqarrabūn*), (2) the Companions of the Right Hand (*aṣḥāb al-maimanah*), that is the righteous generally, and (3) the Companions of the Left Hand (*aṣḥāb al-mashamah*), those who will be placed in the abode of misery because they have rejected God and His Message or have led a wicked and sinful life.[16]

The above references to the Qur'ān and the prophetic traditions, although by no means exhaustive, are sufficient in our view for the purpose we have in mind, namely to demonstrate that the idea of hierarchy is rooted in the Islamic revelation. Since the hierarchic structure of reality extends to all domains of cosmic manifestation, including the realm of human intelligence and cognition, its corollaries in these domains (one of which is the hierarchy of the sciences) are operative in most people's minds when they deal with the place of things in the cosmic order. Thus, in the second sense referred to earlier and in the way of looking at the genesis of philosophical and scientific concepts, al-Fārābī's inquiry into psychology represents no more than an attempt to give a rational dress to the metaphysical idea of the hierarchy of beings and of knowledge. Al-Fārābī certainly did not discover the above metaphysical idea through the process of rational inquiry. Rather, this idea acts as one of the principal guides to that inquiry, while at the same time being confirmed by it.

Although the idea of hierarchy was generally accepted by medieval Muslim thinkers, it was by no means conceptualized and understood in only one way. We may speak of the distinctly Farabian exposition of this idea. Accordingly, in the next three chapters, I will deal with what I consider to be al-Fārābī's treatment and understanding of the specific idea of the unity and the hierarchy of the sciences.

2.2 The Bases of the Hierarchy of the Sciences

To speak of the hierarchy of the sciences is to speak of the reasons why one science or another is accorded a nobler rank or precedence over others. According to al-Fārābī, the true basis of such an ordering consists of one or more of three elements. In his *Risālat fī faḍīlat al-'ulūm wa'l-ṣinā'āt* (*Treatise on the Excellence of the Sciences and the Arts*),[17] he writes:

> The excellence of the sciences and the arts is only by virtue of one of three things: the nobility of the subject matter, the profundity of the proofs, or the immensity of the benefits in that science or art, whether these benefits are anticipated or are already present. As for the (science or art) which excels others because of the immensity of its benefits, it is like the religious sciences (*al-'ulūm al- shar'īyah*) and the crafts needed in every age and by every nation. As for that which excels others because of the profundity of its proofs, it is like geometry (*al- handasah*). As regards that which excels others because of the nobility of its subject matter, it is like astronomy (*'ilm al-nujūm*). However, all these three things or any two of them may well be combined in a single science such as metaphysics (*al-'ilm al-ilāhī*).[18]

In the above passage, al-Fārābī cites three criteria by means of which the hierarchy of the sciences might be established. The first, the nobility of the subject matter (*sharaf al-mawḍū'*), is derived from the fundamental principle in ontology that the world of beings is hierarchically ordered. We may, therefore, speak of the first criterion as constituting the ontological basis of the hierarchy of the sciences. According to al-Fārābī, astronomy fulfills the criterion of having a noble subject matter because it deals with the most perfect of bodies, namely the celestial bodies.[19] The second criterion, the profundity of the proofs (*istiqṣā' al-barāhin*), is based upon the view that systematization of truth claims in the different sciences is characterized by different degrees of clarity and certainty. According to this view, the methods of discovering truth claims and of proving them are more perfect and vigorous in some sciences than in others. On the basis of the second criterion, al-Fārābī considered geometry to be superior to many other sciences. This was, in fact, a prevalent view in his time. The rigor of geometrical proofs was generally admired as perfect. Insofar as the idea of profundity of proofs pertains directly to methodological issues, the second criterion may be regarded as constituting the methodological basis of the hierarchy of the sciences.

As for the third and last criterion, namely the magnitude of the benefits (*'iẓam al-jadwā*) which can be derived from the science in question, it is based upon the fact that both practical and spiritual needs which concern the volitional aspects of the soul are also hierarchically ordered. For al-Fārābī, as well be shown in chapter four, the question of practical human needs and of the benefits or usefulness of objects (including the sciences and arts) and human acts belongs to the ethico-legal domain. It is significant that he mentions the sciences of the *Sharī'ah* as an example of knowledge which is deemed excellent on account of its usefulness. This is because, in Islam, the idea of hierarchy of human needs and of values of human acts in all spheres of life is based upon the ethico-legal teaching of the *Sharī'ah*.[20] In his political philosophy, of which ethics is a part, al-Fārābī defines practical human needs in terms of ethical categories of the useful and the good, which agree with the teaching of the *Sharī'ah*.[21] Since the third criterion pertains directly to ethico-legal issues, it may be described as the ethico-legal basis of the hierarchy of the sciences. Although al-Fārābī dealt with the three above bases for hierarchically ordering the sciences, it is apparent that he is primarily concerned with the methodological basis. That this is so will be made clear in my discussion of his classification of the sciences. One should not lose sight of al-Fārābī's central concern with the theme of the degrees of profundity of proofs. It bears an essential relationship to his scheme of classification of the sciences. However I will refer to his views on all three bases.

A proper treatment of al-Fārābī's theory of methodology entails a discussion of his conception of logic. The latter, in turn, presupposes an understanding of the fundamental elements of his psychology. Al-Fārābī's psychology deals extensively with the theory of the intellect (*'aql*). Included in the theory is the idea of the prophetic intellect which he defines as the vehicle of divine revelation (*waḥy*). This idea of the prophetic intellect is of particular importance to our understanding of his conception of the relationship between revelation, intellect and reason. I will present in this chapter a detailed discussion of al-Fārābī's psychology – that is, his theory of the faculties of the soul. His theory of the relationship between revelation, intellect and reason will be treated in the next chapter. This will be followed in the same chapter by a discussion of his conception of methodology.

2.3 **The Hierarchy of Faculties of the Human Soul**

Al-Fārābī discusses psychology mainly in four of his well known works, the *Risālat fi'l-'aql, al-Madīnat al-fāḍilah, al-Siyāsat al-madanīyah*, and *Fuṣūṣ al-ḥikam*. In the discussions in the four works taken together he explains the hierarchical process of the development of the various faculties of the soul, the nature and functions of each, and the final purpose to which the activities of all of these are directed. In the *Risālat fi'l-'aql*, which is wholly devoted to a discussion of the intellect (*'aql*), al-Fārābī explains the different senses in which the word intellect has been used.[22] He also gives an exposition of the various stages which the development of the human intellect may undergo in the process of actualizing the possibilities latent within it. As has been shown by a number of scholars, the elements of al-Fārābī's psychology (especially the doctrine of the intellect) were drawn from various sources – Aristotelian and Neoplatonic as well as Islamic.[23]

Following Aristotle, al-Fārābī describes the human creature as a rational animal (*al-ḥayawān al-nātiq*) who is superior to all other creatures. Humanity enjoys domination over other species by virtue of having an intelligence (*nuṭq*)[24] and a will (*irādah*), both of which are functions of the rational faculty.[25] According to al-Fārābī, the faculties of the human soul (*al-nafs al-insānīyah*) are five in number. He describes the order of their generation in man as follows:

> When man comes into being, the first faculty to appear is that by which he is nourished, namely the vegetative faculty (*al-quwwat al-ghādhiyah*). After that there develops the faculty with which he perceives the tangible objects such as heat, cold and the rest. This is also the faculty with which he tastes and smells, hears sounds, and sees colors and all other objects of vision such as light rays. Along with the senses is developed (the faculty) with which he yearns for the sensibles; he either likes or dislikes them. Then, after that, appears another faculty with which he retains the impressions of the sensibles upon his soul after the sensible objects have disappeared from his senses. This is the imaginative faculty (*al-quwwat al-mutakhayyilah*). This faculty combines some (of the impressions of the) sensibles with others as well as separates some from others, producing different combinations and separations. Some of these are false and some are true. Associated with this faculty is the power of desire toward the objects of imagination. After that there appears in him the rational faculty (*al-quwwat al-nāṭiqah*) with which he is able to perceive the intelligibles in order to distinguish between the noble and the based and to gain possession of the arts and the sciences. There is also associated with this faculty the desire toward that which has been perceived by the intellect.[26]

The order of development of the faculties of the human soul is the vegetative, the sensitive (*al-quwwat al-ḥāssah*), the appetitive (*al-quwwat al-nuzūʿīyah*), the imaginative, and the rational. Together they constitute the hierarchical order of pre-eminence, since each faculty exists for the sake of the one above it. The highest member of this hierarchy is the rational, for it rules or orders all the others.

The human, says al-Fārābī, gains knowledge of a thing either through the rational faculty, the imaginative faculty, or sensation.[27] In maintaining this tripartite division of the cognitive faculties, he echoed the ancient doctrine that "only the like can know the like." Corresponding to the tripartite structure of body (*corpus*), soul (*anima, psyche*) and spirit (*spiritus*), with which the sensitive, the imaginative and the rational are respectively identified, is the tripartite structure of the corporeal, the psychic, and the spiritual worlds of the cosmos. It is significant that al-Fārābī employed the same set of terms (i.e., *jism*, *nafs*, and *'aql*) for the triadic nature of both the knower and the known. That terminological usage reflects his belief that microcosm and macrocosm correspond to one another. It also reflects his view that the reality of the subject and object of knowledge constitutes an organic whole.

2.3.1 **The Sensitive Faculty**

The sensitive faculty is the lowest of the cognitive faculties since it exists for the sake of both the imaginative and the rational. Al-Fārābī – like many other Islamic philosophers – believed that in the process of development of the human individual the sensitive precedes the imaginative. That is to say, the cognitive power of the human soul is first developed through the external senses. The imaginative is, however, viewed as possessing a higher ontological status than the sensitive. Al-Fārābī expressed the relationship between the two faculties by saying that the imaginative is a form (*ṣūrah*) for the ruling element of the sensitive and that the latter is matter (*māddah*) for the imaginative.[28] The ruling element of the sensitive faculty refers to the common sense (*al-ḥāssat al-mushtarakah*). Al-Fārābī defines it as the power or faculty (*quwwat*) which receives all the impressions of the five external senses. It is necessary to clarify what he means by the imaginative faculty as a form for the common sense and the latter as matter for the former. An important idea in al-Fārābī's psychology is that each lower faculty serves as matter for the higher. It is upon this idea that he constructs his system of the hierarchy of cognitive faculties of the soul. According to him, the common sense is matter for the imaginative in the sense that the existence of the former is the preparatory condition for the coming into being of the latter. The common sense is the instrument through which the imaginative attains perfection.[29]

Viewed as form, the imaginative is the 'reality' for the sake of which the common sense exists. In Aristotelian terms the imaginative is the final cause of the common sense. As will be seen later, al-Fārābī extends or enlarges the above idea of form and matter in an 'upward' direction right to the realm of the universal, supra-individual Intellect. It seems to me that his motive for doing so – which is of

Neoplatonic inspiration – is to establish the pure forms of universals contained in the Intellect as the real content of all knowledge.

Al-Fārābī's above conception of form permits him to speak of the imaginative as form for the sensitive faculty as well. The implication is that the imaginative necessarily imposes a limit to the capacity of the sensitive to know things. The sensitive faculty can only know the external world that is of its own nature, namely, the material world which analogously is matter for the imaginal world (*'ālam al-khayāl*, the Latin *mundus imaginalis*). The last named is the object of the imaginative faculty. Herein lies the significance of the sensitive faculty for man's intellectual realization, despite the fact that as a mode of knowing it is limited by nature. In al- Fārābī's view, the indispensability of sensation for acquiring the arts and the sciences stems from the following consideration: the forms of sensible things perceived first by the external senses and then by the imaginative faculty constitute potential intelligibles. These will become actual intelligibles when man's potential intellect (*'aql bi'l-quwwah*) becomes actual as a result of being illuminated by the active intellect (*al-'aql al-fa''āl*).[30]

According to al-Fārābī, perception, whether by the sensitive or the imaginative faculty, involves some kind of abstraction of the form of the perceived object. By abstraction he means the detachment of form from matter or its material accidents. There are degrees of abstraction. Sense neither abstracts form completely from matter, nor from the accidents of matter. Sensation cannot retain the forms of sensible objects after the absence of their matter. It needs the presence of matter for the presence of form. A higher degree of abstraction of form is effected through the imaginative faculty. This is explained below.

2.3.2 The Imaginative Faculty

In his theory of the imaginative faculty al-Fārābī deals with what he calls the five internal senses (*al-ḥawāss al-bāṭinah*).[31] These are:

(1) the faculty of representation (*al-quwwat al-muṣawwirah*)
(2) the faculty of estimation (*al-quwwat al-wahm*)
(3) the faculty of memory (*al-quwwat al-ḥāfiẓah*)
(4) the faculty of compositive human imagination

(5) the faculty of compositive animal imagination (*al-quwwat al-mutakhayyilah*)

The faculty of common sense, which al-Fārābī describes as the ruling element of the external senses and as the recipient of sensed forms, is excluded from both the external and internal senses. It seems that he makes of it a neutral sense occupying an intermediate position between the two.[32] It is not the function of common sense to preserve the forms it receives. That function belongs to the faculty of representation which is situated in the fore-brain. This faculty retains those forms even after the sensed objects have disappeared. Since this faculty does not need the presence of matter for the presence of form, its power of abstraction is said to be more perfect than that of sensation. However, the forms in the representative faculty are not divested of their material accidents. These forms are perceived with all their material attachments and relationships such as position, time, quality and quantity. It is impossible for a form in the representative faculty to be such as to admit all the individuals of the species to share in it.

The faculty of representation only conserves forms of sensible objects which are perceived by the external senses. There are, however, non-sensible forms connected with the individual sensible objects that cannot be perceived by the external senses. Al-Fārābī attributes the function of perceiving such forms to the faculty of *wahm* (estimation).[33] In illustrating this faculty, he gave the following example: when a sheep sees a wolf it perceives not only the latter's sensible form but also its enmity toward it. The wolf's enmity, which is non-sensible, is perceived directly by the sheep's faculty of *wahm*.

Al-Fārābī gave no other example to illustrate the function of *wahm*. Moreover, in the above example, he offered no further clarification concerning the perceptive power of *wahm*. He does not say whether the sheep's sense of fear at the sight of the wolf is as a result of a previous experience or is an instinctive interpretation by its soul of the latter's image. According to Ibn Sīnā, the operations of *wahm* need not be purely instinctive but may be based on previous experience.[34]

Although al-Fārābī was vague about how *wahm* operates, he was clear enough in spelling out its basic function, namely, the association of non-sensible entities like good and evil with the individual sensible objects. The operation of *wahm* represents another stage of abstraction. Since it abstracts non-material entities from matter, its abstraction is said to be more perfect than that performed by the representative faculty. But insofar as the non-material entities are perceived in their particularity there is no difference between the two abstractions.

Non-material entities perceived by *wahm* are retained in a different faculty, namely the faculty of memory (*al-quwwat al-ḥāfiẓah*). The relation of the faculty of memory to *wahm* is the same as that of the representative faculty to the common sense. Likewise, its relation to the non-material entities perceived by *wahm* is the same as that of the representative faculty to sensed forms.

Another internal faculty is creative in its nature. It possesses a compositive function. It produces new composite images out of the images stored in the representative faculty through what is called the process of combination (*khalṭ* or *tarkīb*) and separation (*tafṣīl*). That is to say, it combines certain images with others and separates some images from others as it chooses. The animal with this faculty performs the above function both in its waking and sleeping states. Some of the newly produced images are objectively true and some false, according as there is, or there is not, a thing in the external world corresponding to them. With reference to humans al-Fārābī calls this compositive faculty *al-mufakkirah* (rational imagination) and with reference to animals *al-mutakhayyilah* (sensitive imagination).

Al-Fārābī explains the distinction between the two kinds of compositive imagination as follows: in compositive human imagination the rational faculty makes use of the created images, with the help of *wahm;* in animal imagination the compositive power is utilized by the faculty of *wahm*. I will deal a little further with the question of how the rational faculty employs the images when discussing next that faculty. Al-Fārābī's explanation of the nature of compositive animal imagination suggests that the supreme internal faculty in the animal is *wahm*. This may be inferred from his description of *wahm* as being served by the rest of the animal faculties. The faculty of memory serves it by conserving its objects of perception. The representative and compositive faculties serve it by letting the images they produce be used by it. Al-Fārābī did not explain the manner in which *wahm* is using the images. On the basis of his own definition of the function of *wahm*, I think what he probably means is that this faculty operates on the images by associating good or evil, pleasure or pain with them.

Viewed as a whole, al-Fārābī's treatment of the imaginative faculty lends itself to the following criticism. First, his terminological usage lacks precision. He is found to use the same term to denote several entities although only one such case is known. *Al-mutakhayyilah*, a key term in traditional Islamic psychology, is used by him in three different senses: (1) in the general sense of a generic noun representing all the internal faculties, which are intermediary between the sensitive and the rational faculties;[35] (2) referring to a specific function of imagination or internal faculty, namely, the faculty of compositive animal imagination, as is the case in the above classification of the internal senses;[36] and (3) referring to the combination of compositive animal imagination and the representative faculty.[37] There is also a case, the only known one, in which al-Fārābī employs different terms to denote the same faculty. He applies two terms, *al-ḥāfiẓah* and *al-dhākirah*, to the faculty of memory which conserves non-material entities perceived by *wahm*.

A second criticism of al-Fārābī's treatment of the imaginative faculty is the lack of concrete examples to illustrate the functions of the different internal faculties he had enumerated. He also left unexplained the manner of operations of some of the faculties. These terminological and conceptual shortcomings notwithstanding, al-Fārābī had clearly defined the basic functions of the imaginative faculty, namely, retention, composition and estimation of images. He made clear that these functions exist for the sake of the rational faculty.

2.3.3 The Rational Faculty

In the *Fuṣūl al-madanī*, al-Fārābī summarizes the constitution and the range of functions of the rational faculty in the following terms:

> The rational faculty is that by which a man understands. By it comes deliberation (*rawīyah*), by it he acquires the sciences (*al-'ulūm*) and arts (*ṣinā'āt*), and by it he distinguishes between the

> fair and the ugly in actions. It is partly practical (*'amalī*) and partly theoretical (*naẓarī*). The practical is partly a matter of skill (*mihnīyah*) and partly reflective (*fikrīyah*). The theoretical is that by which man knows the existents which are not such that we can make them or alter them from one condition to another, e.g. three is an odd and four an even number. For we cannot alter three so that it becomes even, while still remaining three, nor four so that it becomes odd, while still four, as we can alter a piece of wood so that it becomes round after being square, remaining wood in both cases.
>
> The practical is that by which are distinguished the things which are such that we can make them or alter them from one condition to another. What is a matter of skill and art is that by which the skills are acquired, e.g. carpentry, agriculture, medicine, navigation. The reflective is that by which we deliberate on the things which we wish to do, when we wish to know whether to do it is possible or not, and if it is possible, how we must perform the action.[38]

From the above passage we may derive a number of important principles of al-Fārābī's classifications of the sciences. The first of these is that the rational faculty is partly theoretical and partly practical. This principle is the basis for his fundamental division of the sciences into the theoretical and the practical. Furthermore it is the constitution of the rational faculty described above which gives rise to the four elements that constitute human perfection or happiness, namely theoretical virtues, deliberative virtues, moral virtues, and practical arts. These four elements of human happiness in turn serve as a basis for his scheme of classification in the *Iḥṣā' al-'ulūm*. These two principles will be discussed in chapter four. What I wish to consider here are the various degrees of the acquisition of knowledge by the rational faculty.

According to al-Fārābī, the function of the theoretical rational faculty is to receive the forms of intellectual objects. He calls intellectual objects intelligibles (*ma'qūlāt*). The forms of intelligibles are universals. These are immaterial forms which are completely free from matter and material attachments. There are two kinds of intelligibles that are imprinted on the rational soul.[39] Intelligibles of the first kind are forms abstracted from their matters in some manner which I shall presently explain. Before these universal forms were abstracted from their matters they were potential intelligibles. When they were abstracted, they became actual intelligibles. The second kind of intelligibles consist of forms that are always in actuality, that is, forms which are not and never were in matter. These latter intelligibles refer to the First Cause and the whole hierarchy of separate intelligences (*al-'uqūl al-mufāriqah*) situated below it. Placed in the lowest rank in this hierarchy is the active intellect whose function in the acquisition of knowledge by the rational faculty is discussed below. According to al-Fārābī, all human beings share a "certain natural disposition" which he calls the potential intellect (*'aql bi'l-quwwah*).[40] This intellect possesses the capacity to receive intelligible forms or universals. The actualization of this capacity occurs when the potential intellect is illuminated by the active intellect. The potential intellect becomes an actual intellect (*'aql bi'l-fi'l*) in relation to the intelligible forms it has received. It is still a potential intellect in relation to other intelligible forms. The first forms to be imprinted on the rational soul are the intelligibles which the potential intellect abstracts from the matters. What makes possible this simultaneous transformation of the potential intellect and the potential intelligibles into their states of actuality is their illumination by the active intellect.

Al-Fārābī describes the relation of the active intellect to the potential intellect by the analogy of the relation of the sun to the eye in darkness. The eye is only potential sight as long as it is in darkness.[41] It is the sun, insofar as it gives the eye illumination, which makes the eye actual sight and visible things actually visible. Further, the sunlight enables the eye to see not only objects but also the light itself as well as the sun which is the source of that light. In a similar manner, "light" from the active intellect makes the potential intellect an actual intellect and the potential intelligibles actual intelligibles. The potential intellect is then able to perceive that "light" as well as the active intellect.

Al-Fārābī identified the active intellect with the holy spirit (*Rūḥ al-Quds*) or Gabriel, the archangel of divine revelation.[42] He also called it "a separate form of man" or true man (*al-insān 'ala'l-ḥaqīqah*).[43] The active intellect is a perfect repository of intelligible forms. As such, it serves as a model of intellectual perfection. Man attains the highest level of being possible for him when he realizes within himself the being of the true man. That is to say, when man's intellect comes to resemble the active intellect. Al-Fārābī's conception of the active intellect exemplifies how he sought to harmonize Greek philosophical theories with the religious beliefs of Islam.

With the abstraction of universals by the potential intellect, the process of abstraction of forms from matter reaches its most perfect stage. This abstraction presupposes not only a contact between the potential and active intellects but also the presence of images of sensible things in the imaginative faculty. In saying that these imaginative forms emerge into the rational faculty as intelligibles, al-Fārābī did not mean that the source of the latter forms is the imaginative faculty. According to him, the source of intelligible forms is the active intellect. The forms which exist in the material world are bestowed by the active intellect.[44] In this intellect the forms exist as universals and are absolutely separate and simple. When these forms descend into the imaginal and sensible worlds they enter into a plurality. But with the help of the active intellect these particularized forms are once again raised in the intellect of man to the domain of the universal.

There are levels of actualization of all the possibilities latent within the human intellect. The first possibility to be realized by the human intellect is the possession of what al-Fārābī calls primary intelligibles (*ma'qūlāt ūlā*).[45] These intelligibles constitute one of the four classes of indemonstrable premises usually mentioned by Muslim commentators on Aristotle's *Organon*.[46] Indemonstrable premises are those syllogistic premises which are grasped without a syllogism or recourse to a middle term. From them are derived all other syllogistic premises. Of the four kinds of indemonstrable premises, the primary intelligibles are the most worthy of being adopted as the premises of demonstrative syllogisms employed in the philosophical sciences. Al-Fārābī defines demonstration as a syllogism composed of premises which are true, primary, and necessary. I will discuss further his concept of demonstrative syllogism in the next chapter.

Al-Fārābī divides primary intelligibles into two kinds according to the manner in which they arise in the intellect. Of the first kind are intelligibles which occur to an individual without any inquiry or prior desire to know them. The individual is not aware of how and when these intelligibles come to exist in his intellect. These intelligibles appear to be natural (*fiṭrīyah*) to the intellect in the sense that they are common to all who are equipped by natural disposition to receive them. Al-Fārābī calls them "primary principles" (*al-mabādi'al-ūlā*).[47] As examples, he mentions the following: every three is an odd and every four an even number; every part of a thing is smaller than the thing and every whole is greater than its part; two quantities which are equal to a third are equal to one another.[48]

Al-Fārābī calls primary intelligibles of the second kind "axioms of certainty" (*awā'il al-yaqīn*). These intelligibles are deliberately sought by men. They arise in the intellect as a result of inquiry and experience but without recourse to reasoning. The distinction between the two kinds of primary intelligibles may be best understood by referring to al-Fārābī's notions of conception (*taṣawwur*) and assent (*taṣdīq*). In al-Fārābī, conception is variously described as understanding the meaning of a thing, what its name signifies, or what it is.[49] Conception of a thing is free from judgment about the existence of that thing.

There are various levels of conception. The lowest level is the concept of what a name signifies. It is a necessary condition of the higher forms of conception. The highest level of conception is perfect definition (*al-ḥadd al-tamm*) which signifies the essence of the thing defined. As for assent, al-Fārābī defines it as "the belief that a thing about which a judgment has been made exists outside the mind as it is believed to be by the mind."[50] Equivalently, assent is the belief that a judgment made about a thing is true since al-Fārābī understands what is true as a thing's existence outside the mind corresponding to what is believed by the mind. Assent may then be true or false, depending on whether the thing toward which it is directed exists or does not exist. Assent applies to both simple and predicate existence of a thing.

There are also various levels of assent. Since assent is in the nature of belief, it admits of degrees of certainty. Al-Fārābī makes clear that certainty involves only assent to what is true. The notion of certainty is not applicable to an assent to what is false. Each level of assent corresponds to a particular degree of certainty. In one of his works, al-Fārābī classifies assent into certainty, approximate certainty (*muqārib li'l- yaqīn*), and trust (*sukūn al-nafs*).[51] He describes certainty as being composed of three elements: (1) the belief that something is or is not in a specific condition, (2) the belief that that thing cannot be other than it is, and (3) the belief that belief (2) cannot be otherwise.[52] Approximate certainty is composed of the first two beliefs only while trust is merely the first belief.

The class "certainty" is in turn divided into "necessary certainty" and "certainty at times". The former refers to necessary beliefs in what is true and necessary. Primary intelligibles are examples of such beliefs. The latter sub-class refers to beliefs in what is true but not necessary. By non-necessary object of belief al-Fārābī means an object which may undergo changes in its future state of existence. Al-Fārābī's classification of assents is important in understanding his theory of methodology. I will deal with it in greater detail in the next chapter. My present discussion of his notions of conception and assent merely aims at clarifying the distinction between the two kinds of primary intelligibles previously mentioned.

Al-Fārābī maintains that all knowledge is either conception or assent. In general, every assent must be preceded in time by a sufficient amount of conception or by other forms of assent unless it itself is the lowest form of assent.[53] The distinction between primary principles and axioms of certainty lies in the fact that the relations between assent and conception in the two cases are different in nature. The knowledge of primary principles is not sought and is prior in time to all other knowledge. In other words, the assent to a primary principle is not preceded in time either by conception or another form of assent. The knowledge of a primary principle is certainly dependent on the knowledge of the definitions of the terms which constitute the subject and the predicate of the proposition involved. Al-Fārābī explains, however, that in this case the knowledge of the definitions is immediate and is simultaneous with the assent to the proposition containing the terms. For example, when the meanings of the terms "whole" and "part" are understood completely, then the proposition "the whole is greater than its part" is known immediately. Further, the definitions of "whole" and "part" are grasped by most people in an immediate manner from the time their intellects begin to perceive.

In contrast, the knowledge of axioms of certainty is sought. The assent to them is only given after inquiry and experience. By "experience" al-Fārābī means the process of arriving at the perfect definitions of the terms contained in the axioms. In his commentary on Aristotle's *Topics* he says that this process should be called "scientific induction" if at all the term "induction" is to be used.[54] "Scientific induction" resembles ordinary induction in that it produces a judgment about a universal, derived from judgments about particular cases of that subject. However, it differs from the latter in the following respect: "scientific induction" seeks to establish a universal proposition for its own sake. Ordinary induction aims at verifying universal premises to be used in syllogisms. Further, the assent to a universal proposition established by "scientific induction", if it does arise, is of the degree of necessary certainty. In ordinary induction the assent to a universal premise is at best of the degree of approximate certainty, which is the level of dialectical arguments in general.[55]

The sole example which al-Fārābī gave of a universal proposition produced by "scientific induction" is the statement "all teaching and all learning which proceed through reasoning come from prior knowledge." The universal judgment in this example is arrived at only after the definitions of the terms constituting the subject[56] become completely known. The definition of the subject is not known spontaneously but as a result of experience, that is, after an examination of individual instances of learning and teaching which proceed through reasoning. Al-Fārābī asserts, however, that the necessary certainty about the above proposition does not derive from the certainty that all the particular instances of teaching and learning have been exhaustively investigated. For the latter certainty is an impossibility. Rather, the necessary certainty results primarily from the universality of the proposition which is productive of intuition.[57] In some individuals, says al-Fārābī, the intuitive capacity is so strong that they are able to form a universal judgment about a subject on the basis of knowledge of just a few particular instances of the subject.

Al-Fārābī describes both kinds of primary intelligibles as immediate knowledge. In the case of the axioms of certainty, they are said to be immediate in the sense of known without a middle term. The predicate of such an axiom is an inherent part of the subject. Thus knowledge of the predicate existence is known through the definition of the subject itself. Axioms of certainty are not immediate in the temporal sense because the definition of the subject is only known completely through experience. Only the primary principles are immediate in both senses.

Al-Fārābī's notions of conception and assent are also useful in clarifying the manner in which the imaginative serves the rational faculty. Primary intelligibles result from self-evident combinations of single universals either as affirmative or negative propositions. These single universals are conceived in the mind (*nafs*) through the assistance of the faculties of memory and *wahm*. The faculty of rational or deliberative imagination (*al-mufakkirah*) operates on the individual forms stored in the representative faculty and on the individual meanings of sensible objects perceived by *wahm* and stored in memory. It does this by discerning similarities and differences between these individual forms or between the individual meanings. As a result of this discernment the accidental universals come to be conceived. By distinguishing between the essential and the accidental, the essential universals are next to be conceived. It is also the faculty of deliberative imagination which combines these single universal meanings into affirmative and negative propositions.

Al-Fārābī identifies the primary intelligibles with the principles of all the four philosophical sciences which he enumerates in the *Iḥṣā' al-'ulūm* and in some other works.[58] The sciences are mathematics, natural philosophy, political philosophy, and metaphysics. Al-Fārābī maintains that most of the principles of these sciences are acquired gradually through "scientific induction". There is a certain order in which these principles generally come into the possession of the intellect. In al-Fārābī's view, this is the natural order of learning the principles of the sciences. He describes the general rule determining this order as follows: one begins with what is easiest for the intellect to comprehend, followed by the next easiest, and so on.[59] The ease with which principles can be grasped is in direct proportion to the "distance" of their subjects from matter. Metaphysical principles are the exceptions. Although objects of metaphysical inquiry are absolutely free of matter, they are not easily comprehensible unless one has acquired knowledge of the principles of some of the sciences. For example, the knowledge of the four Aristotelian causes[60] in natural philosophy, which al-Fārābī calls the principles of being, is necessary in understanding the idea of God as the first or ultimate cause of all beings.

According to al-Fārābī, the principles of mathematics are the easiest to grasp. In mathematics itself, the easiest to comprehend are the principles of arithmetic, then come those of geometry, optics, astronomy, music, and mechanics, in that order.[61] Mathematics is followed by natural philosophy, then metaphysics, and then political philosophy. In the *Iḥṣā' al-'ulūm,* al-Fārābī enumerates these four philosophical sciences in this order. I will discuss further the principles of these sciences and the significance of their order of enumeration at various places in the next four chapters.

Al-Fārābī refers to the intuitive perception of intelligible forms by the intellect as intellection. There are degrees of intellection. The reception of primary intelligibles constitutes the first degree. Al-Fārābī believes in the doctrine of the identity of the intellect and its intelligible objects in the act of intellection.[62] Thus he says that the illuminated potential intellect, in receiving the primary intelligibles as forms, becomes those forms, just as a piece of wax receives forms not by being imprinted on its surface but by pervading its totality so that the wax is turned into an image.[63] Al-Fārābī also believes that both the primary intelligible forms and the resulting actual intellect, insofar as they are actual, acquire a new ontological status in the totality of beings (*al-mawjūdāt*).[64]

The process of the acquisition of actuality by the potential intellect reaches its perfect stage when that intellect becomes an actual intellect not only in relation to all the primary intelligibles but also to the secondary ones derived from them.[65] When that stage is reached, the resulting actual intellect begins to reflect upon itself and its contents. Al-Fārābī argues that the actual intellect can contemplate every intelligible by receiving its form. The contents of the actual intellect are, in fact, pure intelligibles abstract from matter. Further, the actual intellect can know itself because it is both intellect and intelligible thing. The actual intellect's contemplation of itself and its contents constitutes a second degree of intellection. It is a higher kind of intellection than the first, since its objects are intelligible forms abstract from matter and do not depend on the imaginative and sensitive faculties. When the actual intellect possesses this second intellective power, it becomes what al-Fārābī calls the acquired intellect (*al-'aql al-mustafād*).[66]

The acquired intellect thus refers to the actual intellect when it is both self-intelligible and self-intellective. Al-Fārābī is here giving a technical meaning to the term "acquired intellect". According to him, the "acquired intellect" is the most developed form of the human intellect. However, it admits of degrees of perfection. Of all things in the sublunary world, the "acquired intellect" is the closest in resemblance to the active intellect.[67] Both intellects are forms of form, meaning that both are self-intelligible and self-intellective. Their contents are of the same kind insofar as they are pure intelligibles abstract from matter. Al-Fārābī also says that the acquired intellect does not need a body for its subsistence nor corporeal and animate powers for its thinking activities.[68] This is another respect in which the "acquired intellect" resembles the active intellect.

The two intellects, however, are not of the same rank. It was mentioned earlier that the active intellect is absolutely separate and is the perfect repository of intelligible forms. Although the contents of the two intellects are similar insofar as they are intelligibles abstract from matter, they are not of the same ontological order. The active intellect and its contents never cease being actual whereas the "acquired intellect" represents a stage of the acquisition of actuality by the potential intellect. The two intellects differ in another respect. Intelligible forms contained in them are in an inverse order. The order in which the active intellect contemplates existing things is from the more perfect to the less perfect of them.[69] In the case of the thinking activity of the human intellect, we ascend from that which is best known to us to that which is unknown. That which is more perfect in existence, al-Fārābī insists, is more unknown to us.

When the human intellect becomes the "acquired intellect" it is capable of contemplating the active intellect itself. The degree of perfection of the "acquired intellect" depends on the extent to which it actually acquires intelligible forms from the active intellect. In its highest perfection the "acquired intellect" attains union with the active intellect. By "union" al-Fārābī means that the "acquired intellect" participates in the reality of the active intellect without being essentially identified with the latter. For there is a part of the active intellect's reality which is transcendent to or not participated by the human soul.[70] Through its union with the active intellect, this "prophetic intellect" becomes the human vehicle of divine revelation. In according the highest position to the prophetic intellect in the hierarchy of the faculties of the human soul, al-Fārābī remains faithful to the religious view that revelation is the highest source of knowledge.

His notion of the acquired intellect is the key to his solution to the problem of the relationship between revelation and reason as well as of the relationship between religion and philosophy, which will be examined in the following chapter.

ENDNOTES

Chapter 2

[1]N. Rescher, *al-Fārābī: An Annotated Bibliography*, p. 43.

[2]The term *Divine Principle* used here refers to the realm of the Divine Essence and the Divine Names and Qualities, as distinguished from the realm of manifestation or creation. For a detailed discussion of the doctrine of the Unity of the Divine Principle in Islam, see F. Schuon, *Dimensions of Islam*, London, 1970, chap. 11.

[3]It is true that there was much political strife and decadence during al-Fārābī's time, as we discussed in the last chapter. But religiously, spiritually and intellectually, the Muslim community remained close to the teachings of the Qur'ān. It is true that even on these planes there are tensions within the community generated by the crystallization of the revealed message into its constituent elements through the passage of time. But the community lived under the constant reminder of the ideal set forth in that message, however far short it was of that ideal. As remarked by Sir Hamilton Gibb:

> From the beginning of its existence the Community was made aware (and daily reminded) by the Qur'an of the purpose for which it was created and the destiny to which it was called. It was to represent on earth, before the eyes of all mankind, the principle of Divine Justice, that is to say, of integral Reality, Harmony and Truth. Justice in this sense has little or nothing to do with the political or judicial application of man-made laws. It is a principle of order and wholeness: that all elements, endowments, and activities of life shall be in harmonious relation with one another, each fulfilling its proper purpose and ends in a divinely-appointed system of interlocking obligations and rights.

See his preface to S.H. Nasr, *IICD*, p. xiv.

[4]*The Qur'ān* (II: 255). The verse reads:

> God! There is no god but He – the Living, the Self- Subsisting, Eternal. No slumber can seize Him nor sleep. His are all things in the heavens and

> on earth. Who is there can intercede in His presence except as He permitteth? He knoweth what (appeareth to His creatures as) Before or After or Behind them. Nor shall they compass aught of His knowledge except as He willeth. His throne doth extend over the heavens and the earth, and He feeleth no fatigue in guarding and preserving them for He is the Most High, The Supreme (in glory).

[5]Al-Tirmidhī, *Sunan, Thawāb al-Qur'ān*, 2.
[6]Al-Ghazzālī, *The Jewels of the Qur'ān*, trans. M.A. Quasem, Bangi, Malaysia, 1977, p. 75.
[7]Abū Dāwūd, *Sunan, Witr*, 23.
[8]*The Qur'ān* (XXXVII: 5).
[9]*Ibid* (XXXVII: 6).
[10]*Ibid* (XXI: 17, 19)
[11]*Ibid* (LXXVIII: 38). For a detailed discussion of the Islamic doctrine of the hierarchy of angels, see F. Schuon, *op. cit.*, chap. 8, pp. 102-20.
[12]*The Qur'ān* (IX: 72); (XV: 43-44).
[13]*Ibid* (LVIII: 11).
[14]See al-Ghazzālī, *The Book of Knowledge*, trans. N.A. Faris, Lahore, 1962, p. 10.
[15]*The Qur'ān* (III: 18).
[16]*Ibid* (LVI: 7-56).
[17]See n. 107, chapter 1.
[18]The translation is based on the Arabic text in F. Dieterici, *al-Fārābī's Philosophische Abhandlungen*, p. 105, Concerning subsequent translations, they are mine unless otherwise stated.
[19]Al-Fārābī's explanation of why the celestial bodies are the most perfect of bodies is given in chapter four.
[20]The *Sharī'ah*, the Islamic sacred law, is accepted by Muslims as the concrete embodiment of the divine will. It is concrete and all-embracing in the sense that it includes not only universal moral principles but also details of the way in which man should conduct every facet of his earthly life, both private and social. In other words, in Islam, there is no domain of human life which lies outside the sphere of divine legislation.

The *Sharī'ah* is the source of knowledge of what is right and wrong. According to its teaching, there is a hierarchy of values of human acts and objects in the sight of God. Every object or human act must fall into one of the following categories: (1) obligatory (*wājib*), (2) meritorious or recommended (*mandūb*) (3) forbidden (*ḥaram*), (4) reprehensible (*makrūh*), and (5) indifferent (*mubāḥ*).
[21]Al-Fārābī says:

> Happiness is the good without qualification. Everything useful for the achievement of happiness or by which it is attained, is good too; not for its own sake, however, but because it is useful with respect to happiness; and everything that obstructs the way to happiness in any fashion is unqualified evil.

See F.M. Najjar, trans., "al-Fārābī: The Political Regime," *Medieval Political Philosophy*, p. 33.
[22]In that treatise al-Fārābī offers his critique of the meanings of the term intellect given by the ordinary person, by the *mutakallimūn*, and by Aristotle in a number of his

works. Al-Fārābī's conception of the intellect given below is mainly drawn from his critique of Aristotle's theory contained in that treatise.

[23]See, for example, the study by I. Madkour in his *La place d'al-Fārābī dans l'ecole philosophique musulmane.*

[24]Al-Fārābī, *al-Siyāsat al-madanīyah,* pp. 67-8, The word *nuṭq* is here translated as intelligence so as to embrace the following meanings referred to by al-Fārābī:

> This expression points in the opinion of the ancients to three things: (1) the faculty by which man intellects the intelligibles, that by which the sciences and arts are acquired; (2) secondly, the intelligibles which are produced in the soul of man by the understanding, and are called interior speech; (3) thirdly, the expression by language of what is in the mind and is called exterior speech.

See D.M. Dunlop, trans., "al-Fārābī's Introductory *Risālah* on Logic," *The Islamic Quarterly,* London, 3,4 (Jan. 1957), 233.

That the rational faculty and language together signify the totality of man's intelligence is affirmed by a contemporary thinker:

> Man is distinguished from the animals by the totality – hence the objectivity – of his intelligence, and the sign of this totality is not only the rational faculty, but also language; now the domain of language is that of logic, so much so that the latter concerns all that is expressible.

F. Schuon, *From the Divine to the Human,* World Wisdom Books, Bloomington, 1982, p. 24.

[25]"This liking or disliking, if it is the result of Sensation or Imagination, is known by the generic term Volition. If, on the other hand, it is the result of deliberation or reasoning, it is called choice. Man is the sole possessor of choice, but the inclination resulting from Sensation or Imagination is common to all the animals." See M.Wali-ur-Rahman, "The Psychology of al-Fārābī," *Islamic Culture,* 11 (1937), 245.

[26]Al-Fārābī, *Kitāb ārā' ahl al-madīnat al-fāḍilah,* ed. A.N. Nadir, Beirut, 1968, p. 87. Hereafter, this work will be cited as *al-madīnat al-fāḍilah.*

[27]Al-Fārābī, *ibid,* p. 90.

[28]Al-Fārābī's notions of form and matter are not always the same as those of Aristotle. One instance in which al-Fārābī departs from the Stagirite in the use of the two terms is in his explanation of how the various faculties of the human soul constitute a unity. There (see *al-Madīnat al-fāḍilah,* p. 92), al-Fārābī puts forward the idea that every lower faculty serves as matter for the higher in rank and a higher faculty as form of the lower. The sense in which he uses the two terms there is explained below. His doctrine of form and matter is treated more fully in chapter four in my discussion of the ontological basis of his classification of the sciences.

[29]Common sense thus prepares through its reception of these images the materials upon which the imaginative faculty depends for the actualization of its functions. Prior to al-Fārābī, John Philoponus also defined the relation between the imaginative faculty and common sense in terms of form and matter. See R. Walzer, *Greek into Arabic: Essays on Islamic Philosophy,* p. 208.

[30]According to Muslim philosophers, the active intellect is the last and lowest of a series of ten intellects emanating from God. (See chapter four of this work). It is a separate or angelic substance. With respect to the human intellect, the active intellect is

the external agent of its transformation from potentiality to actuality. The various stages of this transformation are discussed below.

[31]Al-Fārābī's fivefold classification of internal senses is given in the *Fuṣūṣ al-ḥikam*. The text used in this reference is contained in Ibn Buṭlān's *Da'wat al-aṭibbā'*, Beirut (see p. 83). For a detailed and illuminating discussion of this classification by a contemporary scholar, see H.A. Wolfson, "The Internal Senses in Latin, Arabic, and Hebrew Philosophic Texts," *Harvard Theological Review*, 28,2 (April, 1935), 93-5.

[32]A similar treatment of common sense is found in Isaac Israeli. See H.A. Wolfson, *op. cit.*, p. 95. The first Muslim thinker to specifically include common sense in the classification of the internal senses is Ibn Sīnā.

[33]Al-Fārābī was the first to introduce *wahm* into the classification of the internal senses.

[34]That, in Ibn Sīnā, *wahm* operates either instinctively or through the memory-images of past experience is clearly indicated by the varied examples and illustrations which he has given of that faculty. An example of an instinctive operation of *wahm* is the following: a lamb runs to another lamb, even if it has never seen it before; similarly, the instinctive action of a lamb in fleeing from a wolf, even if it has never seen one before.

In illustrating the operation of *wahm* on the basis of a previous experience, Ibn Sīnā mentions the example of a dog which fears a stick because it was once beaten with it. He explains the operation as follows: "If an animal has experienced pleasure or pain or any sensual good or evil associated with a sensible form, the faculty of representation (*al-muṣawwirah*) preserves an imprint of the form and memory conserves the association of good or evil with this sensible form... so that when the same form is presented again to the imagination from without, it stirs in the imagination and so the idea of good or evil associated with it is also moved." See F. Rahman, *Avicenna's Psychology*, Oxford University Press, London, 1952, p. 81.

[35]In both *al-Madīnat al-fāḍilah* and *al-Siyāsat al-madanīyah* the term *al-mutakhayyilah* is mainly used in this general sense. In these works al-Fārābī is mainly interested in distinguishing the imaginative, as an intermediate faculty, from the other cognitive faculties and not in distinguishing the imaginative faculty of man from that of other animals.

[36]*Fuṣūṣ al-ḥikam*, p. 83. Al-Fārābī was the first Muslim philosopher to employ the term *mutakhayyilah* with the specific meaning of compositive animal imagination earlier defined.

[37]See *'Uyūn al-masā'il*, p. 63 of F. Dieterici's *al-Fārābī's Philosophische Abhandlungen* and p. 9 of the Hyderabad edition published under the different title, *al-da'āwī al-qalbiyah*, 1349 A.H.

[38]*Fuṣūl al-madanī*, pp. 30-1

[39]R. Walzer (trans.), *al-Fārābī on the Perfect State: Abū Naṣr al-Fārābī's "Mabādi' arā' ahl al-madīna al-fāḍila,"* pp. 197-9. Hereafter cited as *The Perfect State*.

[40]He also calls it the material intellect (*al-'aql al-hayūlānī*).

[41]*Ibid.*

[42]The position of the active intellect in the hierarchy of separate Intelligences and as one of al-Fārābī's six principles of beings will be discussed in chapter four.

[43]See *'Uyūn al-masā'il*, p. 63.

[44]Al-Fārābī, *Risālat fi'l-'aql*, ed. M. Bouyges, Beirut, 1938, p.29.

[45]*The Perfect State*, p.203

[46]The four classes of known statements not arrived at by syllogistic means are: (1) received opinions (*maqbūlāt*), (2) wellknown and generally accepted opinions (*mashhūrāt*), (3) sense-knowledge (*maḥsūsāt*), and (4) primary intelligibles. See *Short*

Pr. Anal., p. 58; Dunlop, "al-Fārābī's *Introductory Sections* on logic." *The Islamic Quarterly*, 2, 4 (Dec. 1955), 275

A discussion of the nature of each class of statements and of their adoption as premises in the different kinds of syllogisms is given in chapter three.

[47]See M.S. Galston, *Opinion and Knowledge in Fārābī's Understanding of Aristotle's Philosophy*, unpublished PhD thesis, The University of Chicago, 1973, p. 151.

[48]Dunlop, *op. cit.*, p. 276

[49]Al-Fārābī, *Kitāb al-alfāẓ al-musta'malah fi'l-manṭiq* (*The Utterances Employed in Logic*), ed. M.Mahdi, Beirut, 1968, p.87. Hereafter cited as *Kitāb al-alfāẓ*.

[50]*Ibid.*

[51]The classification is given in his *Kitāb al-burhān* (*The Book of Demonstration*). See Galston, *op. cit.*, p. 163

[52]*Ibid*, p. 165.

[53]*Kitāb al-alfāẓ*, p. 87

[54]Galston, *op. cit.*, p. 226

[55]*Short Pr. Anal.*, p. 89

[56]Since al-Fārābī refers to definitions as perfect conceptions, he calls the resulting universal judgment "the primary assent which accompanies perfect conception."

[57]Galston, *op. cit.*, p. 251

[58]Of these other works, one may mention, for example, the previously cited *Introductory, 'Risālah' on Logic* (p. 232); and *The Perfect State*, pp. 203-5.

[59]*The Philosophy of Plato and Aristotle*, p. 18

[60]These are the material, the formal, the efficient and the final causes.

[61]*The Philosophy of Plato and Aristotle*, p. 19

[62]What is meant by this statement is that the intellect as the subject, in the act of cognition, becomes like the intelligible objects it knows.

[63]*Risālah fi'l-'aql*, pp. 13-4; F.Rahman, *Prophecy in Islam*, p. 12.

[64]*Risalah fi'l-aql*, pp. 17-8.

[65]*Ibid*, p. 18

[66]*Ibid*, p. 20

[67]*Ibid*, p. 27

[68]*Ibid*, pp. 31-2. What al-Fārābī means is that when the human intellect reaches the stage of the acquired intellect, it becomes self-operative and pure activity in the sense that it no longer needs the help of the sensitive and the imaginative faculties to know the pure intelligibles.

[69]*Ibid*, p.28

[70]*al-Siyāsat al-madanīyah*, p. 35.

CHAPTER 3

THE METHODOLOGICAL BASIS OF THE HIERARCHY OF THE SCIENCES

The last chapter presented al-Fārābī's view of the hierarchy of the faculties of the human soul. That hierarchy culminates in the prophetic intellect, which he identifies with the active intellect, the archangel of divine revelations. In this chapter three main topics are discussed. The first is al-Fārābī's conception of the relationship between revelation on the one hand and intellect and reason on the other. The second is his theory of the relationship between religion and philosophy. The third is his conception of methodology.

3.1 Revelation, Intellect, and Reason

One of the fundamental articles of faith (*arkān al-īmān*) in Islam is belief in divine revelation (*waḥy*). The human recipient of this revelation is known as a prophet (*nabīy*) or messenger (*rasūl*) of God. Muslims believe that prophets and messengers are the best and the noblest of God's creatures. Al-Fārābī accepted fully this religious tenet. But whereas the ordinary believer is content to accept its truth at the level of faith,[1] al-Fārābī the philosopher sought to understand the reality of divine revelation as a philosophical truth. Thus, he says in his *Kitāb al-millat al-fāḍilah:* "Theoretical opinions in religion have their proofs in theoretical philosophy, while they are taken in religion without proofs."[2]

Armed mainly with the tools of Aristotelian psychology, al-Fārābī set out to expound his doctrine of prophecy and revelation within the framework of a comprehensive theory of the intellect. He describes the nature of the prophetic intellect as the vehicle of revelation as follows:

> The supreme ruler [i.e., the prophet] without qualification is he who does not need anyone to rule him in anything what-

ever but has actually acquired the sciences and every kind of knowledge; he has no need of anyone to guide him in anything. Such a one is able to comprehend each of the particular things he ought to do. He can guide others safely in all matters in which he instructs them, employ all those who do any of the acts for which they are equipped, and determine, define, and direct these acts toward happiness. This is found only in the one who possesses superior natural dispositions when his soul is in union with the active intellect.

He can only attain his [union with the active intellect] by first acquiring the *passive intellect* and then the intellect called *the acquired* (*al-mustafād*); for, as is stated in *On the Soul*, union with the active intellect results from possessing the acquired intellect. This one is the true prince according to the ancients; he is the one of whom it ought to be said that he receives revelation. For a person receives revelation only when attains this rank, that is, when there is no longer any intermediary between him and the active intellect; for the passive intellect is like matter and substratum to the acquired intellect, and the latter is like matter and substratum to the active intellect. It is then that the power that enables one to understand how to define things and actions and how to direct them toward happiness emanates from the active intellect to the passive. This emanation that proceeds (*al-ifādat al-kā'inah*) from the active intellect to the passive through the mediation of the acquired intellect is revelation (*waḥy*).[3]

Al-Fārābī's philosophical explanation of *waḥy* contains reference to the participation of three kinds of intellects. The first is the *active intellect*, a cosmic entity which acts as a transcendent intermediary between God and man. The second is the acquired intellect (*al-'aql al-mustafād*) which the Prophet acquires only insofar as his soul is in union with the active intellect. In this union the acquired intellect receives transcendent knowledge from the active intellect. The third, the passive intellect (*al-'aql al-munfa'il*), refers to the actual receptive intellect of the prophet in general.[4] The acquired intellect is a special intellective power which enters the prophet's mind as the result of his union with the active intellect. The receptacle in the prophet's soul which receives this special intellective power is, according to al-Fārābī, the passive intellect. For this reason the acquired intellect is spoken of as the intermediary between the passive intellect and the active intellect.

The ultimate source of revelation is God: "Since the active intellect emanates from the being of the First Cause,[5] it can be said that it is the First Cause that brings about revelation to this person through the mediation of the active intellect."[6] In the passage cited above, which occurs in the *al-Siyāsat al-madanīyah,* al-Fārābī summarizes all that is essentially contained in the traditional doctrine of revelation as expounded by the *falāsifah.*[7]

According to al-Fārābī, revelation possesses two dimensions corresponding to which are two fundamental prophetic functions. The first is the theoretical dimension. It comprises knowledge of what al-Fārābī calls the natural intelligibles *(al-ma'qūlāt al-ṭabī'īyah).*[8] These are the objects of the theoretical-rational faculty alone. They are defined as those "existents which are not such as that we can make them or alter them from one condition to another."[9] In the above passage, al-Fārābī calls the highest type of this knowledge gnosis (*ma'rifah*). Elsewhere he calls it wisdom (*ḥikmah*), which he defines as "the most excellent knowledge of the most excellent existents."[10] Through wisdom man knows true happiness. Al-Fārābī notes that one who has received this dimension of revelation has acquired the sciences and gnosis. This dimension corresponds to the intellectual function of the prophet in his capacity as a sage and philosopher.[11] For al-Fārābī, every prophet is a sage and a true philosopher but the converse is not necessarily true.

The second dimension is the practical one. It comprises knowledge of what al-Fārābī calls the voluntary intelligibles (*al-ma'qūlāt al-irādīyah*).[12] These are the objects of the practical-rational faculty, more specifically, that part called the deliberative faculty (*al-quwwat al-fikrīyah*). The voluntary intelligibles are those which can be made to exist outside the soul by the will when the accidents and states that accompany them as they come to be have come into actual existence by the will.[13] They are things which are found useful for the attainment of an end which in this case is truly good and virtuous namely the attainment of supreme happiness. This meaning of happiness is perceived by the theoretical intellect. Al-Fārābī calls this second dimension of revealed knowledge "practical wisdom." It acquaints humans with what must be done to attain happiness.[14]

Practical wisdom refers primarily to the divine laws, for he says: "Once the conditions that render their actual existence possible are prescribed, the voluntary intelligibles are embodied in laws."[15] This dimension therefore corresponds to the law-giving and legislative function of the prophet in his capacity as a ruler, king or statesman. It is in referring to this dimension of revelation and this prophetic function that al-Fārābī says that the one who receives revelation is able to "comprehend well each one of the things that he ought to do, to guide well all others in everything in which he instructs them,....and to determine, define and direct these acts toward happiness."

Furthermore, according to al-Fārābī, revelation is a kind of cosmic intellection. He describes it as that objective phenomenon which occurs when the prophet's soul is in intellective union with the supra-individual active intellect. Through the latter, which as the lowest of the cosmic intellects understands the essence of the First Cause and secondary causes (the principles of the heavenly bodies), as well as its own essence,[16] the prophet possesses a supra-rational vision of the One and of the whole realm of the spirit. The prophet's experience of revelation is not limited, however, to having a vision of these spiritual realities in their essence. This can only be experienced by a few. Since the message of revelation is meant for all to whom the prophet is sent, his experience of revelation must also embrace the level of understanding of these spiritual realities that is within the possible reach of the rest of that collectivity.

Al-Fārābī maintains that spiritual or intellectual truths (for example, principles of being by which he means the four Aristotelian causes; the hierarchy of beings; the meaning of supreme happiness) are either understood philosophically or imagined. To understand them is to grasp their essence. To imagine them is to perceive their images. Al-Fārābī uses the following analogy to explain the difference between these two levels of understanding:

> We see a person, we see a representation of him, we see his image reflected in water, and we see the image of a representation of him reflected in water. Our seeing the person himself is like the intellect's cognition of the principles of beings, of happiness, and so forth; while our seeing his reflection in water and our seeing a representation of him is like imagination since we are seeing that which is an imitation of him.[17]

Similarly, says al-Fārābī, when we imagine spiritual realities, we are in fact perceiving imitations of them rather than cognizing the spiritual realities themselves.

According to al-Fārābī, in the case of many people, spiritual truths cannot be grasped in their essence either because of certain impediments in their natural constitutions or because of their habits. Their understanding of these spiritual realities is through images (sing., *khayāl*), symbols (sing., *mithāl*) and imitations or similitudes (sing., *muḥākāh*) provided by revelation. Revelation in its totality is experienced by the prophet not only spiritually and intellectually[18] but also through imagination and sensation. For example, the angel of revelation actually appears to the prophet in visual form and the angel speaks in audible form. It is the prophet's imaginative faculty, described by al-Fārābī as "the highest degree of perfection a person can reach with his imaginative powers,"[19] which transforms the intelligibles bestowed upon it by the active intellect into "vivid and potent symbols capable of impelling to action."[20]

In the *Fuṣūṣ al-ḥikam,* al-Fārābī gives a detailed explanation of the prophet's experience of revelation. What happens when an angel and the prophetic spirit meet is that both the internal and external senses of the prophet get attracted to the world of the spirit. The angel is presented to the senses in accordance with the power of the one who sees the angel not in the absolute but the relative form. The prophet hears the angel's speech as a voice even though it is intrinsically a spiritual communication (*waḥy*). By "spiritual communication" al-Fārābī means that the mind of the angel communicates its content directly to the human mind. According to al-Fārābī, this is what constitutes real "speech," since it conveys the meaning of the addressor's mind (to the addressee's mind) in a direct manner.

In ordinary human communication, the addressor cannot touch the mind of the addressee directly in the manner in which a seal touches a piece of wax and cannot render it like itself. Consequently, the addressor has to adopt an exterior ambassador, like voice, writing or gesticulation. As for the communication between the angel of revelation and the prophet, it is at once spiritual and imaginative-perceptual. It is a direct spiritual communication because the prophetic spirit is pure, so that there is no veil between it and the mind of the angel. The latter shines upon the former as the sun shines upon clear water. The communication is also imaginative-perceptual because

> the prophetic mind is impressed by an impression which then overflows also in the internal sense (i.e., imagination). When the imaginative faculty is strong, the angel's impression on it is such that it is perceived (visibly and audibly). Thus the recipient of revelation contacts the angel by his interior (mind) and receives revelation internally, but the angel also appears to him in a visual form and his speech takes on an audible form. In this way, the angel and the revelation come to his cognitive faculties in both ways (i.e., spiritual and imaginative-perceptual).[21]

Since the prophet's imaginative faculty is strong and perfect, neither the sensations coming from the external world nor its services to the rational soul overpower it to the point of engaging it utterly. On the contrary, despite this engagement it has a superfluity of strength which enables it to perform its proper function. The condition of the prophet's imaginative faculty with all its engagements in waking life is like the condition of other souls when they are disengaged in sleep. It is by virtue of its perfection that the prophet's imaginative soul is able to figurize the intelligibles bestowed upon it by the active intellect in terms of visible symbols. These figurative images, in turn, impress themselves on the perceptual or sensitive faculty.[22]

In explaining the conversion of intelligibles of the higher world into objects of prophetic visions, al-Fārābī introduces an activity or function of the imaginative faculty which was not encountered before in our discussion of that faculty. This activity is termed *muḥākāt*.[23] The term conveys the idea of 'imitating', 'copying', 're-enacting' or 'expressing in symbols.' In its activity of *muḥākāt* the imaginative faculty may concern itself either with sensibles (*maḥsūsāt*), intelligibles or impressions conveyed to it by the appetitive faculty.[24]

Prophetic 'imitations' of revealed truths with images and symbols are necessary because for those devoid of philosophical knowledge they provide the only effective means of understanding those truths. According to al-Fārābī, it is not the prophet himself who creates these images and symbols. The working of 'imitation' within the prophet's imaginative faculty is not the result of his intellectual deliberation. He is completely passive before that activity just as the objects of our dreams are not determined by us. Al-Fārābī maintains that the images and symbols too are revealed to the prophet.

Al-Fārābī's doctrine of the visual and acoustic symbolization of spiritual truths by the imaginative faculty, which was later taken over by Ibn Sīnā, agrees with the views presented in the Qur'ān and the numerous traditional accounts of the prophet's revelation.

For al-Fārābī, religious images and symbols (such as those referring to the angel of revelation) possess an objective validity within the context of the religious tradition in question although they are private to the prophet, hence not empirically verifiable to most people.[25] By objective validity is meant that meaning-wise, the images and symbols admit of a common emotional and rational experience by the members of the religious community. Even the philosophers belonging to that community accept the validity of the symbols seeing that the prophet himself is the source of both the interpretation of their meanings and the philosophical understanding of the truths which these symbols reflect and embody.

The above view of the relation between revelation and the rational faculty underlies al-Fārābī's account of the relationship between revelation, intellect and reason. Revelation understood as the prophet's intellectual or spiritual vision of the essence of the spiritual realities, is of the same nature as the pure intellection (*ta'aqqul*) or intellectual intuition of the philosophers and sages (*ḥukamā'*) or the mystic's experience of gnosis (*ma'rifah*). For this reason, intellection has been described by many traditional authorities, ancient as well as modern, as "Revelation on the scale of the microcosm."[26] It is necessary to stress that, for al-Fārābī, the above similarity pertains only to the nature of the intellectual experience in the two cases.

Revelation, however, is superior to intellection in a number of respects. First, the prophet's intellectual experience of the reality of divine things during revelation is not preceded by his learning about these things from an external human source, or by some kind of reasoning which likewise presupposes the possession of certain data concerning them. As explained, al-Fārābī insists that the prophet has no human teacher or guide and that he becomes the recipient of divine revelation precisely because he possesses superior natural dispositions. A perfect soul likewise makes his intellect the perfect substratum for the angel of revelation.

If al-Fārābī insists that revelation does not occur until and unless the acquired intellect (*al-'aql al-mustafād*) has been attained, this does not in any way contradict his earlier assertion that the prophetic mind does not need external instruction. In his view the attainment of the acquired intellect does not necessarily depend for its realization on the amount of knowledge one receives from external instruction. Al-Fārābī has made it clear that the acquired intellect must precede revelation in time because the former alone can become matter and substratum for the active intellect. In maintaining this view, al-Fārābī was possibly influenced by the popular idea in Islam that prophecy begins at the age of forty. Prior to his first experience of revelation, the prophet already possesses intellectual maturity and has exhibited a certain level of theoretical and practical wisdom, thanks to his superior natural dispositions.

As for intellection or intellectual intuition, says al-Fārābī, its realization in general is only possible within the fold of a religious orthodoxy or revealed tradition. In other words, intellection depends on revelation for its effective realization. In the *Taḥṣīl al-sa'ādah*, for example, al-Fārābī explains clearly the necessary intellectual and moral training which a person must undergo and the qualifications he should possess before he can reach that summit of philosophical experience, which is intellection.[27] He also emphasizes the fact that philosophical training and realization presupposes the existence of a living philosophical and spiritual tradition whose origin is prophetic. He goes so far as to say that the pursuit of philosophy is legitimate only when it is rooted in and never divorced from a revealed tradition, including its religious rites and legal-moral injunctions. Thus:

> A philosopher must perform the external [bodily] acts and observe the duties of the law for if a person disregards a law ordained as incumbent by a prophet and then pursues philosophy, he must be deserted. He should consider unlawful for himself what is unlawful in his millah (i.e., religious community).[28]

In more precise terms, revelation helps to promote intellection in two ways. First, the theoretical dimension of revelation provides a wealth of data for the human mind to reflect upon. Initially, the theoretical revealed data are accepted on the basis of faith or by way of imagining them.[29] Later, in contemplating upon these data, the philosophically-minded person may arrive at an intellection or intuitive understanding of them. Another way in which revelation promotes intellection is by providing knowledge of religious rites and moral injunctions. Al-Fārābī thinks that the observance of these rites and injunctions helps to facilitate the path to intellection.

Revelation is also superior to intellection by virtue of the fact that the former, unlike the latter, is not only experienced by the intellect but also by the other cognitive faculties. Spiritual truths received by the prophet's intellect are transformed into images and symbols by the imaginative faculty. This is not the case with other kinds of intellection.

According to al-Fārābī, there is another lesser level of intellectual intuition which is associated with the acquisition of the first intelligibles. Although they resemble each other in the sense that truths are apprehended in a direct manner in both cases, they refer to qualitatively different experiences. This is because the corresponding objective truths involved are at different levels. As for the remaining ways of knowing objects through the rational faculty, these kinds of knowledge are acquired through the indirect processes of discursive reasoning which employ syllogisms (*qiyās*).[30] Although al-Fārābī uses the same term *'aql* for both intellect and reason, as is true of Islamic intellectual tradition in general, he is clear enough about their distinguishing features and functions. According to him, intellect is the principle of reason in the sense that the latter is the handmaid of the former.

The principles or premises (*muqaddamāh*) upon which the highest kind of reasoning[31] is based come from the intellect. Upon these premises reason produces a body of syllogistic discourse (*mukhāṭabah*).[32] The aim of this discourse is to enable premises derived from supra-rational experience to be apprehended rationally by those devoid of that experience so as to lead them to that kind of experience itself.

The importance in al-Fārābī's view of the above role of reason in relation to the intellect (described by Schuon as the descending or communicating function of reason),[33] becomes apparent when he makes this ability to conduct a rational discourse of supra-rational truths one of his main criteria to determine the true worth of a philosopher. The perfect philosopher is one who not only possesses the highest truths or wisdom but also the ability to conduct a rational discourse concerning these truths for the benefit of others.[34] As for the person who receives the discourse, the hope is that he will be led from the rationally formulated proofs to an intellectual intuition of transcendent truths. Al-Fārābī says that there are individuals who, through a syllogistic discourse, attain an intuitive understanding of intellectual and spiritual truths.[35] In such cases reason becomes an instrument for reaching the divine truths in revelation.

In Islamic intellectual tradition the distinction between intellect and reason is well noted. Some later Muslim thinkers used the term *al-'aql al-kullī* for intellect in contrast to *al-'aql al-juz'ī* for reason. The intellect is compared to a sun that shines within man and reason to the reflection of this sun on the plane of the mind.[36] The intellect is the instrument through which an intuitive understanding of transcendent truths is attained. Reason is the instrument of discursive thought (*fikr*). Al-Fārābī characterizes the intellect as the faculty which "cannot be in error in regard to what occurs to it, but all the species of knowledge which reach it are true and certain and cannot be otherwise."[37] By this statement al-Fārābī means that in contrast to reason which can err, the intellect is by nature beyond error. He seems to argue that reason is not entirely free from the possibility of error inasmuch as it operates by syllogisms. I will deal further with this question in my discussion of al-Fārābī's theory of methodology.

Al-Fārābī's distinction between intellect and reason has an important bearing on his philosophy of science. Reason, says al-Fārābī, operates upon various kinds of data. Some of these are furnished by man's mental faculty. These are the kind of data which are not totally free from the possibility of doubt and error. Through the imaginative power of composition and separation of sensible impressions, the human being, both during waking and sleeping hours, is capable of imagining things which do not have an objective existence outside the soul. Reason may well be fed with this kind of data, the objective truth or falsity of which it has no authority to judge.

For al-Fārābī, all data which are not known to be certain must be judged in the light of those which are.[38] The highest category of certain truths comprises the metaphysical truths furnished by revelation and intellection. This category refers to the secondary intelligibles which are none other than the Platonic ideas or the archetypes of the physical and imaginal worlds. Consequently, for al-

Fārābī the claims of reason concerning the nature and reality of the physical and imaginal worlds ought to be judged in the light of their metaphysical principles.

Al-Fārābī's view of the nature of the relationship between revelation, intellect and reason may be summarized as follows: Reason is not opposed to the intellect or revelation if it is correctly used. Reason, in fact, should serve both the intellect and revelation. Al-Fārābī is a well-known example of those Muslim thinkers who emphasized the positive aspect of reason as a ladder which leads one to the verities of revelation. There is unity of revelation, intellect and reason in the sense that their ultimate source is one, namely the Divine Intellect.

3.2 Religion, Philosophy and the Sciences

In the *Taḥṣīl al-sa'ādah* al-Fārābī has clearly stated his views on the nature of religion and of philosophy and the relationship between them:

> When one acquires knowledge of the beings or receives instruction in them, if he perceives their ideas themselves with his intellect, and his assent to them is by means of certain demonstration, then the science that comprises these cognitions is *philosophy*. But if they are known by imagining them through similitudes that imitate them, and assent to what is magined of them is caused by persuasive methods,[39] then the ancients called what comprises these cognitions *religion*. And if those intelligibles themselves are adopted, and persuasive methods are used, then the religion comprising them is called popular, generally accepted, and external philosophy.[40]

Al-Fārābī revived the ancient claim that religion is an imitation of philosophy. According to him, both religion and philosophy deal with the same reality. Both comprise the same subjects and both give an account of the ultimate principles of the beings (that is, the essence of the First Principle and the essences of the incorporeal second principles). Both give an account of the ultimate end for the sake of which man is made – that is, supreme happiness – and the ultimate end of every one of the other beings. However, says al-Fārābī, in everything of which philosophy gives an account based on intellectual perception, religion gives an account based on imagination. In everything demonstrated by philosophy, religion employs persuasive methods.

The purpose of prophetic 'imitations' of revealed truths with images and symbols has been explained. The nature of these religious images and symbols needs further discussion. According to al-Fārābī, religion draws the similitudes of transcendent truths from the natural world, the world of arts and crafts, or from the realm of socio-political institutions. For example, the intelligibles of utmost perfection, like the First Cause, the angelic beings or the heavens are symbolized by sensibles which are excellent, perfect, and beautiful to look at.[41] Thus, in Islam, the sun symbolizes God, the moon the prophet, and the stars the companions of the prophet.

Functions of political offices such as the kingship with the whole hierarchy of his subordinates and their respective functions provide images and symbols for the understanding of the hierarchy of beings and the divine acts in the creation and administration of the universe. Human production of arts and crafts provides similitudes of the actions of natural powers and principles by means of which natural objects come into being. For example, the four Aristotelian causes which al-Fārābī also terms the four principles of being may be explained by referring to the principles of the production of art objects. In general, says al-Fārābī, religion attempts to bring the similitudes of philosophical truths as close as possible to their essences.

In Islam, the above views concerning the distinction between religion (*millah*) and philosophy (*falsafah*) have generally been identified with the *mashshā'ī* school of philosopher-scientists to which al-Fārābī belongs. Rahman has shown that this distinction was given a detailed formulation in the later developments in Greco-Roman religious philosophy.[42] However, the fundamental idea intended to be conveyed through that distinction is not something foreign to the perspective of Islamic revelation. The same idea is expressed by the Sufis in terms of the exoteric-esoteric distinction. That idea is that truth or reality is one but its comprehension by the human mind is of various degrees of perfection. Although himself a Sufi, al-Fārābī speaks here as a representative of the philosophical tradition.

In the perspective of the *falāsifah* the two fundamental approaches to the truth are philosophy and religion. What is being contrasted here is not philosophy, understood as a rational system formulated independently of intellection and revelation, and religion, understood as a total revealed tradition. This is quite clear from al-Fārābī's language, as well as from his descriptions of philosophy and religion. The term used by him to refer to religion as distinct from philosophy is *millah* and not *dīn*.[43] This shows that al-Fārābī wishes to contrast philosophy not with a revealed tradition in its totality,[44] but with the exoteric dimension of a revealed tradition. Hence he used the term *millah* rather than the term *dīn*. *Millah* is appropriate since it refers to a divinely sanctioned religious community with its body of beliefs and laws or moral-legal injunctions based on revelation. The external dimension of a revealed tradition should be identified with the beliefs and practices of this religious community.

In the above-cited passage al-Fārābī seems to argue that there are two kinds of philosophy. One is what he calls a popular, generally accepted and external philosophy. From his description of the characteristics of this philosophy and those of *kalām*, particularly the one given in the *Iḥṣa' al-'ulūm*, there is no doubt that al-Fārābī regards *kalām* as an example of this kind. The other is an esoteric philosophy meant for the elite, a philosophy into which only those who are intellectually and spiritually prepared may be initiated.[45] This philosophy can best be described as a science of reality based on the method of certain demonstration (*al-burhān al-yaqīnī*), a method which is a combination of intellectual intuition and logical conclusion (*istinbāṭ*) which is certain. It is therefore a superior kind of knowledge to religion (*millah*), since the latter is based on the method of persuasion (*al-iqnā'*).

Furthermore, for al-Fārābī this philosophy refers to that eternal truth or wisdom (*al-ḥikmah*) which lies at the heart of all traditions. This may be identified with the *philosophia perennis* promulgated in the West by Leibniz and comprehensively expounded in this century by Schuon.[46] Speaking of some of the ancient possessors of this traditional wisdom, al-Fārābī writes:

> It is said that this science existed anciently among the Chaldeans, who are the people of al-'Irāq, subsequently reaching the people of Egypt, from there transmitted to the Greeks, where it remained until it was transmitted to the Syrians and then to the Arabs. Everything comprised by this science was

expounded in the Greek language, later in Syriac, and finally in Arabic.[47]

Referring to the Greeks, al-Fārābī says that they call this knowledge of eternal truth "unqualified" wisdom and the highest wisdom. They called the acquisition of it "science," and the scientific state of mind "philosophy". By the latter they meant the quest and the love for the highest wisdom. According to al-Fārābī, the Greeks also held that potentially this wisdom subsumes all the virtues. For this reason, they called it the science of sciences, the mother of sciences, the wisdom of wisdoms and the art of arts. What they meant, says al-Fārābī, is the art that makes use of all the arts, the virtue that makes use of all the virtues, and the wisdom that makes use of all the wisdoms.

Al-Fārābī seems to be fully aware of the fact that while the essence of this eternal wisdom is one and the same in all traditions, it has not found the same modes of expression in these traditions. He did not, however, say what these modes of expression are in the case of pre-Greek traditions. But he refers to the Greeks, more precisely Plato and Aristotle and particularly the latter, as the originators[48] of new forms of expression and exposition of this ancient wisdom, namely the dialectical or the logical. Knowledge of its forms came to be inherited by Islam through the Syriac Christians.

As we have seen, al-Fārābī defines the highest wisdom as "the most excellent knowledge of the One as the First Cause of all existence and as the First Truth which is the source of all truths.[49] Following Aristotle, al-Fārābī uses the term philosophy to refer not only to this metaphysical knowledge expressed in rational forms, but also the sciences which are derived from the former based on the same method of certain demonstration. Thus, al-Fārābī's philosophy comprises four parts: the mathematical sciences, physics (natural philosophy), metaphysics, and the science of society (politics).

The philosophy-religion distinction is thus envisaged by al-Fārābī in the context of one and the same revealed tradition. But the distinction has a universal validity, being applicable to all revealed traditions. By viewing each tradition in terms of its hierarchic divisions into philosophy and religion, al-Fārābī was able to provide a theory to explain the phenomena of the diversity of religions.

According to him, the many religions differ from one another because the same intellectual and spiritual truths can have many different imaginative representations.[50] There is, however, a unity of all revealed traditions at the philosophical level, since the philosophical account of reality is one and the same for all nations and peoples.[51]

At the same time, al-Fārābī entertains the idea of the relative superiority of one religious symbolism over another in the sense that the symbols and images employed in one religion are closer to the spiritual truths they seek to convey, i.e., are more adequate and effective, than those employed in another religion. It is noteworthy that al-Fārābī is not known to have derogated any religion by name, although he does contend that some religious symbols and images are objectionable or harmful. He writes:

> The imitations of those things[52] differ in excellence; some of them are better and more perfect imaginative representations, while others are less perfect; some are closer to, others are more removed, from the truth. In some the points of contention are few or unnoticeable, or it is difficult to contend against them, while in others the points of contention are many or easy to detect, or it is easy to contend against them and to refute them.[53]

The philosophy-religion distinction as al-Fārābī has formulated it again brings into focus the centrality of the hierarchy of the sciences in his thought. When this distinction is applied to both the theoretical and practical dimensions of revelation previously described, we will arrive at a result which throws further light on the way al-Fārābī treats the religious sciences in his classification in contrast to the philosophical sciences. *Kalām* and *fiqh*, the only religious sciences[54] to appear in his classification, are for him the external or exoteric sciences of the theoretical and practical dimensions of revelation respectively. Metaphysics (*al-'ilm al-ilāhī*) and politics (*al-'ilm al-madanī*) are their respective philosophical counterparts.

3.3 Al-Fārābī's Theory of Methodology

A major theme in al-Fārābī's philosophy of science (*'ilm*) is the question of the degrees of profundity of proofs. He maintains that some sciences are more excellent than others because they employ more perfect methods of arriving at truth claims and of proving them than those used in the latter. His conception of methods of proofs is found in his logical theory which deals primarily with syllogisms.

The term *'ilm* (science) is used by al-Fārābī in several senses. This study is primarily interested in his notion of "science" understood as an organized body of knowledge and as a discipline having distinctive goals, basic premises, and objects and methods of study. For it is the "sciences" understood in this sense which al-Fārābī seeks to classify. Every science so defined is comprised of cognitions (*ma'ārif*) which encompass both conception and assent. In chapter two it was noted that the term conception is variously described by al-Fārābī as understanding the meaning of a thing, what its name signifies or what it is. There are levels of conception ranging from what a name signifies to the perfect definition. By assent, al-Fārābī means the belief that a judgment made about a thing is true. This belief admits of degrees of certainty.

Al-Fārābī describes proof (*dalīl*) in its general sense as an argument or reasoning by means of which the assent that is sought is attained. He mentions three types of proofs, namely syllogism (*qiyās*), induction (*taṣaffuḥ*), and rhetorical (*khuṭābīyah*) proof.[55] Since assent admits of degrees of certainty, al-Fārābī identifies each type of proofs with a particular level of certainty the nature of which is discussed below. In fact, one of the main characteristics of his theory of methodology is the idea of the correspondence between methods of proofs and degrees of certainty attained by the human mind.

Al-Fārābī describes syllogism as a form of argument in which two propositions, called premises, are joined together in such a way that a third proposition, called the conclusion, necessarily follows.[56] The two premises share a common element called the middle term (*al-ḥadd al-awsaṭ*). There are said to be three figures of syllogisms (i.e., the first, second and third figures) corresponding to the three positions of occurrence of the middle term in the premises of a syllogism.[57]

Al-Fārābī calls syllogisms of the first figure, that is syllogisms in which the middle term is the predicate of one premise and the subject of the other, perfect syllogisms because it is self-evident in these cases that the conclusions follow necessarily from the premises. In the case of all other syllogisms the conclusion is also necessary but only known after the syllogisms have been reduced by some process[58] to the perfect syllogisms.

The term "syllogism" in the above description is used in its technical sense of "the form of an argument". The term refers to the special way in which *any* premises are arranged. In the context of this technical meaning al-Fārābī is able to speak of the demonstrative, dialectical, and sophistical arguments as syllogisms for they all share the same form of reasoning.[59] In certain of his treatises the term "syllogism" is used in a more general sense to include rhetorical and poetical arguments. These five kinds of syllogism differ from each other in their "matter" (*mawādd*), that is, in the nature of their premises. There is an hierarchy of syllogistic methods or proofs corresponding to an hierarchy of the premises of a syllogism. This correspondence may best be explained by referring to the four classes of indemonstrable premises grasped without a syllogism: received opinions (*maqbūlāt*), generally accepted opinions (*mashhūrāt*), sense-knowledge (*maḥsūsāt*), and primary intelligibles. For all premises used in a syllogism are ultimately reducible to the four classes.

Demonstrative syllogisms are said to be composed of premises that are true, primary and necessary. Al-Fārābī identifies these premises mainly with primary intelligibles and premises derived from them through valid syllogisms. But he also includes certain kinds of sense-knowledge among the demonstrative premises. Demonstrative proof (*burhān*) is the most certain kind of proof since assent to primary intelligibles is of the degree of necessary certainty, the highest possible. The certain and thus excellent nature of demostration is derived from the certain nature of its premises.

Al-Fārābī accepts the validity of some kinds of sense-knowledge as premises for demonstrative syllogisms. The objects of sense-knowledge which he considers valid for demonstration are those things which are such that they have always been observed to be the same everywhere in the past and present and that no syllogism concluding the opposite exists. He contends that sense-knowledge of such kind affords certainty, which he calls certainty at times to distinguish it from necessary certainty.[60] Demonstrative proof is the method employed in the philosophical sciences alone.

Dialectical syllogisms are said to be composed of premises that are only approximately certain. Al-Fārābī equates such premises with the generally accepted opinions. He defines generally accepted opinions as statements which are acknowledged by "all men or the majority of men, or by all scholars (*'ulamā'*) and men of intelligence (*'uqalā'*) or the majority of these," when no one takes exception to those statements.[61] In al-Fārābī, the main difference between demonstrative premises and generally accepted opinions appears to be in the manner in which assent is given to each kind of premise. Whereas assent to demonstrative premises is produced on the basis of intellection and rational examination, generally accepted opinions are acknowledged on the basis of faith or the testimony of other people. But what is generally acknowledged, argues al-Fārābī, is not necessarily true. It is only generally accepted that people should accept such opinions especially if these come from scholars, intellectuals and experts in some field.[62]

To understand why al-Fārābī characterizes generally accepted premises as being only approximately certain, it is useful to recall his classification of assent into certainty, approximate certainty, and trust. Complete certainty is comprised of three types of beliefs: (1) the belief that something is or is not in a specific condition, (2) the belief that that thing cannot be other than it is, and (3) the belief that belief (2) cannot be otherwise. Approximate certainty refers to the first two beliefs only. Likewise, the assent to a generally accepted proposition is in the nature of those two beliefs. For the third belief presupposes a critical examination of the first two beliefs. Yet critical examination is what is lacking in the adoption of generally accepted opinions. In other words, the belief that a generally accepted opinion cannot be otherwise is not scrutinized.

The primary consideration in accepting generally accepted opinions is not that they are true or valid but simply that there is a general consensus concerning them. Moreover, "because fame or popularity and not truth is the standard for their acceptance, opinions which are contrary to or contradict one another may both be generally accepted simultaneously"[63] For this reason, al-Fārābī ranks dialectical proof below demonstration in excellence. He describes it as the kind of proof generally employed in the religious sciences like *kalām* and jurisprudence.

Dialectical proof, however, is ranked above the sophistical, rhetorical and poetical in excellence. Strictly speaking, says al-Fārābī, rhetorical and poetical arguments are not syllogistic. A necessary condition for an argument to be syllogistic is that it should be composed of at least two simple statements connected together in the manner displayed by the three figures of syllogism previously described. Al-Fārābī argues that one of the main characteristics of rhetorical proofs is the omission of one of the two main premises from which the conclusion results.[64] The popular view is that rhetorical proofs are syllogisms. The omission of a premise is claimed to be done for the sake of brevity. Since al-Fārābī grants the fact that demonstrative and dialectical syllogisms may omit a premise for the sake of brevity or because the premise is too obvious, he has to distinguish between the two "types" of omission. He maintains that the real purpose of omitting a premise in rhetorical arguments is to hide the fact that the omitted premise is unsuitable for a syllogism.[65] Rhetorical proofs may then be considered as syllogisms only if the popular understanding or usage of the term is to be accomodated. In certain treatises of his, al-Fārābī does make such an accomodation.

The main aim of rhetorical arguments is to persuade the hearer to accept whatever beliefs by making his soul feel contented with those arguments, without however reaching certainty.[66] Al-Fārābī identifies rhetorical premises generally with received opinions. Of the four indemonstrable premises, received opinions are ranked the lowest in terms of the degree of certainty which the premises afford. Like generally accepted opinions, received opinions secure their adoption on the basis of the testimony of other people. In al-Fārābī's view, the certainty associated with received opinions is only of the level of trust for these opinions are not even generally acknowledged. Rather, these opinions are simply received from a single individual or at best a group of people,[67] without investigating whether or not the thing believed could be otherwise. Thus, trust refers only to the first of three beliefs comprised in complete certainty.

As for poetical arguments, these too do not fulfill the necessary condition for a syllogism. In poetical arguments, the two main premises need not even be connected. Poetical arguments simply seek to imitate objects through speech and to produce imagination in the people's souls so that they may desire or avoid a given object. The premises of poetical arguments are imagined propositions. That is to say, the meanings associated with the words used in the propositions are imitations of the things signified by the words.

Al-Fārābī maintains that, from the point of view of discovering truths or verification (*taṣḥīḥ*) of truth claims, both rhetorical and poetical arguments are of no use to the sciences.

Al-Fārābī appears to distinguish between sophistical arguments that are syllogistic and those which are not. In either case, sophistical reasoning does not lead to the kind of certain knowledge that is sought in the sciences. For when the form of the argument is syllogistic, its premises only appear to be generally accepted when in fact they are false. Or, when the premises are generally accepted, the arguments only appear to be sound syllogisms when in fact they are not.[68] The main aim of sophistical arguments, says al-Fārābī, is on the one hand to lead the hearer into error, falsification (*tamwīh*) and trickery (*makhraqah*) and on the other hand to produce a favorable opinion of the speaker as a man of wisdom and knowledge when in fact he is not.[69]

As for the method of induction, al-Fārābī defines it as the examination of all the particular instances subsumed under a universal subject in order to determine whether a predicate or judgment made about that thing applies or does not apply to it universally.[70] In chapter two, a distinction was made between scientific (i.e. complete) induction and ordinary induction. Al-Fārābī does not recognize the latter as a reliable method of proof to be used in the philosophical sciences since it does not afford certainy. In his view, induction should be identified with the dialectical method because at best it effects an assent of the degree of approximate certainty. Induction, he says, is widely employed in the religious sciences, more so than in the philosophical sciences.

It may be inferred from the above discussion that, from the point of view of the sciences which al-Fārābī seeks to classify, the proofs which really matter are the demonstrative and the dialectical. The philosophical sciences are deemed more excellent than the religious sciences because the former employ the demonstrative method and the latter the dialectical.

There is still a need to explain al-Fārābī's contention that the philosophical sciences themselves are distinguished from each other in excellence on the basis of the methods of proof they employ. Since the central idea in al-Fārābī's conception of methodology is the hierarchy of premises employed in syllogistic proofs, the explanation may be sought in the nature of the premises in the philosophical sciences. He has distinguished between two kinds of demonstrative premises: (1) sense-knowledge which affords "certainty at times" and (2) the primary, necessary premises which afford complete certainty. Metaphysics and mathematical sciences like geometry and arithmetic employ only premises of the second category. Natural philosophy employs premises which are substantially drawn from sense-perceptions. A clearer picture of the hierarchy of the philosophical sciences will emerge in chapter six when I discuss the position accorded by al-Fārābī to each of these sciences in his classification.

ENDNOTES

Chapter 3

[1]Faith, like love, is the participation of man in what he does not know with immediacy but which he yet accepts with his mind and heart. Knowledge which removes the veil of separation does not annul this faith but comprehends it and bestows upon it a contemplative quality." S.H.Nasr, *Knowledge and the Sacred*, Crossroad, New York, 1981, p. 326.

In al-Fārābī's terminology, faith corresponds to acceptance of truths in the form of images (*mutakhayyalah*) while philosophical knowledge corresponds to acceptance of truths as cognitions (*mutasawwarah*). "....the ones who follow after happiness as they cognize it and accept the principles as they cognize them, are the wise men. And the ones in whose souls these things are found in the form of images, and who accept them and follow after them as such, are the believers." *Medieval Political Philosophy*, p. 41.

[2]See M.Mahdi, *Kitāb al-millat*, p. 47.

[3]*Medieval Political Philosophy*, p. 36

[4]Al-Fārābī uses the term *al-'aql al-munfa'il* interchangeably with *'aql bi'l-fi'l* (the actual intellect).

[5]In several of his works, al-Fārābī clearly says that the First Cause is God insofar as He is the Cause of all causes and the Cause of all existence, and as He is envisaged in His ontological aspects. See, for example, *al-Siyāsat al-madaniyah*, p. 31; *Iḥṣā' al-'ulūm*, p. 100; *Fuṣūl al-madanī*, p. 173.

[6]F.Najjar, *op. cit.*, pp. 36-7

[7]His more detailed exposition of this doctrine is given in his *al-madīnat al-fāḍilah*. See *The Perfect State*, especially chapters 13 and 14.

[8]In section 2 of his *Kitāb al-millat al-fāḍilah* (M.Mahdi, *op.cit.*, pp. 44-5), al-Fārābī enumerates the ideas comprised by the theoretical dimension of religion. These encompass the fundamental articles of the Islamic faith (*īmān*). See also his *al-madīnat*

al-fāḍilah (chapter 33, pp. 146-50) where these ideas are dscribed as "the common things which must be known by all the inhabitants of the virtuous city" and where he discusses the different levels of acceptance of these same ideas by different people.
⁹Al-Fārābī mentions mathematical objects as an example of such existents. Man cannot alter, for example, the property of oddness of odd numbers to the property of evenness while those numbers still remain odd.
¹⁰*Fuṣūl al-madanī*, sect. 48, p.48
¹¹With Ibn 'Arabī, for example, we have an analogous division of the prophetic functions: the saintly and the legislative or law-giving functions. Every prophet is a saint but not every saint is a prophet. Moreover, if for al-Fārābī the prophet as sage and philosopher is greater than the prophet as law-giver (*Fuṣūl al-madanī*, sect. 89, pp.72-3), so for Ibn 'Arabī the prophet as saint is greater than the prophet as law-giver.
¹²These embrace the body of religious rites and practices and the domain of moral-legal injunctions which govern the whole sphere of human life. See *al-Madīnat al-fāḍilah*, chaps. 25 and 26); also M.Mahdi, *op. cit.*, p.46).
¹³*The philosophy of Plato and Aristotle*, pp. 26-8.
¹⁴*Fuṣūl al-madanī*, sect. 49, p. 48
¹⁵*The Philosophy of Plato and Aristotle*, p. 45 (sect. 56)
¹⁶*al-Siyāsat al-madanīyah*, p. 34.
¹⁷*Medieval Political Philosophy*, p. 40
¹⁸The respective Arabic terms for intellectual and spiritual are *'aqlīyah* and *rūḥanīyah*. These terms refer to two different aspects of the human soul. Intellectual experience refers to the vision of truth or acquisition of knowledge by the highest cognitive faculty of the human soul, namely *al-'aql*. Spiritual experience refers to the transformation of the very substance of the human soul so that it becomes God-like, made possible by the soul's total submission to, and realization of the truth gained through intellectual experience. See S.H.Nasr, *Knowledge and the Sacred*, p. 311.
¹⁹*al-Madīnat al-fāḍilah*, p. 115; *La Vertueuse*, p. 75
²⁰See F.Rahman, *op. cit.*, p. 36
²¹Al-Fārābī, *Fuṣūṣ al-hikam*, p. 86 (*faṣṣ* 46). The translation is from Rahman, *op. cit.*, p. 74
²²"The imaginative faculty represents many of the things supplied by the active intellect by means of visible sensibles which imitate them. The objects of representation are in turn impressed on the faculty of common sense; once their impressions are present in the faculty of common sense, the faculty of sight is affected by those impressions, and they are impressed on it". See *The Perfect State*, p. 223.
²³This term occurs many times in chapter fourteen of *al-Madīnat al-fāḍilah*. See *ibid*, pp. 211-27.
²⁴Impressions conveyed to the imaginative faculty by the appetitive faculty pertain to qualities or dispositions like wrath, desire or emotion in general. The imaginative faculty imitates the appetitive faculty by putting together the actions which are actually brought about by the instruments of the latter faculty when a desire or emotion occurs within it (i.e., the appetitive). The imagining or 'imitating' of such actions sometimes stir the subordinate limbs and organs to perform in reality those actions. For example, a man sometimes gets up in his sleep and walks away without any exterior cause. This is because that action is imitated by the imaginative faculty as if it had happened in reality. See *The Perfect State*, pp. 217-9.
²⁵According to many hadīths, a number of the companions of the prophet witnessed some of the bodily effects of his experience of revelation. Among these visible signs were his profuse sweating and the tremendous heaviness of his body. The prophet's camel was said to have fallen under the heavy impact of his body.

[26]"Revelation is a kind of cosmic Intellection whereas personal Intellection is comparable to a Revelation on the scale of the microcosm". F.Schuon, *Logic and Transcendence*, Perennial Books, London, 1984, p. 33.

Thus, concerning the various forms of wisdom that are the fruits of this intellection, Nasr writes: "In the more universal sense of 'revelation', they are in fact the fruit of revelation, that is, a knowledge which derives not from a purely human agent but from the Divine Intellect, as in fact they were viewed by the long tradition of Islamic, Jewish, and Christian philosophy before modern times." *Knowledge and the Sacred*, p. 13.

[27]*The Philosophy of Plato and Aristotle*, pp. 35-6.

[28]Al-Fārābī, *Sharḥ risālat zainūn al-kabīr*, p. 9. Translation from Rahman, *op. cit.*, p. 63.

[29]"Until they acquire the theoretical virtues, they ought to be instructed in things theoretical by means of persuasive methods. They should comprehend many theoretical things by way of imagining them. These are the things - the ultimate principles and the incorporeal principles - that a man cannot perceive by his intellect except after knowing many other things." The *Philosophy of Plato and Aristotle*, pp. 35-6.

[30]"A premiss known to be factual is either known by a syllogism or it is not arrived at by a syllogism. Those premises which we have ascertained to be factual but whose truth is not derived from a syllogism, are of four kinds:(l) received, (2) well-known (3) sensory, and (4) intellectual by nature.....The only known statements to be added to these four kinds are those which can be known from a syllogism." *Short Pr. Anal.*, pp. 58-9. As noted in chapter 1 (n. 38), this work is in fact a treatise on syllogism.

[31]The different kinds of reasoning are discussed below under the section dealing with al-Fārābī's theory of methodology.

[32]The nature of this syllogistic discourse as seen by al-Fārābī is discussed under his theory of methodology.

[33]F.Schuon, *op. cit.*, p. 37.

[34]*The Philosophy of Plato and Aristotle*, sect. 54, p. 43.

[35]This is true of Islamic intellectual tradition in general:

> "The link between intellect and reason is never broken, except in the individual ventures of a handful of thinkers, among whom there are few that could properly be called scientists. The intellect remains the principle of reason; and the exercise of reason, if it is healthy and normal, should naturally lead to the intellect. That is why Muslim metaphysicians say that rational knowledge leads naturally to the affirmation of the Divine Unity".

S.H.Nasr, *Science and Civilization in Islam*, p. 26.

[36]S.H.Nasr, *Sufi Essays*, SUNY Press, Albany, 1972, p. 54.

[37]*Fuṣūl al-madanī*, p. 42 (sect. 31).

[38]This is clearly implied by al-Fārābī when he classifies the indemonstrable premises of syllogisms into four classes according to the degree of certainty which they afford. These premises are discussed in detail under al-Fārābī's theory of methodology.

[39]The term *iqnāʻ* here translated as "persuasive" is used in a more general sense than persuasion which al-Fārābī identifies with rhetorical arguments. See my discussion below (p. 133-4).

[40]*The Philosophy of Plato and Aristotle*, pp. 44-5.

[41]*The Perfect State*, p. 219.

[42]F.Rahman, *op. cit.*, pp. 61-3.

[43]Both these terms are Qur'anic and are usually rendered as community and religion respectively. I cite here one verse in which the two terms appear together: Say: Lo! As for me, my lord hath guided me unto a straight path, a right religion (*dīn*), the community of Abraham (*millat Ibrāhīm*), the upright who was no idolator (VI:161).

For a good discussion of the etymological meanings of the word *dīn* and its derivative concepts, see S.M.N.al-'Attas, *Islam: The Concept of Religion and the Foundation of Ethics and Morality*, Kuala Lumpur, 1976.

[44]'....the Arabic term *al-dīn* means at once tradition and religion in its most universal sense, while religion as used in its widest sense is understood by some to include the application of its revealed principles and its later historical unfolding so that it would in turn embrace what we mean by tradition...." S.H.Nasr, *Knowledge and the Sacred*, p. 73.

For al-Fārābī, as we shall see, his philosophy refers to the inner or esoteric dimension of *al-dīn* understood in the sense indicated above by Nasr.

[45]Al-Fārābī considers Aristotle's philosophy to be such a philosophy. Its complete realization demands the fulfillment of certain conditions in the manner prescribed by the traditional methods of its study. Thus, in his treatise entitled *Fī mā yanbaghī an yuqaddam qabla ta'allum falsafah Aristū* (*On What Must Precede the Study of Aristotle's Philosophy*), Cairo (1910), al-Fārābī lists nine different conditions which ought to be fulfilled by anyone who aspires to embark upon the philosophical path. These include the knowledge of the different schools of philosophy, the ultimate goal of philosophy and knowledge of the moral and spiritual state which the student of philosophy should acquire (see p. 2). The power of the rational faculty becomes sharpened when man purifies his soul and directs his desire toward the Truth instead of the sensual pleasures (p.15).

[46]See for example his *The Transcendent Unity of Religions, Stations of Wisdom, Light on the Ancient Worlds, Esoterism as Principle and Way, Gnosis: Divine Wisdom*, besides the works that I have cited previously.

[47]*The Philosophy of Plato and Aristotle. p. 43.*

[48]"The philosophy that answers to this description was handed down to us by the Greeks from Plato and Aristotle only. Both have given us an account of philosophy, but not without giving us also an account of the ways to it and of the ways to re-establish it when it becomes confused or extinct." *Ibid*, p.50.

Although al-Fārābī's view of Plato and Aristotle is that "their purpose is the same and that they intended to offer one and the same philosophy," he is also clear about the fact that their methods of expounding that philosophy are different. Thus, he says at the beginning of his *The Philosophy of Aristotle:* "Aristotle sees the perfection of man as Plato sees it and more. However, because man's perfection is not self-evident or easy to explain by a demonstration leading to certainty, he saw fit to start from a position anterior to that from which Plato started." *Ibid*, p. 71. In his *Harmonization of the Opinions of Plato and Aristotle*, he attributes Aristotle's methodological departure from Plato to the former's "excess of natural power". What al-Fārābī means is that Aristotle in formulating his metaphysical truths does not begin from those truths but from the world of the senses and common experience and then goes to ascend to those truths.

[49]*Fusūl al-madanī*, sect. 34, pp. 43-4

[50]*Medieval Political Philosophy*, p. 41.

[51]Al-Fārābī's view of the philosophical unity of religions is almost identical to Schuon's idea of the transcendent unity of religions. See F.Schuon, *The Transcendent Unity of Religions*, Wheaton-Madras-London, 1984. In his introduction to this edition

of the work, Huston Smith writes: "There is unity at the heart of religions. More than moral it is theological, but more than theological it is metaphysical in the precise sense of the word.... that which transcends the manifested world." (p. xxiii).

[52]These refer to the principles of beings, their ranks of order, happiness, and the rulership of the virtuous cities.

[53]*Medieval Political Philosophy*, p. 41.

[54]*Religious* as this term is commonly understood and not in the sense of having the monopoly or exclusivity to the knowledge of God and of that related domain which is usually considered as being the concern of religion. For, in the universal sense of the term "religious," al-Fārābī's metaphysics and politics are also of a religious character.

[55]Galston (*op. cit.*, pp. 62-96) has given a good analytical account of al-Fārābī's views concerning these three types of proofs.

[56]*Short Pr. Anal.*, p. 59.

[57]In syllogisms of the first figure the middle term is the predicate of one premise and the subject of the other; in the second the middle term constitutes the predicate of both premises; in the third the middle term is the subject of both premises.

[58]*Short Pr. Anal.*, p. 62. The process of reducing "imperfect" syllogisms into the perfect ones involves the addition of one or more propositions. Al-Farabi mentions two methods by which the reduction process could be realized. One is the method of conversion (*'aks*). In this case the additional propositions are implicit in the original premises. The other method is called *al-iftirād*, translated by Rescher as *ecthesis*. In this method the additional propositions are laid down as hypotheses. See pp. 66-72 of the above treatise for al-Fārābī's discussion of the two methods.

[59]See *Introductory 'Risālah' on Logic*, pp. 255-6.

[60]He calls the certainty derived from sense-knowledge "certainty at times" because its duration may not extend necessarily to the future.

[61]*Short Pr. Anal.*, p. 58.

[62]Galston, *op. cit.*, p. 137.

[63]*Ibid*, p. 139.

[64]Al-Fārābī, *Kitāb al-khaṭābah* (*The Book of Rhetoric*), in J.Langhade (ed.), *al-Fārābī: deux ouvrages inedits sur la rhetorique*, Beirut, 1971, p. 63.

[65]*Ibid*, p. 69

[66]*Introductory 'Risālah' on Logic*, p. 231.

[67]*Short Pr. Anal.*, p. 58.

[68]Galston, *op. cit.*, pp. 67-8.

[69]*Introductory 'Risālah' on Logic*, p. 231.

[70]*Short Pr. Anal.*, 88.

CHAPTER 4

THE ONTOLOGICAL AND THE ETHICAL BASES OF THE HIERARCHY OF THE SCIENCES

In chapter two, I asserted that the ontological basis of the hierarchy of the sciences is derived from al-Fārābī's hierarchically ordered view of the world,[1] and the ethical basis is derived from his hierarchical ordering of human practical and spiritual needs. In this chapter, I will examine how each basis is conceptually related to al-Fārābī's classification of the sciences.

4.1 The Ontological Basis

The idea of a hierarchy of beings, expressed by al-Fārābī by the term *marātib al-mawjūdāt*, found rich formulations and expositions in the writings of many of the great thinkers throughout the ages.[2] As pointed out by Arthur O.Lovejoy,[3] this idea "has been one of the half-dozen most potent and persistent presuppositions in Western thought" and the phrase *The Great Chain of Being* used to express it "was long one of the most famous in the vocabulary of Occidental philosophy, science, and reflective poetry." Until a little more than a century ago when the modern idea of an evolutionary chain, which is its temporalized, linear version, began to gain currency in Western thought, it was, Lovejoy says, "probably the most widely familiar conception of the general scheme of things, of the constitutive pattern of the universe."[4]

In the Islamic philosophical tradition itself, al-Fārābī appears to have been the first to give a systematic treatment of the hierarchy of beings in terms of a hierarchy of intelligences and souls and their effusion or emanation (*fayḍ*) from God. This theory, which was without doubt influenced by the Plotinian cosmological scheme, was inherited by Ibn Sīnā with further elaboration and certain modifications.[5] In his overall exposition of the idea, however, which includes some other schemes of classification of beings, al-Fārābī has drawn his materials from diverse sources, Aristotelian as well as Neoplatonic. But the final synthesis is characteristically Farabian.[6]

Al-Fārābī discusses the doctrine of hierarchy of beings mainly in his two major works, namely *al-Siyāsat al-madanīyah* and *al-Madīnat al-fāḍilah*. The term he uses for being or existent is *mawjud*. There are many beings (*mawjūdāt*) and they vary in excellence. He offers several schemes of hierarchically ordering the beings. In one scheme, the hierarchy, in a descending order of perfection, is given as follows:[7]

(1) God who is the cause of the existence of all other beings
(2) The angels which are completely immaterial beings
(3) The celestial bodies
(4) The terrestrial bodies

In another scheme, the hierarchy of beings is described in terms of the six grades (*marātib*) of incorporeal principles (*mabādi'*) which govern the constitution of bodies and their accidents.[8] In the descending order of their ranks, these "principles" are: (1) The First Cause (*al-sabab al-awwal*), (2) The Second Causes (*al-asbāb al-thawānī*), (3) the active intellect (*al-'aql al-fa''al*), (4) soul (*nafs*), (5) form (*ṣūrah*), and (6) matter (*māddah*). It is necessary to explain al-Fārābī's usage of the term "principle" here since the Arabic term *mabda'* of which it is a rendering is used with a wide variety of meanings. Etymologically, the term *mabda'* conveys the basic idea of that which is primary or essential and that which is anterior to something else.[9] Al-Fārābī dwells on this idea of anteriority in his logical works.

According to him, a thing is said to be anterior to another in five ways: (1) in time (*zaman*), (2) by nature (*bi'l-ṭab'*), (3) in rank (*martabah*), (4) in excellence (*faḍl*), nobility (*sharaf*) and perfection (*kamāl*), and (5) as cause (*sabab*) of the existence of the other thing.[10] Al-Fārābī understands "cause" in its Aristotelian sense as referring to the material, formal, efficient or final cause of the existence of a particular being. A cause may be proximate or remote, essential or accidental, universal or particular, and actual or potential.[11] His description of the six listed "principles" shows that each of these principles may be said to be anterior to either celestial or terrestrial bodies or both in at least one of the five ways. But the only common element of anteriority with respect to these bodies shared by the six principles is the idea of anteriority as cause understood in one or more of the senses indicated above. For example, prime matter (*māddah ūlā*) is anterior to terrestrial bodies in time and as cause but not so in excellence.[12]

By "principle" (*mabda'*) al-Fārābī usually means "cause."[13] The six listed principles refer to causes of the existence of bodies. Thus, for example, he says that the soul is a principle of the animate substance[14] as an agent, as a form and as an end.[15] The idea of *mabda'* as something primary and essential appears in al-Fārābī's discussion of the nature and goal of the philosophical sciences. For him, knowledge of the causes of things constitutes essential elements of definitions and serves as a source of middle terms to be employed in demonstrative proofs. The primary goal of the sciences is to gain knowledge of the causes of all things.

In yet another related scheme of classifying beings, al-Fārābī divides the genera of beings into three kinds, according to the number of their causes.[15] The first admits of having no cause at all for its existence. Al-Fārābī is here referring to the First Being (*al-mawjūd al-awwal*) or the First Cause which he says is the ultimate principle for the existence of all other beings. Regarding this ultimate principle, we can have only the principles of our knowledge of it and not the principles of its being. The second possesses all the four causes, that is the material, the formal, the efficient, and the final. This second kind refers to the genera of sensible bodies, including the celestial ones. The third admits of having only three causes. Beings belonging to this kind do not possess the material cause. These are the completely immaterial beings other than the First Being and beings which in their essence are not bodies but inhere or dwell in bodies.

It is possible on the basis of these three closely related schemes of classification of beings to establish the ontological basis of al-Fārābī's hierarchy of the sciences.

4.1.1 **The Subject-Matter of Metaphysics**

The highest philosophical science is metaphysics (*al-'ilm al-ilāhī*) because its subject-matter is comprised of the absolute incorporeal beings which occupy the highest rank in the hierarchy of beings. In religious terminology, these incorporeal beings refer to God and the angels. In philosophical terminology, they refer to the First Cause, the second causes, and the active intellect.

Absolutely incorporeal beings are beings which are neither bodies nor in bodies. They have never been nor will they ever be in bodies.[16] They are intellects and intelligibles in act. The First Cause is the most perfect Being, having no associate or contrary, is necessary, self-sufficient, and beyond definition,[17] and completely transcendent with respect to all other beings.

The First Cause is the immediate cause (*al-sabab al-qarīb*) of the second causes and the active intellect.[18] These latter are often grouped by al-Fārābī under the same class of existents that he calls "second beings" (*al-mawjūdāt al-thawānī*). In one of the above classifications of beings, the active intellect has been mentioned separately as an ontological principle. I will explain later why al-Fārābī has done so. He also calls the second beings the separate intelligences (*al-'uqūl al-mufāriqah*).

The separate intelligences come into existence from First Being through the metaphysical process of 'emanation' (*fayḍ*).[19] They are arranged in a fixed order of rank and merit. The order of their generation is the order of their ranks within the hierarchy of second beings. As a result of the self-thought of the First Being, a second existent called the 'second intellect' permanently emanates from the former. The 'second intellect' thinks of its own essence and also thinks the First Being. As a result of the latter kind of thinking, the 'third intellect' necessarily comes into existence; and as a result of the former intellectual activity, the First Heaven follows necessarily. This process of generation of intellects and the heavens, whereby each intellect thinks the First Being to produce a lower intellect and thinks of its own essence to produce a celestial body continues until it reaches its termination with the eleventh intellect and the ninth heaven.[20] The last and lowest separate intellect is the active intellect; the last and lowest heaven, the heaven of the moon.

The ten separate intellects are, next to the First Cause, the most excellent of beings, since they are free from all matter and potentiality. Each one of these intellects is unique in its species. The active intellect alone among them is not a generator of another separate substance or of a heavenly body. It is not a principle of the celestial world, but of the sublunar world. As the last of the heavenly intellects, it serves, together with the sphere of the moon, as the link between heaven and earth.[21] The beings of the terrestrial world owe their perfections to both the active intellect and the celestial bodies.[22]

In its capacity as an agency which governs the sublunar world, the active intellect depends on the celestial bodies for the substrata upon which it acts.[23] With respect to humanity, the active intellect acts as a principle of the human intellect, that is as an end, as a form, and as an agent.[24] In my view, it is because of this special role and function of the active intellect as a principle of terrestrial beings that al-Fārābī mentions it as a separate ontological principle from the rest of the second beings in his hierarchy of the principles of bodies.

4.1.2 The Subject-Matter of Natural Science

The lowest of the philosophical sciences is natural science (*al-'ilm al-ṭabī'ī*) because its subject-matter is comprised of terrestrial bodies, which possess the lowest rank in the hierarchy of beings. More precisely, natural science deals with natural bodies for, as will be explained in section 4.1.4, al-Fārābī distinguishes between terrestrial bodies which come into existence by nature and those whose existence is brought about by human will. Al-Fārābī divides natural, terrestrial bodies into the following different grades: (1) rational animals, (2) non-rational animals, (3) plants, (4) minerals, and (5) the four elements. These natural bodies are described as belonging to the world of generation and corruption in contrast to the incorruptible and eternal nature of the celestial world.

Each natural body is made up of form (*ṣūrah*) and matter (*māddah*).[25] Al-Fārābī maintains two kinds of existence of bodies, namely potential existence and actual existence. A body is merely a potential entity as long as its matter continues to exist without its form. A body becomes actually existing only when its form is present. None of the natural bodies is actual "from the very outset." 'In the beginning' all natural bodies exist only potentially in their 'common prime matter' (*al-māddat al-ūlā al-mushtarikah*),[26] an incorporeal existent which al-Fārābī says is the eternal outcome of 'celestial matter.'[27]

Out of this prime matter arise the first natural bodies, namely the four elements (*usṭuqussāt*): fire, air, water, and earth. As a result of the mixing of the four elements in various proportions and degrees of complexity, the other kinds of natural bodies come into actual existence. The source of the forms by which the natural bodies become actual may be traced to the active intellect.

Qualitative differences in the mixtures of the elements in the different species of natural bodies are related to the presence of entities called souls within these bodies. Al-Fārābī accepts the Aristotelian definition of the soul as the form and entelechy of the body. There are different kinds of souls depending on the kinds of bodies with which they enter into relations. Each kind of soul is described in terms of its powers or faculties. The most perfect kind is the human soul which in addition to having the faculties of the plant and non-rational animal souls possesses the rational faculty.[28]

The different branches of natural science deal with the generation and properties of the different mixtures of the four elements in natural bodies and in the case of ensouled bodies with the nature and functions of the faculties of their respective souls. Thus, in al-Fārābī, psychology is a branch of natural science.

4.1.3 The Subject-Matter of Mathematics

The mathematical sciences (*'ulūm al-ta'ālīm*) and political science (*al-'ilm al-madanī*) appear to occupy a kind of intermediate position between metaphysics and natural science. To see that this is so, it is necessary to identify the subject-matter of each of these sciences and to show that it is comprised of beings which ontologically are situated between the eternal beings studied by metaphysics and the natural bodies studied by natural science.

According to al-Fārābī, the subject-matter of mathematics is comprised of numbers and magnitudes (*a'ẓam*).[29] By magnitudes he means lines, surfaces, and solids, which are said to be continuous quantities (*al-kam al-muttaṣil*).[30] Numbers are discreet quantities (*al-kam al-munfaṣil*).[31] Numbers and magnitudes exist either as abstract or concrete quantities. As concrete quantities, they exist in material objects in such forms as weight, shape, color, and motion. As abstract quantities, that is, as pure numbers and magnitudes, they exist in the human mind as intelligibles which have been stripped off their accidental attributes and material attachments.

In chapter two, it was asserted that al-Fārābī confers an ontological status to the intelligibles acquired by the actual human intellect in the totality of beings. Moreover, these mathematical intelligibles exist in the active intellect. These mathematical intelligibles are said to be a kind of 'intermediate' beings for they are neither like the metaphysical intelligibles which have never been in matter nor like the natural intelligibles which are not completely free of mater.

Al-Fārābī's mathematics also deals with numbers and magnitudes as these entities inhere in other beings. These "other beings" range from the celestial bodies, which fall outside the domain of natural science, to the natural objects studied by natural science. Consequently, some branches of his mathematics, like the study of weights and mechanics, are found to be closer to natural science than are the other branches. But mathematics studies natural bodies only insofar as these bodies possess the mathematical "properties of measurement and orderly proportions, composition and symmetry"[32] by virtue of the fact that either numbers or magnitudes or both are inherent in them.

Al-Fārābī maintains that there are beings in which number and magnitude are inherent essentially. What he means is that number and magnitude enter into the very definitions of these beings. These beings refer to the celestial bodies, including light which al-Fārābī makes the subject-matter of his mathematical science called *'ilm al-manāẓir* (optics). Al-Fārābī regards the celestial bodies as mathematical entities. It is pertinent to mention here the fact that the arabic term *hay'ah* used by al-Fārābī to refer to the state of a thing also means the mathematical forms of the heavens. Thus, in Islamic science, astronomy is often called *'ilm al-hay'ah* (the science of the shape of the heavens).[33]

It is necessary to make a brief reference to the position of the celestial bodies within al-Fārābī's hierarchy of beings. In *al-Madīnat al-fāḍilah,* al-Fārābī presents the view that the celestial bodies occupy an intermediate ontological position between the separate intellects and terrestrial bodies.[34] According to him, the celestial bodies belong to the same genus as the terrestrial bodies because they have matters and forms by which they become substances. However, celestial matter and celestial forms are ontologically more excellent than their terrestrial counterparts in the sense explained below.

The form of each celestial body cannot have contraries and the substratum of this form cannot receive any other form than the one it has and cannot be without it.[35] This means that, in contrast to sublunar matter which can be linked to contrary forms in turn, the matter of each celestial body is permanently linked to the same form. As a result, each body is the only representative of its species. Moreover, celestial matter is eternally actualized and provided with form, unlike the common prime matter of terrestrial bodies which 'in the beginning' was merely potential existence.

Al-Fārābī maintains that the form of each celestial body is actual intellect (*'aql bi'l-fi'l*).[36] But this intellect is inferior to the separate intellects in that it is not completely free from matter and substratum and in that the objects of its thought are not entirely intellects. In this respect, a celestial body has something in common with man. However, unlike the human soul, the soul of a celestial body[37] has neither the sensitive nor the imaginative faculty nor earthly desires or emotions found in man.[38] The soul of a celestial body has only the intellectual faculty and the power of desire associated with this faculty. Its joy and pride in its own essence and its love for the First Being are all intellectual in nature.

Al-Fārābī also maintains that celestial bodies have the finest and most excellent of whatever they have in common with terrestrial bodies. They have the best of shape, which is the spherical and they have the best of visible quality, which is light. And their motion is the best of possible motions, which is circular.[39]

It is clear from the above explanation that al-Fārābī has subordinated the celestial bodies to the separate intellects in his hierarchy of beings and at the same time views them as ontologically superior to terrestrial bodies. He states explicitly that the celestial bodies constitute an intermediate domain of inquiry between natural science and metaphysics.[40] Correspondingly, astronomy, a major branch of al-Fārābī's mathematics, is counted among the intermediate sciences. So is his optics which deals with the related subject of light. Al-Fārābī believes in the incorporeal nature of light.[41]

Mathematical objects are also said to be intermediate between metaphysical and natural objects in the following sense. The genus of beings studied by natural science possess all the four Aristotelian causes that are related to the same genus; metaphysical beings (excluding the First Being), three causes related to the same genus: they do not have the material cause. As for the causes of numbers and magnitudes, there are two possibilities. Pure numbers and magnitudes possess only one cause related to the same genus, namely the formal cause. Their efficient and final causes belong to a different genus, namely the separate intellects. Numbers and magnitudes possess all the four causes only when they exist in material objects.[42]

To summarize, we may assert that the main branches of mathematics like arithmetic, geometry and astronomy deal with entities which occupy an intermediate ontological position between metaphysical beings and natural bodies. Other branches of mathematics, like the science of weights, mechanics, and engineering, deal with objects that are closely related to the subject-matter of natural science.

4.1.4 The Subject-Matter of Political Science

According to al-Fārābī, political science (*al-'ilm al-madanī*) investigates the various kinds of voluntary actions and ways of life; human tendencies, morals and states of character that lead to these actions and ways of life; the ends for the sake of which they are performed, and how they must exist in man.[43] Political science also distinguishes between ends which are true happiness and those which are presumed to be so although they are not. On the basis of this description, al-Fārābī's political science appears as a very comprehensive discipline. In short, this science deals with that vast domain called man and human society.

Natural science too deals with man but as a natural body. Political science studies man insofar as he possesses will (*irādat*) and choice (*ikhtiyār*) by which he brings about the existence of many kinds of beings, which al-Fārābī calls voluntary beings (*mawjūdāt irādīyah*). Man is studied in this science both as an individual and in relation to his fellow men. In my discussion of al-Fārābī's psychology (chapter two, pp.72-4), I refer to his notion of the appetitive faculty or the power of desire. This faculty is one of the five faculties of the human soul, the other four being the vegetative, the sensitive, the imaginative, and the rational. According to him, human desire is of three kinds. The first is the desire toward the objects perceived by the sensitive faculty; the second the desire toward the objects of imagination; and the third the desire toward the objects perceived by the rational faculty.

Al-Fārābī uses the same term 'will' (*irādat*) to mean both desire in its general sense of all the three kinds and desire of the first two kinds only.[44] What he terms 'choice' refers to desire that is the outcome of deliberation or rational thought alone, and so is unique to man.

Al-Fārābī divides terrestrial beings into three groups:(1) natural beings, (2) voluntary beings, and (3) beings which are both natural and voluntary.[45] Natural beings are those whose existence is brought about neither by art (*ṣinā'at*) nor by human will.[46] Examples are the minerals, plants, and animals. Voluntary beings are those whose existence is brought about by human will; for example, human crafts, and human actions that are the product of choice. The existence of the third group of beings is brought about by the combination of nature and human will, as for example in agricultural activity which al-Fārābī calls an art.

The realm of voluntary beings, as conceived by al-Fārābī, embraces the habitual states of the human soul, man's spiritual, mental (*fikrīyah*) and physical (*badanīyah*) activities[47] as well as the products of these activities. By habitual states (*malakāt*) of the soul, al-Fārābī means the virtues (*faḍā'il*) and the vices (*radhā'il*). Virtues are the states of the soul by which a person does good deeds (*khayrāt*) and fair actions (*af'āl jamīlah*).[48] Vices are the states of the soul by which a person does wicked deeds (*shurūr*) and ugly actions (*qabīḥah*). Al-Fārābī's notions of virtue and vice will be discussed further in the next section (4.2). The existence of voluntary beings presupposes the existence of spiritual and natural beings. This fact perhaps explains why al-Fārābī's political treatises always begin with an exposition of the nature of spiritual and natural beings and their respective ontological positions in the universe.

Al-Fārābī's political science deals with the entire realm of voluntary beings. It should be noted, however, that this science does not seek to study the different kinds of voluntary actions and their products in all their aspects. It is merely concerned with their ethical and societal aspects.[49] The intermediate nature of political science, in relation to natural science and metaphysics, stems from the corresponding nature of the human soul and its activities. Human activities range from the lowest kind, which man shares with the non-rational animals, to the highest kind, which he shares with the separate intelligences. According to al-Fārābī, the most perfect state of the human soul is attained when his theoretical intellect becomes like the separate intellects.[50] For then the human intellect would be able to contemplate God in the most perfect manner possible, without the help of the lower faculties of the soul. Man should emulate the pure intellectual activity of metaphysical beings.

In general, human activities are ontologically inferior to the intellectual activity of metaphysical beings for two main reasons. First, the objects toward which human activities are directed are inferior. It is sufficient to consider the objects of the highest kind of human activity, namely rational activity. Like the intellects of celestial bodies, the human intellect does not think of the separate beings alone. It also thinks of inferior objects like celestial bodies, and, inferior still, terrestrial bodies. Second, human activities, apart from the pure thinking activity realized in certain individuals, can only be performed with the help of various faculties of the soul and their instruments. In performing these activities, the organs of the human body need matter to act upon. In contrast, the thinking activity of metaphysical beings has no need of auxiliaries.

As beings, the different states of the human soul are inferior to the metaphysical states of the separate intelligences. Again, it is sufficient to consider just the virtuous states of the soul. The states of perfection of the human soul are not immutable by nature in the way that the separate intelligences are so. Virtuous states of the soul need to be sustained through the performance of virtuous actions, which need the support of matter for their existence. At the societal level, virtuous ways of life can only be preserved through political actions, which again need the support of material accidents and states.[51]

Human activities, however, are generally superior to the activities of non-rational animals. Spiritual and rational activities of man are the most excellent activities of terrestrial beings because they are the activities of the most excellent faculty of the soul, namely the rational faculty. Even in those kinds of activities which are shared by both man and the non-rational animals, the human ones are deemed superior because, in general, they are not indiscriminate actions but defined and determined actions[52] which result from choice and man's moral character. Human activities, whether individual or collective, reflect a higher level of organization and purposiveness than what is indicated by the activities of the non-rational animals. In general, al-Fārābī, would say, human activities are superior to those of non-rational animals because man possesses a superior kind of will called choice, which arises out of rational deliberation that is unique to him.

In al-Fārābī's account of natural science and political science, there is some overlapping between the two disciplines with respect to their subject-matters. This overlapping pertains mainly to psychology. Al-Fārābī's natural sciences discusses the different faculties of the human soul and establishes the conclusion that man's final perfection is intellectual in nature, namely the perfection of the theoretical intellect. Political science incorporates this idea of man's final perfection into its body of knowledge and makes it the central theme of its inquiry. For natural science does not deal with the question of how man, in the context of his terrestrial existence, may attain that perfection.

To conclude, it may be asserted that the subject-matter of al-Fārābī's political science occupies a kind of intermediate ontological position between the subject-matters of natural science and metaphysics. Being an intermediate science, political science shares certain things in common with the highest science (metaphysics) and with the lowest science (natural science). However, the greater part of the subject-matter of political science clearly lies, ontologically speaking, between the subject-matters of natural science and metaphysics.

4.2 The Ethical Basis

To order the sciences hierarchically on an ethical basis is to order them according to their degrees of usefulness. But usefulness has a meaning only in relation to some final goal. For al-Fārābī – and in this he saw no disagreement between Plato and Aristotle[53] – the ultimate goal of human existence is to attain supreme happiness (*al-sa'ādat al-quswā*). Al-Fārābī equates supreme happiness with the absolute good (*al-khayr 'alā al-iṭlāq*), namely

> that which is chosen and desired for itself and is not chosen, at any time whatever, for the sake of anything else. All else is chosen for its use in the attainment of happiness.[54]

This absolute good is God, the First Cause, since He is "the end beyond which there is no other end to be sought by [means of] happiness."[55] What al-Fārābī means is that in man's final perfection in the life hereafter, man is able to have the vision of God, which constitutes his esternal bliss.[56] This idea of the vision of God as the highest felicity of man has its basis for Muslims in the Qur'ān.[57] It receives a more detailed treatment in the writings of al-Ghazzālī. As al-Fārābī explains, the nature of this vision is intellectual. The perfected intellect of man, which survives after his death, is able to "seize upon the essence of the First Principle without the need for its representation by analogy or example."[58]

Al-Fārābī makes it very clear, however, that this supreme happiness in a life hereafter is conditional upon happiness in the present life *(al-sa'ādat al-dunyā)*, which he calls man's first perfection.[59] In his view, the soul is immortal. Its state of happiness in the afterlife is a consequence of its state of health in the present life. By the health of the soul, al-Fārābī means its states and the states of its parts by which it always does good and noble deeds and fair actions.[60]Likewise, the misery of the soul in the next life is a consequence of its state of sickness in this life. By the "sickness of the soul," he means its states and the states of its parts "by which it always does wicked and evil deeds and ugly actions."[61] If man were to attain earthly happiness or the first prefection, as well as supreme happiness, then it is necessary that all the parts of his soul be in the state of perfect health. In other words, it is necessary that man acquire all the virtues and be rid of all the vices.

4.2.1 **Al-Fārābī's Theory of Virtue**

It is in relation to man's final goal of attaining perfection in this life and supreme happiness in the next that the usefulness of each science should be measured. Since, in al-Fārābī's view, human perfection in the present life results from the acquisition of the virtues, a discussion of some aspects of his theory of virtue is necessary. Moreover, as noted in chapter two (p. 83), al-Fārābī's classification of virtues bears a close relation to his scheme of classification of the sciences. This is because, in his ethical philosophy, virtue is closely related to the acquisition of the different sciences and arts.

4.2.1.1 Classification of Virtues

Like Aristotle, al-Fārābī divides virtues into two fundamental categories: rational (*nuṭuqīyah*) and ethical (*khuluqīyah*).[62] These correspond respectively to the two fundamental parts of the human soul, namely, the rational faculty and the faculty of choice. The rational virtues are of three kinds: the theoretical (*naẓarīyah*), the deliberative (*fikrīyah*), and the artistic (*ṣināʿīyah*). These correspond to the three constitutive parts of the rational faculty. Thus, there are four principal virtues in al-Fārābī's ethical philosophy, corresponding to to the four principal parts of the human soul. The correspondence is summarized below:

Faculties of the Human Soul	**Virtues**
A. The rational faculty (*al-quwwat al-nāṭiqah*)	A. Rational virtues (*al-faḍāʾil al-nuṭuqīyah*)
1. Theoretical intellect (*al-ʿaql al-naẓarī*)	1. Theoretical virtues (*al-faḍāʾil al-naẓarīyah*)
2. Practical intellect (*al-ʿaql al-ʿamalī*)	2. Practical virtues (*al-faḍāʾil al-ʿamalīyah*)
(a) Deliberative faculty. (*al-quwwat al-fikrīyah*)	(a) Deliberative virtues (*al-faḍāʾil al-fikrīyah*)
(b) Faculty of skill (*al-quwwat al-mihnīyah*)	(b) Artistic virtues (*al-faḍāʾil al-ṣināʿīyah*)

B. The appetitive faculty (*al-quwwat al-nuzū'īyah*) associated with the rational faculty, namely, the faculty of choice (*ikhtiyār*).

B. Ethical virtues (*al-faḍā'il al-al-khuluqīyah*)

Since al-Fārābī defines the virtues as the states of the soul by which a person does good deeds, it is necessary to explain what he means by the *good* itself. According to him "good is of two sorts; (a) a sort of which no evil at all is the opposite, and (b) a sort of which evils are the opposite."[63] The first category refers to God alone who is the Supreme Good. He is in fact the ultimate source of all that is good. The second kind of good, which is relative, pertains to both the created and the moral orders. The world, that is to say the created order, is good insofar as it manifests the Supreme Good. Insofar as it implies separation or remoteness from its Source, it possesses a contingent aspect of evil. This is because the world possesses the possibility to set itself against God or as a would-be equal to God.[64] The world is, therefore, evil insofar as it lends itself to the possibility of polytheism and idolatry or leads man away from God.

In the sphere of human life, the good refers to that which brings about participation in the Supreme Good. The moral good is symbolized and actualized by obedience to Divine Law. Its opposite, moral evil, results from man's misuse of his will by disobeying the law. Disobedience to the law constitutes evil because God willed the law in conformity with His nature, which is goodness and perfection.

The good, natural, intellectual as well as moral,[65] contributes to the attainment of man's final perfection, namely his perfect knowledge of God in this life and his vision of God in the hereafter. Al-Fārābī also describes the ultimate perfection of the human soul as the state of its being God-like (*al-tashabbuhu bi'Llāh*) which, he says, is the final goal of the study of philosophy.[66]

4.2.1.2 **Virtues and Their Corresponding Knowledge**

Closely related to the correspondence between the fundamental parts of the human soul and the principal virtues is the correspondence between rational virtues and the sciences. This latter correspondence is illustrated below:

The rational faculty and its parts	**Corresponding knowledge**
A. Theoretical intellect	Theoretical knowledge (1) theoretical parts of the philosophical sciences (2) wisdom (metaphysical knowledge)
B. Practical intellect	Practical wisdom
(1) the deliberative faculty	(a) practical parts of the philosophical sciences possessing such parts (b) practical wisdom in general
(2) the faculty of skill	the practical arts and sciences, and the crafts, for example, carpentary, agriculture, medicine, and navigation

4.2.1.3 **Hierarchy of Virtues**

According to al-Fārābī, the most excellent of the virtues are the theoretical virtues. These he classifies into three types: (1) the excellence of the theoretical intellect, (2) knowledge (*al-'ilm*), and (3) wisdom (*al-ḥikmah*).[67] This knowledge and wisdom are the fruit of the theoretical intellect.

What al-Fārābī means by "the theoretical intellect as a theoretical virtue" is that this intellect has actually acquired knowledge of the primary intelligibles and is in a sound state to gain certain knowledge of the rest of the theoretical intelligibles or existents (*al-mawjūdāt al-naẓarīyah*).[68] The knowledge that is a theoretical virtue is certain knowledge of the existence of theoretical beings and certain knowledge of the cause of their existence. This knowledge is arrived at through demonstrative proofs, based on the primary intelligibles acquired by the theoretical intellect.[69] This knowledge, therefore, refers to the theoretical parts of the philosophical sciences.[70]

Wisdom as a theoretical virtue refers to the highest form of theoretical knowledge. This is the knowledge of the One and of Its relationship with the many. It is the knowledge of the divine essence, attributes, and acts.[71] Quite often, al-Fārābī uses the term wisdom in a more general sense to include knowledge of metaphysical beings other than God. This would mean that in the above division of theoretical virtues, he has placed a part of metaphysics under the category of *al-'ilm* rather than of *al-ḥikmah*.

Al-Fārābī's deliberative virtues are comprised of (1) the excellence of the practical intellect,[72] and (2) practical wisdom (*ta'aqqul*) which he defines as

> the power of excellence of deliberation (*rawīyah*) and production (*istinbāṭ*) of the things which are most excellent and best in what is done to procure for a man a really great good and an excellent and noble end, whether that is happiness or something which is indispensable for obtaining happiness.[73]

According to him, the practical intellect is a deliberative virtue when it enables a person to formulate premises concerning the usefulness of things in general on the basis of experience and personal observation.[74] The excellence of this intellect increases with the increase of experiences in various situations of a man's life. The practical wisdom that is a deliberative virtue covers many types of human activities, including the political, the economic, and the military.[75] Further, this practical wisdom consists of many kinds of practical, rational virtues like cleverness (*kays*), discernment (*dhihnī*), readiness of wit (*dhakā'*), excellence of idea (*jūdat al-ra'y*), and excellence of understanding (*jūdat al-fahm*).[76]

In al-Fārābī's view, the theoretical virtues are superior to the deliberative virtues for the following reason. The excellence of the theoretical intellect is superior to the excellence of the practical intellect. The practical intellect exists to serve the theoretical intellect.[77] In fact, we saw that, for al-Fārābī, man's final perfection is the perfection of the theoretical intellect whose fruit he variously describes as wisdom, the absolute good, and true happiness. It is the theoretical intellect which establishes for man both his immediate and final goals of life that are truly good and virtuous. For example, the meaning of true happiness is perceived by the theoretical intellect. By "immediate goal" al-Fārābī means the things which are indispensable for attaining the final goal.

The practical intellect is indifferent to ends. The deliberative faculty discovers things only insofar as they are found useful for the attainment of an end, regardless of whether this end is truly good, evil, or only believed to be good.[78] In other words, the practical intellect may be employed to deliberate on and discover the means for attaining evil ends. Al-Fārābī considers the practical intellect to be a deliberative virtue only when it is used to discover and produce things that are useful for some virtuous end. Otherwise, it should be called by other names such as deceit, artifice and strategem.[79] For this reason, al-Fārābī identifies the fruit of the excellent practical intellect with practical wisdom earlier defined.

The superiority of the theoretical over the deliberative virtues implies that the theoretical part of a science is more excellent than its practical part. Pure mathematics, that is, arithmetic and geometry, would then be viewed as more excellent than applied mathematics like mechanics and engineering. Similarly, the theoretical part of political science is more excellent than its practical part.

Metaphysics is clearly established as the most useful science from the point of view of man's final perfection. As for the other philosophical sciences, it is not yet possible to establish the degrees of their usefulness in relation to man's final goal on the sole basis of the superiority of the theoretical over the deliberative virtues. It is necessary to refer to al-Fārābī's conception of ethical virtues. The artistic virtues will not be discussed since the practical arts and crafts are excluded from his classification of the sciences.

The ethical virtues, says al-Fārābī, are the virtues of the appetitive faculty.[80] They are qualities of the soul, like temperance (*'iffah*), generosity (*sakhā'*), courage (*shajā'ah*) justice (*'adālah*), humility (*tawāḍu'*), and forbearance (*ḥilm*). Just as good actions are "moderate, mean actions between two extremes, both of which are bad, the one excess and the other defect," ethical virtues are likewise "mean states and qualities of the soul between two other states, both of which are vices, the one excessive and the other defective."[81]

Theoretical, deliberative, and ethical virtues, as well as the practical arts are discussed in al-Fārābī's political science, or more precisely its theoretical part. This is because his political science concerns itself with human goals, the highest of which is happiness. In dealing with these virtues, and by implication with the sciences and arts, political science does not seek to usurp the functions of the other sciences and of the arts. Al-Fārābī's "political scientist" is interested in the virtues only insofar as they are related to the organization and life of the political community.[82] He seeks to study the virtues as rational principles by which people living in political associations attain happiness, everyone according to his natural disposition.

In al-Fārābī's political science, the moral life is not the highest type of life. The ethical virtues are means for realizing intellectual perfection. The ethical virtues, together with the deliberative virtues and the practical arts, are to prepare man for the acquisition of the theoretical virtues, which alone enable him to have an intellectual vision (*naẓar; theoria*) of Reality.[83] Moral life is necessary for the acquisition of theoretical virtues because of the peculiar nature of man as a terrestrial being. Man's intellectual activity of the nature envisaged by al-Fārābī depends on the well-being of his body and soul.[84] Moreover, man is by nature a political animal who can satisfy his basic needs and realize his perfections only in political association. Moral life, says al-Fārābī, guarantees the well-being of both his physical body and the body politic. It is significant, in this respect, that al-Fārābī compares the morally excellent society with the perfect and healthy body of man.[85]

From the point of view of the individual quest for the attainment of theoretical perfection, al-Fārābī considers the ethical virtues to be more important than the deliberative virtues. Following Plato, al-Fārābī imposes the following stringent conditions on the student of the philosophical sciences:

> He should excel in comprehending and conceiving that which is essential. He should have good memory and be able to endure the toil of study. He should love truthfulness and truthful people, and justice and just people... He should not be gluttonous for food and drink, and should by natural disposition disdain the appetites, the *dirhem*, the *dīnār*, and the like. He should be high-minded and avoid what is disgraceful in people. He should be pious, yield easily to goodness and justice, and be stubborn in yielding to evil and injustice.... Moreover, he should be brought up according to laws and habits that resemble his innate disposition. He should have sound conviction about the

> opinions of the religion in which he is reared, hold fast to the virtuous acts in his religion, and not forsake all or most of them. Furthermore, he should hold fast to the generally accepted virtues and not forsake the generally accepted noble acts.[86]

However, says al-Fārābī, in the person of the perfect philosopher who has attained theoretical perfection, the deliberative and the ethical virtues are inter-dependent and equally important.[87] It is perhaps for this reason that al-Fārābī does not separate ethics from politics and economics. The foregoing discussion clearly shows that, for al-Fārābī, the most useful sciences are not those which supply the greatest material benefits but those which are indispensable for and contribute most to spiritual and intellectual perfection. Mathematics and political science are, next to metaphysics, the most useful. Mathematics is the science best suited to contribute to the perfection of the theoretical intellect by enhancing its power to know metaphysical beings with demonstrative certainty. This is because (pure) mathematics deals with numbers and magnitudes, entities which are the easiest to be grasped independent of matter. Further, the exact nature of this science and the profundity of demonstrative proofs it employs resemble those of metaphysics. For this reason, mathematics is regarded as the best intellectual discipline to prepare the theoretical intellect for the highest kind of metaphysical speculation.[88]

Political science, however, appears to be a more useful and important science than mathematics. Political science deals with man's final perfection. It seeks to distinguish between true and presumed happiness. Then, it seeks to identify and explain the ways and means through which true happiness is attained, and to distinguish them from elements that obstruct his achieveing this happiness. Consequently, this science deals with the four principal virtues by which citizens attain happiness. It possesses an almost "architectonic" position in relation to all the sciences and arts in the sense that it exercises supervision and control over them, insofar as they are relevant to the life of the society.[89]

Political science prescribes the kind of opinions citizens of the virtuous state should possess in common. These opinions include knowledge of the First Cause, the separate intelligences, the celestial substances, man and the rest of the terrestrial beings, happiness, and life after death. Political science also prescribes the kind of actions which citizens should perform. These include the actions and practices by which God and the angels are glorified.[90] This science also gains its eminent position from the fact that its subject-matter includes prophecy and the *Sharī'ah*, the source of moral life for both the individual and the community. The position of metaphysics as the most useful science is, however, secure. Despite its eminent position, political science remains basically a "prescriptive" science. It points the way to true happiness but does not directly deliver that happiness.

Conclusion

Al-Fārābī's ethical basis for the hierarchy of the sciences is, therefore, to be sought in his theory of the good, virtue, and happiness, which is subsumed under his political science. On this basis, the theoretical part of a science is deemed more excellent than its corresponding practical part. The place of each science in this hierarchy is determined by its degree of contribution to the perfection of man's knowledge of God and the perfection of the soul to the point of being God-like, either directly or indirectly.

Metaphysics is the most useful science because it deals with God, the Supreme Good, and with spiritual beings close to God. The knowledge of God is sought for its own sake since it constitutes man's true happiness. The knowledge of the separate intelligences is the next most useful knowledge since, next to God, they possess the highest degree of goodness. The knowledge which the separate intelligences have of the First Cause is the prototype of all gnosis.[91] Man's intellectual perfection is modeled on that of the separate intelligences.

The least useful of the philosophical sciences is natural science. Natural science deals with natural things, not all of which are good. The greater part of natural things are good, not in themselves, but because they are useful for the attainment of happiness.[92] A part of natural science deals with what man possesses by nature. The greater part of this science deals with minerals, plants and non-rational animals. All these natural things, including man's rational powers that are given him by nature, are considered the natural good only if they are used for the sake of the higher goods, namely, the moral good and the other principal virtues. But the knowledge of natural things is also used to serve evil ends as clearly demonstrated by the people of the ignorant states (*ahl al-jāhilīyah*).[93]

The intermediate sciences with respect to their degrees of usefulness are mathematics and political science. Applied mathematics deals with the material goods. In general, it is, therefore, no more useful than natural science. But insofar as it is needed in many of the practical arts, it is perhaps more useful than natural science. Pure mathematics is more useful than natural science with respect to the highest intellectual good. Knowledge of the properties of numbers and magnitudes is the key to understanding God's attributes as these attributes are manifested in creation. According to al-Fārābī, the order pervading the whole cosmos is a manifestation of divine justice.[94] The fundamental nature of this order and justice is mathematical. Thus, in *al-Madīnat al-fāḍilah* and *Taḥṣīl al-sa'ādah,* he gives great emphasis to what Walzer calls the "geometrical" justice pervading the heavens and the terrestrial world.[95]

A part of political science deals with voluntary goods that are material and cannot, therefore, be considered as more useful than natural science. The greater part of this science, however, deals with the moral and spiritual good, the things that are most indispensable for the attainment of happiness.

Endnotes

Chapter 4

[1]For al-Fārābī's detailed account of the hierarchy of being, see his *al-Madīnat al-fāḍilah* (chapters 1-19) and *al-Siyāsat al-madanīyah* (pp. 31-69).

[2]For a good account of how late-medieval and Renaissance philosophers dealt with the idea of the hierarchy of being, see E.P.Mahoney, "Metaphysical Foundations of the Hierarchy of Being According to Some Late-Medieval and Renaissance Philosophers," *Philosophies of Existence: Ancient and Medieval*, ed. P.Morewedge, New York, 1982, pp. 167-257..

[3]See his *The Great Chain of Being,* Harvard University Press, 1982, Preface, p.vii.

[4]*Ibid.*

[5]For a comprehensive treatment of Ibn Sīnā's cosmological doctrines, see S.H.Nasr, *IICD*, chapters 12-14.

[6]"Some of the elements of the Theory of the Ten Intelligences can be traced to the different sources they have been derived from. Its astronomical aspect is closely identical with Aristotle's interpretation of the movement of the spheres. The Theory of Emanation has been borrowed from Plotinus and the school of Alexandria. But, in its

entirety, it is a Farabian theory dictated and formulated by his desire for showing the unity of truth and his method of grouping and synthesis. He reconciles Plato and Aristotle and religion and philosophy." I. Madkour, "al-Fārābī," *A History of Muslim Philosophy*, pp. 459-60.

[7] *The Perfect State*, chapters 1-9.

[8] *al-Siyāsat al-madaniyah*, p. 31.

[9] E.W.Lane, *Arabic-English Lexicon*, The Islamic Texts Society, Cambridge (England), 1984, vol.I, p.165.

[10] *Introductory Sections on Logic*, pp. 276-7.

[11] Al-Fārābī, *Kitāb al-burhān* (The Book of Demonstration), quoted by Galston, *op. cit.*, p. 247.

[12] *The Perfect State* chapter 6, sect. 1, p.113.

[13] In *Tahṣīl al-sa'ādah*, al-Fārābī identifies the principles of being with the four Aristotelian causes. See *The Philosophy of Plato and Aristotle*, p. 15.

[14] By animate substances, al-Fārābī means bodies which admit of life, namely, plants and animals.

[15] *The Philosophy of Plato and Aristotle*, pp. 15-16.

[16] *Risālat al-'aql*, p. 23; *al-siyāsat al-madaniyah*, p. 31.

[17] See the first four chapters of *al-Madīnat al-fāḍilah* for al-Fārābī's exposition of these different aspects of divine nature and attributes (the first two chapters in *The Perfect State*).

[18] *al-Siyāsat al-madaniyah*, p. 31.

[19] This process is described in detail in *al-Madīnat al-fāḍilah*. See *The Perfect State*, chapter 3, pp.101-5.

[20] The universe consists of nine spheres: (1) the First Heaven, which is starless, (2) the sphere of the fixed stars, (3) Saturn, (4) Jupiter, (5) Mars, (6) Sun, (7) Venus, (8) Mercury, (9) Moon. According to the most recent discovery, Ptolemy was the first to introduce the starless heaven above the sphere of the fixed stars. See B.R.Goldstein (ed.), *The Arabic Version of Ptolemy's 'Planetary Hypotheses'*, Transactions of the American Philosophical Society, N.S., Philadelphia, 1967, vol.57, pt.4; see also *The Perfect State*, p.364.

Al-Fārābī was apparently the first Muslim to believe in this astronomical system of nine spheres. He was followed by Ibn Sīnā and al-Ghazzālī, but rejected by Ibn Rushd.

[21] The role of the active intellect as the link between heaven and earth is treated by al-Fārābī in a number of his treatises. See *Risālah fi'l-'aql*, p.33; *al-Siyāsat al-madanīyah*, pp. 35-6; F.Najjar, *The Political Regime*, pp.33-4 ; *The Perfect State*, pp.199-203 ; and De Boer, *The History of Philosophy in Islam*, p. 115.

[22] F.Najjar, *op. cit.*, p.33

[23] *Risālah fi'l-'aql*, pp. 33-4.

[24] *The Philosophy of Plato and Aristotle*, p. 127.

[25] For al-Fārābī's discussion of form and matter, see *The Perfect State*, p. 109-13; and *al-Siyāsat al-madanīyah*, pp.36-40. Al-Fārābī views form as a higher ontological principle than matter because what gives a body its actual existence is its form. "Matter exists for the sake of form, and if there were no form in existence, matter could not exist." *The Perfect State*, p. 109.

[26] *Ibid*, p. 112.

[27] Al-Fārābī differs from Aristotle in his notions of prime matter and celestial matter. According to him, Aristotle did not postulate a terrestrial prime matter out of which were to arise the four elements (*The Philosophy of Plato and Aristotle*, p.104). Aristotle, he says, simply refers to a fifth element, ether, the 'stuff' from which the heavenly

bodies are made, as the celestial cause of the existence of the four elements. For al-Fārābī, celestial matter first gives rise to terrestrial prime matter, then only to the four elements.

No explanation is offered by al-Fārābī, at least in his presently known works, as to how prime matter originates from celestial matter. He is also vague about the exact nature of 'celestial matter.' According to Walzer, al-Fārābī's 'celestial matter' is not the Peripatetic 'ether' but the neo-Platonic *neote hylē* (a 'spiritual, intelligible matter'). See *The Perfect State*, p. 370.

[28]The faculties of the plant soul are the nutritive, the reproductive, and the faculty of growth. The non-rational animal soul possesses the additional faculties of motion, the sensitive, the imaginative, and the appetitive.

[29]*The Philosophy of Plato and Aristotle*, p. 18.

[30]Al-Fārābī defines quantity as "everything of which the totality can be measured by a part of it." Dunlop, "al-Fārābī's Paraphrase of the 'Categories' of Aristotle," *The Islamic Quarterly*, 4:3 (Oct. 1957), p. 186. A continuous quantity is any quantity whose middle part can be posited as a boundary or limit common to its two equal parts. Magnitudes are a particular kind of continuous quantity, i.e., that quantity whose parts have position in it or have position of mutual relation in one or more direction or dimension. For example, a line consists of parts having position of mutual relationship in one dimension. *Ibid*, p.189.

Another kind of continuous quantity is time. Its parts have no 'position' in relation to one another.

[31]A discrete quantity is that quantity whose two equal parts do not possess a common limit. *Ibid*, p. 188.

[32]*The Philosophy of Plato and Aristotle*, p. 18.

[33]Al-Fārābī's use of the term *hay'ah* to mean a particular kind of shape or form, namely mathematical shape, as distinguished from shape (*khilqah*) and form (*ṣūrah*) in general, is found in his discussion of form and matter in *al-Madīnat al-fāḍilah*. See *The Perfect state*, p. 109.

[34]*Ibid*, chapter 7.

[35]*Ibid*, p.121

[36]Al-Fārābī says: "Because the substrata of their forms have no privation (*'adam*) whatsoever, and their forms have no privations opposed to them, their substrata, consequently, do not prevent their forms from thinking and from being intellect in their essences. Hence the form of each of these celestial bodies is 'actual intellect'." *The Perfect State*, pp. 122-3.

[37]For al-Fārābī's discussion of the souls of celestial bodies, see *al-Siyāsat al-madanīyah*, pp.33-4.

[38]*Ibid*, p. 34.

[39]*The Perfect State*, p. 125.

[40]*The Philosophy of Plato and Aristotle*, sec.16, pp.21-2.

[41]According to al-Fārābī, light (*ḍiyā'*) is not a bodily thing. It is incorporeal but possesses a visible quality. In his incorporeal theory of light, al-Fārābī is following the tradition of Alexander of Aphrodisias and Plotinus. See S.Sambursky, *The Physical World of Late Antiquity*, London, 1962, pp.113 ff.

Al-Fārābī's optics or the mathematical properties of some aspects of the phenomena of light will be treated in chapter 6.

[42]*The Philosophy of Plato and Aristotle*, sec. 11, p.19.

[43]R.Lerner and M.Mahdi, *Medieval Political Philosophy*, p.24; *Iḥṣā' al-'ulūm*, p.102.

[44]*The Perfect State*, p.205.

[45]*Ibid*, p. 107

[46]*Iḥṣā'al-'ulūm* p. 91.
[47]*The Perfect State*, p.207; in *Kitāb al-millat al-fāḍilah* (pp.45-6), al-Fārābī enumerates some of the spiritual activities which should be performed by citizens of the virtuous state. The most excellent of these activities are the practices by which God and the angels are glorified.
[48]*Fuṣūl al-madanī*, p.27
[49]For example, political science is interested in agriculture, not as a scientific or artistic activity, but insofar as it possesses economic, political, and moral dimensions.
[50]*The Perfect State*, pp.205-7
[51]Since ways of life belong to the category of voluntary intelligibles, they can only be made to exist outside the soul (i.e., in society) after the "states and accidents that must accompany them when they exist" have been known and willed. *The Philosophy of Plato and Aristotle*, p.26
[52]*The Perfect State*, p.207
[53]Al-Fārābī judged the views of these Greek philosophers to be in harmony with the religious view of life: "But Socrates, Plato and Aristotle taught that man has two lives. The continuance of the first is due to nourishment and the other external things which we need today for our continued existence. It is our first life. The other is that of which the continuance is in its essence without its requiring for the continuance of its essence things external to it, but it is sufficient in itself for its continued preservation. It is the after-life. Man, according to them, has a first and a last perfection. The last results to us not in this life but in the after-life, when there has preceded it the first perfection in this life of ours." *Fuṣūl al-madanī*, p.39
[54]*Ibid*, pp.39-40
[55]*Ibid*, p.60
[56]"This is the after-life, in which a man sees his Lord and is not defrauded in the seeing of Him." *Ibid*, p.65
[57]The phrase which occurs frequently in the Qur'ān is, the face of Allāh' (*wajh Allāh*) which is interpreted by many commentators as the sight or vision of God. One verse reads: "And who, from desire to see the face of their Lord, are constant amid trials, and observe prayer and give alms in secret and openly, out of what we have bestowed upon them, and turn aside evil by good......" (XIII:22)
[58]*Fuṣūl al-madanī*, sec.76, p.65
[59]According to al-Fārābī, the first perfection consists in acquiring all the virtues as well as doing the actions of these virtues. See *Ibid*, p.39
[60]*Ibid*, sec.l, p.27
[61]*Ibid*, p.27
[62]*Ibid*, sec.7, p.31
[63]*Ibid*, sec.69, p.60
[64]See F.Schuon, *Islam and the Perennial Philosophy*, London, 1976, p.166
[65]"Happiness is the good without qualification. Everything useful for the achievement of happiness or by which it is attained, is good too, not for its own sake, but because it is useful with respect to happiness.......The good that is useful for the achievement of happiness may be something that exists by nature or that comes into being by the will." F.Najjar, *The Political Regime*, p.34
[66]"As for the goal which is aimed at in the study of philosophy, it is the knowledge of the Creator Most High, that He is One, the unmoved, the real cause of all things and that He orders this universe with excellence, wisdom and justice. As for the actions which the philosopher performs, they are aimed at attaining likeness to the Creator so far as this lies within the human power." Al-Fārābī, *Fī mā yanbaghī an yuqaddam ta'allum falsafah arisṭū*, p.13

[67] *Fusūl al-madanī*, sec.30, p.42
[68] *Ibid*
[69] *Ibid*, p.43
[70] For, in *Tahṣīl al-sa'ādah*, al-Fārābī says that the ultimate purpose of the theoretical sciences is only to gain certain knowledge of the essences or principles of beings. *The Philosophy of Plato and Aristotle*, sec.2, p.13
[71] *Fusūl al-madanī*, pp.43-4
[72] *Ibid*, p.42
[73] *Ibid*, p.45
[74] *Ibid*
[75] *The Philosophy of Plato and Aristotle*, sec.38, p.46
[76] For al-Fārābī's discussion of these rational practical virtues, see *Fusūl al-madanī*, pp.46-7
[77] *The Perfect State*, p.209
[78] *The Philosophy of Plato and Aristotle*, sec.27, p.28
[79] *Fusūl al-madanī*, sec.88, p.72
[80] *Ibid*, p.31
[81] *Ibid*, sec.16, p.34
[82] F.Najjar, "al-Fārābī on Political Science," *The Muslim World*, vol.48 (1958), p.97
[83] E.I.J.Rosenthal, "The Place of Politics in the Philosophy of al-Fārābi," *Islamic Culture*, 29:3(1955), p.161
[84] This "fact" was made quite clear in my discussion of al-Fārābī's view of man's intellectual development in chapter 2.
[85] *The Perfect State*, p.231
[86] *The Philosophy of Plato and Aristotle*, sec.60, p.48
[87] *Ibid*, sec.34, p.32
[88] *Fusūl al-madanī*, sec.89, p.73. According to the Ikhwān al-Ṣafā', number "is a sample from the superior world, and through knowledge of it the disciple is led to the other mathematical sciences, and to physics and metaphysics. The science of number is the 'root' of the sciences, the foundation of wisdom, the source of knowledge, and the pillar of meaning." Translation by S.H.Nasr in *Science and Civilization in Islam*, p.153.

"Mathematics in the Islamic perspective is regarded as the gateway leading from the sensible to the intelligible world, the ladder between the world of change and the heaven of archetypes." *Ibid*, p.146
[89] F.Najjar, *op. cit.*, p.97
[90] *Kitāb al-millat al-fāḍilah*, pp.44-6
[91] *IICD*, p. 204
[92] *The Political Regime*, p.34 (see note 65 above)
[93] *The Perfect State*, chapter 18. For al-Fārābī's classification of societies and socio-political systems and detailed discussion of them, see chapters 15-19.
[94] "Inasmuch as the substance of the First is a substance from which the existents emanate, while it does not neglect any existence beneath its existence, it is generous, and its generosity is in its substance, and inasmuch as all the existents receive their order of rank from it, and each existent receives from the First its alloted share of existence in accordance with its rank, the First is just, and its justice is in its substance." *The Perfect State*, pp.95-7
[95] *Ibid*

CHAPTER 5

CLASSIFICATION AND DESCRIPTION OF LINGUISTIC SCIENCE AND LOGIC

This chapter consists of three main parts. In the first, I present al-Fārābī's classification and enumeration of the sciences as given in the *Iḥṣā' al-'ulūm*. In the second, I discuss some of the major features of this classification. And in the third and final part, I explain the specific positions of linguistic science and logic in al-Fārābī's classification.

5.1 Classification and Enumeration of the Sciences

In his *Iḥṣā' al-'ulūm*, al-Fārābī presents the following classification and enumeration:[1]

I. Science of language (*'ilm al-lisān*); its seven subdivisions are the following sciences:
   (1) simple expressions (*alfāẓ mufradah*)
   (2) composite expressions (*alfāẓ murakkabah*)
   (3) the rules (*qawānīn*) governing simple expressions
   (4) the rules governing composite expressions
   (5) correct writing
   (6) the rules governing correct reading (*qirā'ah*)
   (7) the rules of poetry (*shi'r*)

II. Logic (*'ilm al-manṭiq*). This is divided into eight parts that deal with the following:
   (1) Rules governing simple intelligibles or ideas and simple expressions which signify these intelligibles, corresponding to Aristotle's *Categories*
   (2) Rules governing simple statements or propositions composed of two or more simple intelligibles; and

composite expressions signifying the composite intelligibles, corresponding to Aristotle's *On Interpretation.*

(3) Rules of the syllogisms which are common to the five syllogistic arts – the demonstrative, the dialectical, the sophistical, the rhetorical, and the poetical, corresponding to Aristotle's *Prior Analytics*

(4) Rules of demonstrative proof and the special rules by which the philosophic art is constituted, corresponding to Aristotle's *Posterior Analytics*

(5) The means of discovering dialectical proofs, questions and answers, and the rules by which the art of dialectic is constituted, corresponding to Aristotle's *Topics*

(6) Rules governing matters which are such as to turn man away from truth to error and to lead him to deception, corresponding to Aristotle's *On Sophistic Refutations*

(7) The art of rhetoric. It deals with the rules by which rhetorical statements may be examined and evaluated, corresponding to Aristotle's *Rhetoric*

(8) The art of poetry, corresponding to Aristotle's *Poetics*

III. The mathematical or propaedeutic sciences (*'ulūm al-ta'ālīm,*) which consist of the following:

(1) Arithmetic (*'ilm al-'adad*) comprised of
  (a) The theoretical science of numbers
  (b) The practical science of numbers

(2) Geometry (*'ilm al-handasah*), comprised of
  (a) Theoretical geometry ·
  (b) Practical geometry

(3) Optics (*'ilm al-manāẓir*), which includes study of
  (a) What is observed by means of straight rays
  (b) What is observed by means of other rays

(4) Science of the heavens (*'ilm al-nujūm*) which is divided into
  (a) judicial astrology (*'ilm aḥkām al-nujūm*)
  (b) astronomy (*'ilm al-nujūm al-ta'līmī*), which includes study of
    (i) Figures, masses, and relative positions of the heavenly bodies

   (ii) Motions of the heavenly bodies and their conjunctions
   (iii) The earth's climatic zones

(5) Music (*'ilm al-mūsīqā*), which is comprised of
   (a) Practical music (*'ilm al-mūsīqā al-'amalīyah*)
   (b) Theoretical music (*'ilm al-mūsīqā al-naẓarīyah*)

(6) Science of weights (*'ilm al-athqāl*)

(7) Engineering or science of ingenious devices (*'ilm al-ḥiyal*) such as
   (a) Arithmetical devices
   (b) Mechanical devices
   (c) Devices for making astronomical, musical and other instruments for use in various practical arts, including weapons
   (d) Optical devices

IV. Physics or natural science (*al-'ilm al-ṭabī'ī*). This is divided into eight main parts that deal with the following:
   (1) Principles of natural bodies
   (2) Principles of the elements and simple bodies
   (3) Generation and corruption of natural bodies
   (4) Reactions which the elements undergo in order to form compound bodies
   (5) Properties of compound bodies
   (6) Minerals
   (7) Plants
   (8) Animals, including man

V. Metaphysics (*al-'ilm al-ilāhī*). Its three parts deal with the following:
   (1) Beings and their essential attributes insofar as they are beings
   (2) The principles of demonstration (*mabādi' al-barāhin*) in the particular theoretical sciences
   (3) Absolute incorporeal beings

VI. Political science (*al-'ilm al-madanī*), jurisprudence (*'ilm al-fiqh*), and dialectical theology (*'ilm al-kalām*)

(A) *Political Science.* Its two parts deal with the following:
  (1) Happiness and human virtues
  (2) Ethics and political theory
(B) *Jurisprudence.* Its two parts deal with
  (1) Articles of faith
  (2) Religious rites, practices, and moral-legal injunctions
(C) *Dialectical Theology.* Its two parts deal with
  (1) Articles of faith
  (2) Religious acts

5.2 **Characteristics of al-Fārābī's Classification**

According to al-Fārābī, he had composed the above classification, with its detailed subdivisions, with several objectives in mind.[2] First, the classification is intended as a general guide to the different sciences so that students choose only to study the.subjects that are really beneficial to them. Second, the classification is to enable a person to learn about the hierarchy of the sciences. Third, its various divisions and subdivisions provides a useful means of determining the extent to which specialization may be legitimately pursued. And fourth, the classification informs the student of what must be studied before one can claim expertise in a particular science.

The detailed subdivisions of the sciences show that al-Fārābī considered the pursuit of specialization to be a legitimate activity, provided that this does not destroy the unity and hierarchy of the sciences. The more excellent and useful sciences may not be sacrificed for the sake of greater quantitative accumulation of knowledge within a single, narrow discipline. The classification as a whole provides a constant reminder of the unity and hierarchy of the sciences.

Al-Fārābī himself informs us that his classification is not complete, but is limited solely to the "generally known sciences."[3] The sciences that are clearly given prominence in his classification are the sciences associated with the pre-Islamic philosophical tradition. Considering the fact that some of these philosophical sciences were still new to the Muslims of al-Fārābī's time, the use of the phrase "generally known sciences" is somewhat misleading. Al-Fārābī was the real founder of logic and political science in Islam. Opposition to *manṭiq* came from among both the grammarians and religious scholars and it was serious enough to compel him to write a special treatise defending logic on religious ground. Al-Fārābī himself claimed that he was heir to a logical tradition that was almost extinct.

It would be more appropriate to say that al-Fārābī had composed the classification to make logic and the philosophical sciences better known and more generally accepted among Muslims. The classification constitutes an ingenious attempt at projecting a superior image of the philosophical sciences in relation to the religious sciences. The only sciences which could pose a serious challenge to his metaphysics and political science, namely, *kalām* and *fiqh,* were included on the list that he discussed but only received brief treatment. Further, by showing that the science of language and logic constitute two related but independent disciplines, al-Fārābī's classification helps to secure a place for logic among the generally accepted sciences. When al-Fārābī declares that his classification helps a person "to find out which science is better or more useful or more accurate, more reliable and effective," the science he has in mind could only be one or more of the philosophical sciences.

The best way, in al-Fārābī's view, to compare and contrast the philosophical sciences and religious sciences, is to appeal to methodological ground. This would mean for him an appeal to logic or syllogism. This "methodological motive" of al-Fārābī explains why his classification is limited to the syllogistic arts or sciences. In the *Iḥṣā' al-'ulūm,* al-Fārābī presents a powerful case for the necessity of logic. He asserts that logic is necessary for whoever does not wish to base his conviction on mere opinion.[4] In an obvious reference to the critics of logic among the grammarians, he argues that just as experience and perfect innate disposition is no substitute for the rules of grammar as means for testing correct language, likewise experience or perfect innate disposition alone is insufficient to guarantee correct reasoning. Rules of logic are necessary for correctness and reliability of knowledge in the sciences.

Al-Fārābī's discussion of logic in the above treatise presents the position that the philosophical sciences are methodologically superior to the religious sciences. His divine science, for example, is said to offer certain knowledge of God and the other spiritual beings whereas *kalām* or *fiqh* offers at best approximate certainty in that knowledge. Later Muslim philosophers like Mullā Ṣadrā (d. 1050/1641) took a more critical stance toward *kalām* by even questioning its legitimacy to concern itself with the knowledge of God.[5]

None of the hidden (*khafīyah*) or occult (*gharībah*) sciences, including alchemy and the interpretation of dreams,[6] is included in al-Fārābī's classification. These sciences fulfill the condition of not being generally known or generally accepted sciences. However, in my view, al-Fārābī's omission of these sciences from his classification is dictated not so much by this consideration as by the fact that they are non-syllogistic sciences. The practical arts like medicine, architecture, agriculture, and navigation are likewise omitted because these are non-syllogistic arts.[7]

In limiting his classification to the syllogistic arts and sciences, he has to exclude many disciplines which he himself, in his other writings, considers useful. This exclusion of many useful arts and sciences is a major defect of his classification. This defect was overlooked in his lifetime. Scholars were impressed by his originality.[8] His work was the most influential of the early classifications.

Why was al-Fārābī's classification so influential? According to Nasr, al-Fārābī "molded and formulated the various branches of knowledge in a complete and permanent form within Islamic civilization."[9] Al-Fārābī's scheme and structure also was original. It is more congenial to Islam's epistemology than al-Kindī's Aristotelian classification scheme.[10] What made al-Fārābī's classification at once original and appealing was his integration of the Greek sciences and the Islamic disciplines into an organic unity based on his idea of the hierarchy of the sciences, although within this scheme *kalām* and *fiqh* are subordinate to the philosophical sciences.

With only minor changes, later Muslim philosophers like Ibn Sīnā and Ibn Rushd could accommodate al-Fārābī's classification to new disciplines as well as those disciplines that al-Fārābī omitted.

5.3 Divisions of Linguistic Science and Logic

Let us examine in the final part of this chapter each of the main divisions of the science of language and of logic, that is, the two sciences that precede his discussion of philosophy proper.

5.3.1 The Science of Language

Al-Fārābī begins his classification with the science of language, divided into seven parts. He makes it clear that the division itself is universal in the sense that it applies to every language of the human race (*'ilm-al-lisān 'ind kull ummat*).[11] He distinguishes between two fundamental functions of this science.[12] The first is to preserve significant expressions (*al-alfāẓ al-dāllah*) which are either simple or composite. Simple expressions belong to one of three genera: noun (*ism*), verb (*fi'l*), or particle (*ḥarf*).[13] The name is a simple expression, signifying a meaning which can be understood alone and by itself, without signifying, by its essence, structure and form, the time of the meaning in question.[14] The verb is a simple expression, signifying a meaning which can be understood alone and by itself, and at the same time it signifies, by its structure and essence, the time in which that meaning exists.[15] The particle is a simple expression, signifying a meaning which cannot be understood alone and by itself, except when it is joined to a noun or verb, or to both together.[16] Each of the three kinds of simple expression is said to signify a single notion.

Composite expressions signify either single[17] or compound notions. They are composed of two or all of the above genera of simple expressions. Expressions which signify compound notions are called phrases. Phrases may be classified into a number of categories.[18]

The second function of the science of language is to formulate rules and conventions governing significant expressions. Every art, whether theoretical or practical, is characterized by a set of rules. By "rule" al-Fārābī means a universal statement that embraces many individual things belonging to the art in question. The rules of every art are aimed at three things.[19] First, to attain a complete and sound understanding of everything comprised in an art so that the elements of this art may be clearly distinguished from the elements of another art. Second, to provide means of checking errors that may have been committed by someone ignorant of the rules. Third, to facilitate the learning of what is comprised in the art. Al-Fārābī's emphasis on the need to distinguish between the elements of one art from those of another art, in the midst of discussing the rules of the science of language, suggests that he was sensitive to the claim of prominent grammarians that their grammar encompasses the logic of the *manṭiqīyūn*.

Al-Fārābī's sevenfold division of linguistic science is claimed to be based on the "anatomy" of human language and the above two functions. The first major branch, called the science of simple expressions, is lexicography and lexicology.[20] The second branch, the science of composite expressions, categorizes composite phrases. This branch includes questions about their preservation and transmission in a nation's literary history.

The third branch is the science of rules governing simple expressions. It deals with morphology, phonetics and orthography.[21] The fourth branch is the science of rules of compound expressions. It deals with two different kinds of rules. First, rules concerning signs of inflexion of nouns (including pronouns), verbs and particles when they are combined or ordered according to established usage.[22] Second, rules that combine and arrange words into meaningful phrases.[23] The science of the first al-Fārābī calls grammar.

The fifth branch is the science of rules of correct writing. It distinguishes letters written in lines from those which are not, and explains how writing in lines is to be done. The sixth branch is the science of the rules of correct reading. It deals with convention ways to ensure the correct reding or pronounciation of words and phrases.[24]

The seventh branch is prosody, which deals with forms of metric composition. It is itself divided into three parts.[25] The first enumerates the metres used in poetic verses.[26] The second deals with the different kinds of rhymes used in each metre. The third investigates the kinds of expressions that are appropriate for poetic verses.

Al-Fārābī depicts linguistic science, especially grammar, as a necessary science. However, grammar is not as important as the art of logic although the significance of the former is closely tied up to that of the latter.[27] Grammar and logic are both indispensable to constructing a rational philosophical system. They are necessary tools for studying all other sciences. Linguistic science is listed first because learning language necessarily precedes mastering logic.

5.3.2. **Logic**

According to al-Fārābī, logic deals with "intelligibles insofar as they are signified by expressions and with expressions insofar as they signify intelligibles."[28] At another place, he says that logic is about "thoughts as signified by expressions and as somehow related to things."[29] Logic does not deal with things as such or with intelligibles in the mind in themselves. Entities outside the human soul and intelligibles in the mind as signifying the natures of things perceived by the intellect are treated in the philosophical sciences. Logic concerns itself with the mental states and accidents of intelligibles as exemplified in being subject or predicate, universality or particularity of predication, and essentiality or accidentality of predication.[30] Similarly, logic does not deal with linguistic expressions in themselves for these would be the concern of linguistic science. Logic deals with expressions insofar as they are common to all languages.[31] Al-Fārābī emphasizes, however, that logic has less to do with the ordering and arrangement of expressions than with the ordering and arrangement in the mind of the intelligibles which the expressions signify.[32]

Logic is comprised of rules governing intelligibles and expressions in their mutual relations that I have just described. For this reason, each part of logic is described by al-Fārābī as "the science of the rules of." He divides logic into eight parts, corresponding to the eight books of the Aristotelian *Organon*. To understand this division and the ordering of its parts, it is necessary to refer to the aims of logic as conceived by al-Fārābī. According to him, the general aims of logic are the following. First, to regulate (*tuqawwim*) and guide reason toward right thinking with regard to all intelligibles that admit of error. Second, to provide safeguards against error in regard to those intelligibles. Third, to provide means of testing the intelligibles that admit of error.[33] The means by which these aims may be achieved are the "rules" (*qawānīn*). The relation of these logical rules to the intellect and the intelligibles is analogous to the relation of grammatical rules to language and the expressions.[34]

The ordering of the eight parts of logic is determined by what al-Fārābī perceives to be the primary intention (*al-qaṣd al-awwal*) of logic. He considers the fourth part of his logic, corresponding to Aristotle's posterior analytics, as the first in nobility and authority, as well as its primary intention.[35] The fourth part is comprised of the rules of demonstrative syllogism and the special rules by which the demonstrative or philosophic art is constituted. According to al-Fārābī, the primary aim of the student of logic is to study the demonstrative art because this art best fulfills the goal of attaining certain knowledge in the philosophical sciences.

The order in which al-Fārābī has arranged the eight parts of logic is said to be the order of learning or instruction (*tartīb al-ta'līm*).[36] The first three parts are described variously as preludes (*tawṭi'āt*), introductions (*madākhil*) and means (*ṭuruq*) to demonstrative syllogisms.[37] These parts contain elements that are common to all the five remaining parts.

Logic as a science must embrace all forms of reasoning. Although demonstrative reasoning constitutes the primary aim of logical study, the study of the non-demonstrative syllogistic arts is also necessary. This latter study helps one to know what the art of demonstration is not, which in turn helps him to guard against falling into the use of methods that lead to error, mere opinion or an image of the truth.[38]

In order to have a perfect knowledge of the demonstrative art, one must also have a knowledge and a sound one of its contrasts and opposites. To know what leads to truth and certainty is to know at the same time what leads to error and doubt. But one does not begin with error and then gradually discovers the truth. One should first learn what leads to the truth and then seeks to know all the possible avenues to error. Moreover, says al-Fārābī, understanding of the demonstrative art facilitates the learning of the imperfect syllogistic arts.[39] For this reason, the learning of demonstrative art precedes that of the other syllogistic arts (corresponding to the last four parts). The last four parts are themselves ordered according to an hierarchy of syllogistic proofs discussed in the third chapter.

Al-Fārābī affirmed the traditional ordering of the eight books of the *Organon* established by the Alexandrian school of philosophy.[40] Although the arts of rhetoric (the seventh part) and poetics (the eighth part) are not strictly speaking syllogistic arts, they are incorporated into logic to serve as auxiliaries of the demonstrative art and as tools of comparison and contrast with the syllogistic arts proper.

It is possible, says al-Fārābī, to associate each syllogistic art with a particular class or school of thinkers. He associates demonstrative art with philosopher-scientists, dialectical with theologians, rhetorical with politicians,[41] poetical with poets, and sophistic with sophists.[42]

Logic, in al-Fārābī's classification, is not a part of any philosophical science. It is an instrument or tool of the philosophical sciences.[43] But it is also a science (*'ilm*). Al-Fārābī seems to maintain that logic as a science is intermediate between the science of language and the philosophical sciences. In his view, the science of language, the science of logic, and the philosophical sciences are all concerned with meanings of things, but in different forms and at different levels.[44] Intelligibles or the meanings of things studied under the philosophical sciences manifest themselves at the level of interior speech (*al-nuṭq al-dākhil*)[45] and at the "lower" level of linguistic expression or exterior speech (*al-nuṭq al-khārij*). Logic is concerned with both kinds of speech, but primarily with the first kind. Linguistic science is primarily concerned with exterior speech.

Al-Fārābī, unlike his teacher Abū Bishr, did not make the strict distinction that "logic enquires into the meaning (*ma'anā*), whereas grammar enquires into the sound (*lafẓ*) ."[46] For al-Fārābī, the science of language does not ignore meanings totally in its study of speech. However, interior speech signifies meanings of things more directly than does exterior speech. It is in this sense that al-Fārābī understands logic as an intermediate science between linguistic science and the philosophical sciences.

Al-Fārābī presents logic and linguistic science as two very closely related sciences. He notes that this close relationship between them is reflected in the Arabic language itself. The word for logic in Arabic, *manṭiq,* is etymologically related to the word for speech, *nuṭq*.[47] Al-Fārābī considers logic to be a kind of universal grammar[48] whose validity extends to all humanity. He gives two reasons for his view. First, logic is concerned with thoughts or interior speech, which belongs to all humanity. Second, logic is only interested in expressions to the extent that they are common to the languages of all communities. As for grammar, it does take into account features that the language in question has in common with those of other communities, but it does not study them insofar as they are common features. The main concern of grammar is the rules peculiar to the language of a given community.

What the boundary of separation is between the two disciplines was debated in al-Fārābī's time, when logic was a newcomer to the Islamic intellectual scene. Al-Fārābī's accounts of linguistic science and the science of logic in the *Iḥṣā' al-'ulūm* represents an attempt at delineating their respective boundaries.

ENDNOTES

Chapter 5

[1]*Iḥsā' al-'ulūm*, pp. 45-113
[2]*Ibid*, pp.43-4. See also, F.Rosenthal, *The Classical Heritage in Islam*, p. 55
[3]The term used is *al-'ulūm al-mashhūrāt*. *Iḥṣā' al-'ulūm*, p.43
[4]"If the logical arts are not in practice distinguished from one another and the rules which pertain to each isolated, men may employ less than certain methods without recognizing them as such and thus be led unaware to mere opinion (*ẓunūn*), persuasion, error and perplexity, or imagining." Galston, *Opinion and Knowledge in Fārābī's Understanding of Aristotle's Philosophy*, p.41 (cf. *Iḥṣā' al-'ulūm*, p.63).
[5]See J.W.Morris, *The Wisdom of the Throne: An Introduction to the Philosophy of Mullā Ṣadrā*, pp. 31-2, and the section concerning "Knowledge of God, of His Attributes, His Names, and His Signs", pp. 94ff.
[6]Other than alchemy and the science of dream interpretation, we may mention as belonging to this category of sciences *līmiyā'* (magic), *hīmiyā'* (the subjugating of souls); *sīmiyā'* (the producing of visions), *raml* (geomancy), *jafr* (numerical symbolism of the letters of the Arabic alphabet), and *'ilm al-firāsah* (physiognomy). *Jafr*, said to have been first cultivated by 'Alī, the fourth Caliph of Islam and the cousin and son-in-law of the Prophet, was probably the most popular of the occult sciences. For a detailed treatment of the occult sciences in Islam, see S.H.Taqizadeh, "The Open and Secret Sciences," *Mélanges d'Orientalisme offerts à Henri Massé*. Tehran, 1963, pp. 383-7; and S.H.Nasr, *Islamic Science*, chapter IX.

Al-Fārābī defended the legitimacy of alchemy and the interpretation of dreams as sciences. His attitude toward the other occult sciences is unknown. According to him, alchemy should only be studied by those who possess the necessary intellectual and spiritual qualifications. Intellectually, the student of alchemy should have mastered at least logic, mathematics, and mineralogy. In addition, he should be a lover of wisdom. His aim in studying alchemy should be to acquire that wisdom by which man gains a comprehension of the inner reality of things through the interiorization of the alchemical art. Spiritually, his intention should be to purify his soul. Without these conditions, alchemy as a science and an art poses grave danger to its student and the community in general. See his *Fī wujūb ṣanā'at al-kīmiyā'*, pp. 76-8.

Al-Fārābī's treatise on dream interpretation has not survived to enable us to know his views in more detail. He considered this science to be a symbolic one, outside the scope of philosophy. The principles of the natural sciences cannot adequately deal with the total phenomena of dreams. According to Muslim philosophers, there are three kinds of dreams: (1) false dreams, (2) patho-genetic dreams, and (3) true dreams. For al-Fārābī, dreams of intellectual objects, which are represented by the faculty of imagination in the form of sensible objects whose perfection and beauty is directly proportional to the perfection and beauty of the intellectual objects, are really the dreams that stand in need of a symbolic interpretation. See Wali-ur-Rahman, "al-Fārābī and His Theory of Dreams," *Islamic Culture* , 10(1936), pp. 148-9

[7]In *Introductory 'Risālah' on Logic* (pp.231-2), al-Fārābī explains the distinction between syllogistic and non-syllogistic arts. "The syllogistic arts are those which, when their parts are integrated and perfected, have as their action thereafter the employment of syllogism, while the non-syllogistic are those which, when their parts are integrated and perfected, have as their action and end the doing of some particular work, such as medicine, agriculture, carpentary, building, and the other arts which are designed to produce some work and some actions."

The syllogistic arts are said to be five in number: philosophy (demonstrative), dialectical, sophistical, rhetorical, and poetic arts. Al-Fārābī granted the fact that a part of the knowledge embodied in each of the non-syllogistic arts may have been produced through the employment of syllogism.

[8]In the words of Ṣā'id ibn Aḥmad al-Andalusī, "the classification, the like of which had never before been composed and the scheme of which had never been adopted by any other author, is an indispensable guide to students in the sciences." *Ṭabaqāt al-umam*, p.53

[9]See chapter 1, note 126

[10]Al-Kindī's work on the classification of the sciences is called *Fī aqsām al-'ulūm* (On the Divisions of the Sciences). Unlike that of al-Fārābī, it made no impact upon the main fields of Islamic scholarship. For a detailed discussion of al-Kindī's classification, see G.N.Atiyeh, *al-Kindī: The Philosopher of the Arabs*, Islamic Research Institute, Rawalpindi, 1966, pp. 32-40; see also L.Gardet, "Le probleme de la philosophie musulmane," in *Mélanges offerts à Etienne Gilson*, Paris, 1959, pp. 261-84.

Al-Kindī's Aristotelian division of philosophy into the theoretical and the practical is not the only feature of al-Fārābī's classification. Al-Fārābī's original features include the incorporation of linguistic science, logic, *kalām*, *fiqh* and political science. Linguistic science occupies an important place in religious scholarship, because it is indispensable to the study of the Qur'ān and to prophecy and the *Sharī'ah*. All these features contribute to the favorable respone shown to his classification even by religious scholars.

[11]*Iḥṣā' al-'ulūm*, p. 46

[12]*Ibid*, p. 45. In Islam, up to the times of al-Fārābī, the question of the nature and origin of language was generally investigated in *kalām*, *fiqh* and philosophy, rather than in grammar or philology. It was only in the latter part of the tenth century A.D. that Arabic philology began to show interest in this question. Even then philology was greatly dependent on those disciplines for general theories of language. See M.Mahdi, "Language and Logic in Classical Islam," *Logic in Classical Islamic Culture*, p.53, n.6

[13]Al-Fārābī also made frequent use of the Arabic renderings of the corresponding concepts in Greek grammar, namely *kalimah* (vocable), *adāt* (instrument), and *ism* (name) itself. For al-Fārābī's discussion of these grammatical concepts and also his reference to their Greek equivalents, see his *Introductory Sections on Logic*, pp.278-82; *Iḥṣā' al-'ulūm*, pp.49-50; and Zimmermann, *al-Fārābī's Commentary and Short Treatise on Aristotle's 'De Interpretatione'*, pp. 220-5.

[14]*Introductory Sections on Logic*, p.278

[15]For example, 'he walked', 'he walks', 'he will walk' all signify a meaning and at the same time signify, essentially and not accidentally, the times in which the meaning exists. *Ibid*, p.278. In contrast, 'yesterday', 'tomorrow', and the like are not vocables, since each of these in the first place signifies a time, without signifying a meaning in that time.

[16]*Ibid.*

[17]An example of compound expressions which signify a single notion, given by al-Fārābī, is 'Abd al-Malik (servant of the king) when it is used as a person's proper name.

It is not a phrase since part of it does not signify part of the person. See Zimmermann, *op. cit.*, p.225.

[18]Al-Fārābī says, "There are complete and incomplete phrases. The genera of complete phrases are, according to many of the ancients, five in number, namely, statement, imperative, entreaty, request, and vocative. Stating sentences are those that are true or false, being composed of a predicate and subject. The remaining four genera are neither true nor false, except by accident." *Ibid*, p.226.

[19]*Iḥṣā' al-'ulūm*, p.45

[20]*Ibid*, p.47

[21]In the short section on morphology in *Iḥṣā' al-'ulūm* (p.48), al-Fārābī uses the term *mithāl awwal* ("prototype") to denote a word which is not derived from another. Morphology seeks, among other things, to distinguish between prototypes that are maṣdars, that is, have verbs derived from them, and prototypes that are not maṣdars.

In his *Kitāb al-ḥurūf* (p.71), where he adopts a semantic approach to Arabic morphology, he draws attention to the need to distinguish between what is prototypal in form and what is prototypal in meaning.

[22]*Iḥṣā' al-'ulūm*, p. 49

[23]*Ibid*, pp.50-1

[24]*Ibid*, p.51

[25]*Ibid*, pp. 51-2

[26]This part also seeks to distinguish which of the metres are more perfect, beautiful, and more pleasant to hear than others. In his *Paraphrase of the 'Categories' of Aristotle*, al-Fārābī speaks of the cultivation of this part of linguistic science in all languages. "It is possible to find the like of it in the existing Arabic language. For the experts in Arabic call the short syllables 'movable' letters, and the long syllables and what resembles them they call *asbāb* or cords. What can be combined in their language of both kinds of syllables they call *awtād* (pegs). Then they combine some of these with others and make of them measures greater than these, by which they measure their metrical expressions and discourses, e.g., *fa'ūlun, mafā'ilun, mustaf'ilun*. If this is so, then every expression can be measured by a long or short syllable or a combination of both." Dunlop, "al-Fārābī's Paraphrase of the 'Categories' of Aristotle," *The Islamic Quarterly*, p.187 (cf. *Iḥṣā' al-'ulūm*, p.52, where he gives the Arabic renderings of the Greek terms for syllables and metrical feet as *maqāṭi'* and *arjul* respectively).

[27]Zimmermann, *op. cit.*, p.cxviii

[28]*Iḥṣā' al-'ulūm*, p.59

[29]Zimmermann, *op. cit.*, p.2 (18.5f) and p.ll (24.23).

[30]Ibn Sīnā, *Remarks and Admonitions, Part One: Logic*, trans. Shams C.Inati, p.10

[31]*Introductory 'Risālah' on Logic*, p.233; *Iḥṣā' al-'ulūm*, pp. 60-61

[32]*Kitāb al-alfāẓ*, p.102

[33]*Iḥṣā' al-'ulūm*, p. 53; and *Introductory Risalah on Logic*, p.231.

[34]*Ibid*; and *Iḥṣā' al-'ulūm*, p.54

[35]*Ibid*, p.72

[36]*Ibid*.

[37]Domingo Gundisalvo (fl. 1140), whose *De Divisione Philosophiae* contains many direct quotations from al-Fārābī's *Iḥṣā' al-'ulūm*, especially the part on logic, explains why the first three parts should precede the study of demonstration as follows:

> The sure cognition of truth is not obtained except through demonstration. Therefore, it was necessary that a book be composed in which would be taught how and out of what things demonstration is

made.... But since demonstration is made only by means of syllogism and syllogism in truth consists of propositions, it was necessary to have a book in which would be taught the number and kind of propositions, and how according to mode and figure syllogisms should be constructed. For this reason, the *Prior Analytics* was written.

But propositions cannot compose a syllogism unless they first have been composed out of their own terms. Therefore, it was necessary to have a book which would teach the number of terms and which terms go to make up a proposition. This is fully taught in... *On Interpretation*. Further, a proposition is never well composed from terms unless the signification of each term is first recognized. Therefore, the *Book of Categories* were written to teach how many kinds of terms there are and what is the signification of each of them..... See D.Gundisalvo, 'Classification of the Sciences,' trans. N.Clagett and E.Grant, in E.Grant, ed., *A Source Book in Medieval Science*, Harvard University Press, 1974, p.67

[38]*Ihsā' al-'ulūm*, p.73
[39]Gālston, *op-cit.*, p. 38
[40]In his *Rhetoric*, Aristotle speaks of the rhetorical art as being "a branch of dialectic and similar to it" (Barnes, ed., *The Complete Works of Aristotle*, p.2156). And in his *Poetics*, he describes the art of poetry as an imitative mode of expressing things that is rooted in human nature (*ibid* p.2318). Thus, although the Rhetoric and the Poetics did not form parts of the original *Organon*, the rhetorical and poetical arts were accepted by Aristotle as possible modes of reasoning or discourse. The later incorporation by the Commentators of these two works of Aristotle into the original *Organon* was not an error, as claimed by some modern historians of logic, but the result of a conscious attempt to formulate in an explicit manner the whole Aristotelian logical system.
[41]See I.Madkour, "al-Fārābī," in *A History of Muslim Philosophy*, p.456
[42]For al-Fārābī's discussion of the characteristics of the sophists and the sophistic art, see *Ihsā' al-'ulūm*, pp.65-6; and *The Philosophy of Plato and Aistotle*, pp. 88-92
[43]"The art of logic is an instrument (*ālāt*) by which, when it is employed in the several parts of philosophy, certain knowledge is obtained of all which the several theoretical and practical arts include, and there is no way to certainty of the truth in anything of which is sought save the art of logic." *Introductory 'Risalah' on Logic*, p.234
[44]Zimmermann, *op. cit.*, p.11
[45]*Introductory 'Risalah' on Logic*, p.233; *Ihsā' al-'ulūm*, p.60
[46]D.S.Margoliouth, "The Discussion between Abū Bishr Mattā and Abū Sa'id al-Sīrāfī on the Merits of Logic and Grammar," *JRAS* (1905), p.97.
[47]*Introductory 'Risalah' on Logic*, p.232
[48]This view of al-Fārābī is discussed by Zimmermann, *op. cit.*, p. xliii.

CHAPTER 6

CLASSIFICATION AND DESCRIPTION OF THE PHILOSOPHICAL SCIENCES

According to al-Fārābī, the domain of mathematics extends well beyond the Latin *Quadrivium* (arithmetic, geometry, astronomy, and music) to include optics, weights, and "ingenious devices",[1] which, in modern science, are parts of physics. These seven divisions comprise the entire world of numbers and figures,[2] including their properties or attributes as inherent in various classes of beings.

Al-Fārābī subdivides three of them – arithmetic, geometry, and music – into theoretical and practical parts.[3] The theoretical part considers mathematical forms independent of the materials in which they inhere, whereas the practical part considers the forms insofar as they enter into relations with concrete things.[4] It should be noted that, in al-Fārābī, the above distinction between "theoretical" and "practical" is relative in the sense that the distinction only applies to the parts of one and the same science. For example, mathematical astronomy deals with mathematical forms in relation to celestial bodies without, on that account, being called practical science.

Similarly, he subdivides astronomy (*'ilm al-nujūm*) into two parts, namely judicial astrology and mathematical astronomy. However, he did not explicitly distinguish them as the practical and the theoretical. In the case of optics too, described as both a science and an art,[5] there is no theoretical-practical division. As for the science of weights and the science of ingenious devices, they do not admit of such a division, since they are entirely applied mathematics.

Al-Fārābī considers optics a mathematical science, because it deals with the geometrical properties of light, especially as related to the phenomena of vision. Optics investigates the geometrical nature of visual images, some of which are said to be real, others apparent.[6]

It seeks to demonstrate[7] what causes these visual images. It also deals with the problem of measurement of large and distant magnitudes, like the depths of valleys and rivers, the altitudes of mountains and clouds, and the elongations of celestial bodies that are observable from earth.[8] It classifies the different kinds of light rays.[9] Knowledge obtained in optics finds its partial application in the production of optical devices. The science of optical devices, however, is not considered a part of optics. Rather, it is a part of the science of "ingenious devices." Al-Fārābī judges optics to be a subdivision of theoretical geometry.[10] He does not distinguish the theoretical foundation of optics from geometry. Consequently, he does not divide optics into a theoretical and a practical part.

Al-Fārābī does not distinguish between theoretical and practical astronomy because mathematical astronomy is a mathematical science whereas judicial astrology is not.[11] The latter is not even part of natural science. In fact, he considers astrology an interpretative or occult science.[12] His treatise *On the True and the Untrue in Judicial Astrology* shows that he was critical of astrology but did not reject it altogether. Since astrology was popular during his time, and its practitioners made all kinds of spurious claims, he felt obliged to mention it in his classification. He wished to draw attention to the fact that astrology, contrary to many claims, is not an exact and certain science in the sense that mathematical astronomy is one.

Mathematical astronomy includes the study of the planet earth: its shape, mass, motion, its position in relation to the heavenly bodies as well as its climatic conditions. For al-Fārābī, the study of the earth's climatic zones falls under mathematical astronomy, because these zones are the effects of the motions of the heavenly bodies and of their conjunctions.

Al-Fārābī describes both the science of weights and the science of ingenious devices as applied mathematics. The former investigates "the principles of doctrine on weights" and "the principles of instruments by which heavy bodies are lifted and on which they are moved from place to place."[13] The latter science is what is today called the engineering sciences. This science devises "ways to make all the things happen whose "modes of existence" were stated and demonstrated in the theoretical sciences."[14] It includes "principles of the civil practical arts employed in bodies, figures, positions, arrangements and measurement as in the arts of building construction, carpentary and others."[15]

Of the mathematical sciences listed by al-Fārābī, only theoretical arithmetic and theoretical geometry are truly "pure" mathematical sciences in the sense that they deal with numbers and magnitudes absolutely independent of material objects. Al-Fārābī also considers these two sciences to be the roots and foundations of all the sciences.[16] The other mathematical sciences are ordered in the classification according to the following principle. The study of optics, mathematical astronomy, music, weights, and ingenious devices in this very order is the study of the properties of numbers and magnitudes in progressively more complex relations and greater reference to material things.[17] Although in the hierarchy of the sciences mathematics is an intermediate science between natural science and metaphysics, it precedes natural science in the order of instruction. According to al-Fārābī, one should first study the things which are easier for the human mind to grasp and in which less confusion is likely to occur. These things are numbers and magnitudes.[18]

6.2 Natural Science

Al-Fārābī introduces his division of natural science with a definition of the discipline. It is that science which "inquires into natural bodies and the accidents inherent in them" and which deals with their causes in terms of the four Aristotelian causes.[19] To illustrate effectively the subject-matter of this science, al-Fārābī resorts to examples drawn from the world of the practical arts, that is, by referring to artificial bodies. He maintains that the principles of artificial bodies and of their accidents are generally better known than the principles of natural bodies and their accidents. Most of the principles of artificial bodies can be known through sense-perception[20] whereas most of the principles of natural bodies "can only be verified by reasoning and certain demonstrations."[21] Al-Fārābī believes that such a method of illustration is valid because there is a complete analogy between human production of artificial bodies and the divine creation of natural bodies.

Al-Fārābī divides natural science into eight parts. As in logic, each part of natural science corresponds to a book of Aristotle or some parts of it. The first part of natural science, which al-Fārābī says is treated in the book called *Natural Things That are Heard,*[22] inquires into "what all natural bodies, simple as well as composite,[23] share with respect to principles and the accidents following upon those principles."[24] In the *Iḥṣā'* itself, al-Fārābī does not explain what these "principles" mean. But on the basis of his treatise, *The Philosophy of Aristotle,* we know that these principles refer to the four principles of being.[25] This part of natural science contains an account of universal propositions, premises, and rules covering all natural bodies.[26] It presupposes a knowledge of the *Categories* of the science of logic.[27]

The second part of natural science deals with simple bodies in three main respects. First, it inquires into the existence of the different kinds of simple bodies and into their common nature. The cause of existence of the four elements and their common nature form a part of this inquiry. Second, it inquires into simple bodies insofar as they are elements of composite bodies. It seeks to determine whether or not all simple bodies constitute parts of composite bodies.[28] It shows that some simple bodies are not elements and principles of composite bodies. Third, it examines the common properties of simple bodies that form parts of composite bodies.

The third part of natural science deals with the generation and corruption of natural bodies generally and of the things of which they are composed. It explains the principles of the generation of composite bodies from the elements.[29] The fourth part deals with the principles of the reactions which the four elements undergo in order to form compounds without, however, considering the bodies of which they are composed.[30] No further explanation is given of this part. The third and the fourth parts appear to be very closely related.

The fifth part of natural science deals with the different kinds of bodies composed of the four elements. It classifies these bodies into two main groups.[31] The first group consists of the homogeneous bodies,[32] and the other of heterogeneous bodies.[33] The first group is itself divisible into two subgroups: (1) those bodies that only form parts of a heterogeneous body,[34] and (2) bodies which cannot be part of a natural body of diverse parts, but which are generated only to form a part of "the sum of the world, the sum of the generated bodies, or the sum of a certain genus or species."[35] The main function of the fifth part of natural science is to investigate the common physical properties of composite bodies and in particular of homogeneous bodies that form parts of heterogeneous bodies.

The sixth part of natural science inquires into homogeneous bodies that are not parts of heterogeneous bodies. Al-Fārābī identifies this part with mineralogy in its comprehensive sense. The seventh part deals with all the plant species, their common properties, and properties that are specific to each species.[36] According to al-Fārābī, plants and animals are the two kinds of heterogeneous natural bodies. The eight and last division of natural science deals with the different animal species, their common properties as well as properties that are specific to each species.[37]

Al-Fārābī, evoking the hierarchic reality of the cosmos of which the natural world is a part, maintains that the sciences of plants and animals, including the human body, cannot be adequately established on the basis of natural or physical properties alone.[38] What al-Fārābī means is that principles discovered in the first six parts of natural science are insufficient to deal with "most matters relating to animals and in many matters relating to plants. Sciences of plants and animals need to be founded on higher principles, which he calls animate or psychical principles. As we saw, by this kind of principle al-Fārābī means the soul. The plant soul with its faculties of nutrition, growth, and reproduction accounts for many of the properties that are characteristic of plant species. Similarly, the animal soul with its greater number of and superior faculties accounts for most of the animal properties.

Al-Fārābī's natural science is primarily a science of principles of the different species of natural bodies. His natural science would include many of the sciences that fall under the broad disciplines of mineralogy, botany, and zoology. Of the sciences related to mineralogy, chemistry and geology are certainly included, but others like alchemy and metallurgy, being practical arts, are not. Agriculture, the most important science related to botany, is similarly excluded. I have also referred to the omission of medicine, one of the most important sciences and arts related to zoology. One important feature of al-Fārābī's natural science, however, is the inclusion of psychology. His psychology, like that of the Peripatetic philosophers generally, falls under the heading of natural philosophy, because it is concerned with the various faculties of the souls which dominate the three kingdoms.

6.3 Metaphysics

Al-Fārābī divides metaphysics into three parts. The first part is ontology, that is, the science which deals with "beings (*mawjūdāt*) and their attributes, insofar as they are beings."[39] Al-Fārābī says practically nothing about this science in the *Iḥṣā'*. The second part of metaphysics seeks to classify the different kinds of beings with a view of establishing the subject-matters of the theoretical sciences. This part establishes the principles of demonstrations (*mabādi' al-barāhin*) in the science of logic, the mathematical sciences and natural science. That is to say, this part of metaphysics provides the proofs of legitimacy of the subject-matter of each particular theoretical science by making known the true nature and characteristics of the things comprised in that subject-matter.[40] For example, the proof that the point, the unit, lines and surfaces, studied in mathematics, are not separate substances.

The third and last part of metaphysics deals with "beings that are neither bodies nor in bodies."[41] It demonstrates that these absolutely incorporeal beings exist, are many, finite in their number, and can be ranked hierarchically. The hierarchy culminates in God.[42] This metaphysical science also demonstrates that the universe is a unity and hierarchically ordered.[43] This science includes angelology, cosmology, and cosmogony.[44] In addition, it introduces us to an important function of metaphysics. It acts as judge of all the false notions and theories concerning God in His relation to the Universe.[45]

Al-Fārābī himself informed us that he followed Aristotle in his division of metaphysics. The first part of al-Fārābī's metaphysics corresponds to Aristotle's science of "being *qua* being."[46] This science differs from the particular (*juz'īyah*) theoretical sciences investigated in the second part of metaphysics in that none of the latter deals generally with being as being. As Aristotle says it, the particular sciences "cut off a part of being and investigates the attributes of this part." The first part of al-Fārābī's metaphysics includes Aristotle's "second substances," which refer to universals[47] like existence, quiddity, unity, necessity, contingency, substance and accident.

The second part of al-Fārābī's metaphysics corresponds to Aristotle's treatment of the particular sciences of different genera of beings.[48] Al-Fārābī's third metaphysical science corresponds to Aristotle's theology.[49] It is significant that in his account of metaphysics al-Fārābī presents this science as the most "demonstrative" of all the sciences that he enumerates.[50]

6.4 Political Science

The central theme of al-Fārābī's political science, as we saw, is happiness. This theme determines the nature, scope, functions and aims of his political science. Al-Fārābī divides this science into two parts. The first part deals with the various kinds of human actions and ways of life with a view of understanding their goals and man's moral character. It judges these goals in the light of its presupposition that the ultimate goal of human life is supreme happiness. It explains that true happiness is attainable only through the virtues and the good and noble things. Such things as wealth, honor, and sensual pleasures, when these are made the only ends in this life, do not constitute true happiness but are only presumed to be so. The first part of political science, therefore, deals with the theory of happiness and human virtue.

Al-Fārābī's second subdivision of political science is what

> comprises the way of ordering the virtuous states of character and ways of life in the cities and nations; and making known the royal functions by which the virtuous ways of life and actions are established and ordered among the citizens of the cities, and the activities by which to preserve what has been ordered and established among them. It then enumerates the various kinds of the non-virtuous royal crafts – how many they are, and what each one of them is; and it enumerates the functions each one of them performs, and the ways of life and the positive dispositions that each seeks to establish in the cities and nations under its rulership (*ri'āsah*).[51]

The operation of the royal craft al-Fārābī calls politics (*siyāsah*). Politics thus occupies an important place in his political science. He calls political science practical or human philosophy (*al-falsafat al-'amalīyah*) distinct from theoretical philosophy (*al-falsafat al-naẓarīyah*) which consists of mathematics, natural science, and metaphysics.[52] We encounter here another usage by al-Fārābī of the distinction between "theoretical" and "practical." For, as we saw, mathematics, now a part of theoretical philosophy, possesses branches that are divisible into theoretical and practical parts; and, as we shall see, political science, here described as political philosophy, possesses theoretical and practical parts. According to al-Fārābī, practical philosophy differs from theoretical philosophy in three respects. First, the subject matter of political science consists of voluntary intelligibles and that of theoretical philosophy natural intelligibles.[53] Second, the primary principle of political science is human will or choice, while that of theoretical philosophy Nature. Third, the aim of theoretical philosophy is theoretical knowledge alone, whereas that of political science is action that leads to the realization of happiness.

Al-Fārābī does not explicitly distinguish between theoretical and practical parts of political science. But Ibn Rushd, who knew al-Fārābī's political works well, states at the beginning of his *Paraphrase of Plato's Republic* that the first subdivision of al-Fārābī's political science is the theoretical part of this science, since it investigates the general rules that are far removed from action.[54] According to Ibn Rushd, the two subdivisions of political science differ from each other in degree in bringing about action. The more general the themes treated and the rules laid down in this science, the further away are they from bringing about action. Clearly, the first part of al-Fārābī's political science deals more with theoretical knowledge than with action, while the reverse is true in the second part. On this basis, the two parts may be called respectively the theoretical and the practical dimensions of political science.[55]

Theoretical political science's role is to ascertain the extent to which the different sciences and arts are necessary to people achieving their ultimate perfection within the organized life of a community. It must also determine how best the cultivation of the excellent sciences and arts could be carried out. In undertaking the above role, theoretical political science presupposes the knowledge furnished by each science or art. As such, political science is what Aristotle calls "the most authoritative art and that which is most truly the master art."[56]

The practical part of al-Fārābī's political science refers primarily to politics proper (*siyāsah*). It is concerned with both the virtuous and the non-virtuous kinds of rulership or government. It describes the factors which may transform virtuous governments and ways of life into corrupt governments and ignorant ways of life.[57] It prescribes the practical measures of preventing this transformation from taking place.[58] It also prescribes the measures to be used to restore virtuous governments and ways of life to their previous state once they have been transformed into corrupt rulerships and ignorant ways of life. Practical political science also deals with the fundamental elements which constitute the virtuous government[59] and with the kind of political education and succession to guarantee the virtuous character of a human community through laws.[60]

Al-Fārābī's political science is concerned as well with social change, transformation and decadence. These phenomena are understood in the light of a clearly defined concept of a perfect society and socio-political order, identified with the one originally founded by a prophet.[61] Another important feature of al-Fārābī's political science concerns the place of the science of ethics in relation to that science. Al-Fārābī, unlike Ibn Sīnā and al-Ghazzālī, does not adopt in the *Iḥṣā' al-'ulūm* Aristotle's threefold division of practical wisdom into politics, economics, and ethics. He conceives ethics to be subordinate to political science.[62] Al-Fārābī does not mention ethics by name in his divisions of political science, but that science clearly forms a part of the first division. His definition of politics proper (*al-falsafah al-siyāsīyah*)[63] shows that this science ought to be identified with his second division of political science.

In general, al-Fārābī's political science (*al-falsafah al-madanīyah*) embraces anthropology, sociology, philosophy of law, practical psychology, ethics, and public administration. As such, it is the most comprehensive branch of the humanities.

6.5 **Jurisprudence and Dialectical Theology**

Al-Fārābī's account of political science is immediately followed by that of the science of jurisprudence (*'ilm al-fiqh*). According to him, jurisprudence is the art that enables man to infer the determination of whatever was not explicitly specified by the Lawgiver, on the basis of things that were explicitly specified and determined by him.[64] A jurist should strive to infer correctly by taking into account the Lawgiver's purpose with the religion he had legislated for the nation to which he gave that religion. Al-Fārābī says that every religion (*millah*) consists of both opinions (*ārā'*) and actions (*af'āl*), which constitute the two parts of this science.[65] He offers no further explanation about this science.

Finally, al-Fārābī describes the science of dialectical theology (*kalām*).[66] Kalām, he says, is a religious science which arose in a religious tradition at some point in its history out of a need to formulate a systematic defense of the tenets of that religion against attacks from various sources such as from the followers of other religions.[67] He divides the *mutakallimūn* into five different groups according to the nature and kinds of arguments they employ.[68] The first group seeks to defend religion by claiming that divine revelation is superior even to the knowledge gained by the best of human intellects.[69] The second group seeks to prove that religion is true by showing that scientific knowledge is in harmony with the religious texts.

The remaining three groups cannot be said to be concerned with the intellectual defense of religion. The third group attacks its adversary's beliefs. The fourth group tries to silence its adversary through shame and fear of bodily harm. And the fifth group accuses their opponents of being either an enemy or ignorant. In all three cases, the faith is defended by using rhetorical or sophistical methods rather than logical-demonstrative arguments.

Al-Fārābī's political science, *fiqh,* and *kalām* are conceptually related insofar as they all are based on revelation. Political science, for al-Fārābī, is essentially a science of revealed doctrines and practices or the Divine Law (*Sharī'ah*) understood at the level of philosophy, while *fiqh* and *kalām* are two sciences of the same doctrines and practices understood at the level of *religion.*[70]

It is also true that al-Fārābī's political science deals with laws of human rather than divine origin. However, even these laws are for al-Fārābī ultimately rooted in revelation. He argues that in virtue of his knowledge of the inner meaning of the *Sharī'ah,* the philosopher should play a leading role in administering the religious political state. However, al-Fārābī seemed resigned to the fact that most jurists and dialectical theologians could not concede such a role to philosophers.[71]

The philosopher considers *kalām* to be a necessity for the protection of the non-philosophic masses of the community. Likewise, *fiqh* has value for the philosopher because it is a necessary basis for that part of political science concerned with legal philosophy. It is significant that al-Fārābī did not treat the science of *uṣūl al-fiqh* (principles of jurisprudence) as a separate science in his classification when this science, generally attributed to al-Shāfi'ī (d.820), already was established by the time al-Fārābī composed his classification. Instead, he subsumed it under political science.[72]

In conclusion, al-Fārābī attempted to secure a predominant place within the Islamic intellectual universe for his new discipline of political science. He advocated it as a comprehensive and supreme science of man and of society to which the sciences of *fiqh* and *kalām* would be subordinate. However, later thinkers, like Ibn Sīnā and al-Ghazzālī, modified his claim in their classifications of the sciences.

ENDNOTES

Chapter 6

[1]I have followed the usual practice of rendering *'ilm al-ḥiyal* as "the science of ingenious devices." On the basis of al-Fārābī's own description, this science may be identified with engineering, especially mechanical engineering.

[2]In Islamic philosophy of mathematics generally, numbers and figures are considered to exist on three levels of reality: (1) as archetypes in the Divine Intellect, (2) as abstract or 'scientific' entities in the human mind, and (3) as concrete quantities in material things. See S.H.Nasr, *Islamic Science*, p.88

Al-Fārābī only deals with the last two levels of reality of numbers and figures. However, he is known to have asserted that perfect knowledge of mathematical entities exists in the active intellect.

[3]*Iḥṣā' al-'ulūm*, pp.76, 78, 86 (see the relevant parts in my summary of the classification in chapter 5).

[4]Arithmetic deals with numbers; geometry with lines, surfaces, and solids; and music with melodies.

[5]*Ihsa' al-'ulūm*, p.80

[6]*Ibid.*

[7]*Ibid*, pp. 78, 80. Optics employs the demonstrative method of theoretical geometry. Al-Fārābī maintains that ancient geometers, with the exception of Euclid, employed both the method of analysis (*taḥlīl*) and the method of synthesis (*tarkīb*) in their study. Euclid organized his *Elements* according to the method of synthesis alone (p.79).

[8]*Ibid*, pp.80-1

[9]Al-Fārābī listed four kinds of light rays: the straight, the reflected, the reversed, and the refracted. Optics studies the mathematical properties of these light rays independently of material things.

[10]"Optics inquires into the same things as does geometry, namely figures, magnitudes, order, positions, equality, inequality, and other things but only insofar as they are in lines, surfaces, and solids in abstraction. The study of geometry is thus more general than that of optics." *Ibid*, pp.79-80. A.G.Palencia's text (*al-Fārābī: Catalogo de las Ciencias*, p. 36, Spanish trans., p.43) would make optics a banch of practical geometry.
[11]Judicial astrology, says al-Fārābī, deals with the planets insofar as they serve as indices and clues to the understanding of future events and past and present occurrences in the world. It is basically concerned with predictions the nature of which is uncertain. *Iḥṣā' al-'ulūm*, p. 84
[12]Al-Fārābī is not alone in placing judicial astrology among the occult sciences. Ibn Ḥazm (d.1064) also treated that discipline in his classification in this way. See A.G.Chejne, *Ibn Ḥazm*, Kazi Publications, Chicago, 1982, pp.180-4.
[13]*Iḥṣā' al-'ulūm*, p.88
[14]*Ibid.*
[15]*Ibid*, p.90
[16]With reference to theoretical arithmetic, see *ibid*, p.75; for theoretical geometry, p.78.
[17]*The Philosophy of Plato and Aristotle*, pp.19-20.
[18]*Ibid*, p.18
[19]In al-Fārābī's terminology, these are: (1) *allatī 'anhā* (that from which) corresponding to both the material and efficient causes; (2) *allatī bihā* (that by which) referring to the formal cause; (3) *allatī lahā* (that for which) corresponding to the final cause. See *ibid*, p.15; and *Iḥṣā' al-'ulūm*, p.91.
[20]Either directly, as in the case of, for example, the principles of a garment, or indirectly, as in the case, for example, of the principles of the healing power of medicine. *Ibid*, p.94
[21]*Ibid.*
[22]This is an alternative title for Aristotle's *Physics*. Al-Fārābī's Arabic title is *al-samā' al-ṭabī'ī* (lit: what is heard concerning natural things). Scholars cannot agree why such a name has been given to this part of natural philosophy. One explanation is that it concerns what the student 'hears' before everything else when studying natural philosophy, hence the term *samā'*.
[23]"Natural bodies are either simple (*basīṭah*) or composite (*murakkabah*). The simple bodies are those whose existence is not generated from other bodies distinct from them. Composite bodies are those whose existence is generated from other bodies distinct from them. Examples of the latter are plants and animals." *Iḥṣā' al-'ulūm*, pp.95-6.

Al-Fārābī identifies the simple bodies not only with the four elements but also with "bodies of their genus and near to them," like vapours ahd winds in the air. *The Perfect State*, p.137.
[24]*Iḥṣā' al-'ulūm*, p.96.
[25]*The Philosophy of Plato and Aristotle*, p.99. The aim of the first part of natural science is to define clearly the subject-matter of that science. Al-Fārābī's lengthy discussion of the four principles or causes in the *Iḥṣā'* is in fact to make known (but not to prove) that subject-matter before actually undertaking a division of natural science into its various branches.
[26]*The Philosophy of Plato and Aristotle*, p.98.
[27]*Ibid*, pp.82-3. The ten categories are substance, quantity, quality, relation, time, place, situation, state, action, and passion. For al-Fārābī's discussion of these categories, see his previously cited *Paraphrase of the 'Categories' of Aristotle*.
[28]The whole of the second part of al-Fārābī's natural science corresponds to the four books of Aristotle's *On the Heavens* (al-Fārābī's Arabic title: *al-samā' wa'l-'ālam*).

[29]*Ihsā' al-'ulūm*, p.97. The third branch of al-Fārābī's natural science corresponds to Aristotle's *On Generation and Corruption* (Arabic title: *al-kawn wa'l-fasād)*.

Al-Fārābī speaks of four qualitatively different kinds of mixtures (*ikhtilāṭ*) of the elements, corresponding to the four classes of composite, natural (terrestrial) bodies mentioned in his hierarchy of being. In their order of greater complexity, the mixtures are of minerals, plants, non-rational animals, and rational animals. *The Perfect State*, p.141.

[30]The fourth branch corresponds to the first three books of Aristotle's *Meteorology* also known as *On Phenomena of the Upper Regions* (Arabic title: *al-āthār al-'alawīyah*).

[31]*Ihsā' al-'ulūm*, p.97; *The Philosophy of Plato and Aristotle*, p.112. The fifth branch, which is wholly devoted to homogeneous and heterogenous bodies, corresponds to the fourth book of Aristotle's *Meteorology*.

[32]Homogeneous bodies are bodies all of whose parts are similar in nature. Al-Fārābī gives the examples of flesh and bone.

[33]"Heterogeneous bodies originate only from that of combination of homogeneous bodies in which the essence of every one of the latter bodies is preserved; it is the combination of being together and in contact." *The Philosophy of Plato and Aristotle*, p.112.

[34]For example, flesh and bones are both homogeneous but each can become part of a heterogenous body, such as the arm.

[35]Al-Fārābī mentions gold and silver as example of homogeneous bodies which do not form part of another natural body.

[36]*Ihṣā' al-'ulūm*, p.98. The seventh part corresponds to Aristotle's *On Plants* (generally regarded by modern scholars as pseudo-Aristotelian). The Arabic title is *Kitāb al-nabāt*.

[37]*Ibid*. The eight branch of natural science corresponds to Aristotle's *On Animals* (*Kitāb al-hayawān*) and *On the Soul* (*Kitāb al-nafs*).

[38]*The Philosophy of Plato and Aristotle*, pp.115-6. According to al-Fārābī, an inquiry into the principles of human bodies, their natural powers and acts, shows that the animate or psychical principles themselves are not adequate to explain the causes of those powers and acts (p. 112). A higher principle, is needed, namely, the intellect. His inquiry into this principle terminates with the active intellect, the intermediary between human intellect and the transcendent intellects.

[39]*Ihsā' al-'ulūm*, p.99

[40]*Ibid*. Related to this function, the second part of metaphysics seeks to criticize the false views concerning the principles of the particular sciences.

[41]*Ibid*.

[42]*Ibid*, p.100. Al-Fārābī's more detailed discussion of Divine Attributes and Qualities may be found in *al-madīnat al-fāḍilah* and *al-siyāsat al-madanīyah*.

[43]*Ihṣā' al-'ulūm*, p.101

[44]As shown in chapter four, al-Fārābī's cosmology and cosmogony are very closely related to angelology.

[45]*Ihsā' al-'ulūm*, p.101

[46]Says Aristotle, "There is a science which investigates being as being and the attributes which belong to this in virtue of its own nature. This is not the same as any of the so-called special sciences." *Metaphysics*, Book IV in Barnes, *op. cit.*, p.1584.

[47]Al-Fārābī refers to Aristotle's distinction between "first substances" and "second substances" (cf. Aristotle's *Categories* in Barnes, *op. cit.*, p.4; and *Metaphysics* in *ibid*, Book V, sec.8, pp.1606-7) in his *Kitāb al-ḥurūf* (sec.68, p.102) as well as in his *Paraphrase of the 'Categories' of Aristotle* (sec.4, p.185).

[48]See Aristotle's *Metaphysics*, 1025-30.

[49] *Ibid*, 1071-6
[50] Al-Fārābī's metaphysics not only provides demonstrative knowledge of the unity of God, the hierarchy of being, and the principles of the sciences, but also provides demonstrative refutation of all contrary views.
[51] Lerner and Mahdi, *Medieval Political philosophy*, pp.25-6
[52] Al-Fārābī, *Kitāb al-tanbīh 'alā sabīl al-sa'ādah*, p.20
[53] The term "natural intelligibles" is used here with a wider connotation to include celestial and divine beings; in fact, all things which man cannot make or change.
[54] See Averroes, *Commentary on Plato's Republic*, trans. E.I.J.Rosenthal, Cambridge University Press, 1956, p.112.
[55] Ibn Rushd compares political science in this respect to medicine. Medicine is known as a practical science or art. But physicians divide it into its theoretical and practical parts. The theoretical and the practical parts of political science, says Ibn Rushd, stand in the same relationship to each other as do the books of *Health and Illness* and the *Preservation of Health and the Removal of Illness* in medicine. *Ibid.*

In *Taḥṣīl al-sa'ādah*, al-Fārābī confirms that the first part of his political science is its theoretical part. There he states clearly that giving an account of happiness and the principal virtues belongs to the "theoretical affairs." *The Philosophy of Plato and Aristotle*, sec.21, p.25.
[56] See his *Nicomachean Ethics* in Barnes, *op. cit.*, p.1729
[57] Lerner and Mahdi, *op. cit.*, p.26; for detailed discussion of these different kinds of ways of life, see *The Perfect State*, chaps. 15-19.
[58] Lerner and Mahdi, *op. cit.* This prescriptive role of practical political science with respect to human society is analogous to that of medical art with respect to the human body (*Iḥṣā' al-'ulūm*, p.104). In his political works, al-Fārābī resorts to the symbolism of human anatomy to explain his theory of human society. See *The Perfect State*, pp.231-7; also H.K.Sherwani, "al-Fārābī's Political Theories," *Islamic Culture*, 12:3 (July, 1938), p.300.
[59] The most important element in al-Fārābī's virtuous government is virtuous leadership in the person of the *Imām* or ruler in whom are combined many leadership qualities. For a discussion of these qualities, see *The Perfect State*, pp.247-53.
[60] *Ibid*, p.251
[61] Lerner and Mahdi, *op. cit.*, p.37. Maimonides likewise views the founding of a perfect nation and the proclamation of a perfect law to serve as a constitution for the perfect nation as the very *raison d'etre* of prophecy. See L.Strauss, "Quelques remarques sur la science politique de Maimonide et de Farabi," in *Revue des Etudes Juives*, 100 (1936), p.20.
[62] Lerner and Mahdi, *op. cit.*, p.95
[63] In his *Kitāb al-tanbīh 'alā sabīl al-sa'ādah*, al-Fārābī divides political philosophy (*al-falsafat al-madaniyah*) into two branches: (1) the ethical art (*al-ṣanā'at al-khuluqiyah*) and (2) politics proper (*al-falsafat al-siyāsiyah*). The latter is defined as that discipline which "comprises knowledge of the affairs by which the noble things and the means of acquiring and preserving them for the inhabitants of the city are realized." See pp.20-1.
[64] *Iḥṣā' al-'ulūm*, p.107; Lerner and Mahdi, *op. cit.*, p.27.
[65] *Ibid.*
[66] *Ibid.* pp.107-8. This science, likewise, deals with both opinions and actions of a religion.
[67] Al-Fārābī sees the *kalām* of his time as mainly an apologetic discipline.
[68] *Iḥṣā' al-'ulūm*, pp.109-13.
[69] *Ibid*, p.109

[70]In his *Kitāb al-millat* (sec.5, pp.46-7), for example, al-Fārābī mentions explicitly that political philosophy deals with the doctrines and practices comprised by the *Sharī'ah*. For al-Fārābī and other Muslim philosophers, philosophy in general and political philosophy in particular "derive from the niche of prophecy." See S.H.Nasr, *Knowledge and the Sacred*, p.35.
[71]*Kitāb al-ḥurūf*, sec.149, p.155.
[72]My inference is based on the following passage: "Regarding the voluntary actions, ways of life, positive dispositions, and so forth, that it investigates, political philosophy gives an account of the general rules. It gives an account of the patterns according to which they should be determined with due regard to particular states and times: how, with what, and by how many things, they are to be determined. Beyond this, it leaves them undetermined, because actual determination belongs to another faculty, with a different function, which should be joined to this one." Lerner and Mahdi, *op. cit.*, p.25; *Iḥṣā' al-'ulūm*, p.104.

God raises in degrees those of you who believe
and those to whom knowledge is given.
The Qurān: Repentance, Verse 72

PART II

Al-Ghazzālī's Classification

CHAPTER 7

THE LIFE, WORKS AND SIGNIFICANCE OF AL-GHAZZĀLĪ

7.1 Religious and Political Background of al-Ghazzālī's Period

Abū Ḥāmid Muḥammad ibn Muḥammad ibn Muḥammad al-Ṭūsī al-Ghazzālī,[1] an outstanding jurist, theologian and Sufi, was born in 450/1058 at Ṭūs, near the modern Mashhad, in Khurāsān which, prior to his time, had already produced so many eminent Sufis that Hujwīrī (d.464/1071) called it the land "where the shadow of God's favour rested" and where "the sun of love and the fortune of the Sufi Path is in the ascendant".[2] The district of Ṭūs[3] itself was the birthplace of many outstanding personalities and men of learning in Islam, including the poet Firdawsī (d.416/1025) and the statesman Niẓām al-Mulk (d.485/1092), who was destined to play a significant role in the intellectual life of al-Ghazzālī.[4] Among its distinguished religious scholars, we should specifically mention al-Ghazzālī's own uncle who was reported to have taught *fiqh* to al-Fārmadhī (d.477/1084), another famous native of Ṭūs and one of al-Ghazzālī's teachers in Sufism.[5]

Since al-Ghazzālī's life[6] and works were intertwined with the various contending religious and political movements prevailing in the Islamic world of his times, it is necessary to make a brief reference to some of them.[7] In 447/1055, three years before al-Ghazzālī's birth, the Shī'ite Būyid (Buwaihid) domination over the Sunni caliphate in Baghdad came to an end when the Turkish Seljūqs, under their leader Ṭughrul-Beg (d.455/1063), entered the city and deposed the Būyid regime. Prior to this historic event, Ṭughrul-Beg, who first rose to prominence in 429/1038 when he proclaimed himself Sulṭān of Naishapur,had already brought under his rule most of the eastern provinces of the 'Abbāsid Empire. Among these were eastern Persia which he captured from the Turkish Ghaznavids and western Persia from the Būyids themselves. With Baghdad, which was still the center of the Islamic world, now under his political and military control, Ṭughrul-Beg was conferred the title "King of the East and of the West,"[8] by the reigning Caliph al-Qā'im (d. 467/1075). When Ṭughrul-Beg died in 455/1063, his nephew Alp-Arslān succeeded him to become the first Great Seljūq.

The Seljūqs thus once again brought the eastern provinces of the Islamic world under Sunni rule after more than a century of domination by Shī'ite rulers. The only serious challenge to the Seljūqs in their quest for supremacy came from the Fāṭimid dynasty in Egypt, which at this time also held sway over much of North Africa and Syria. But Alp-Arslān further consolidated and even extended the dominion of Seljūq power and jurisdiction. He conquered new territories in Asia Minor from the Byzantines and forced the emir of Aleppo to abandon the suzerainty of the Fāṭimids, who were Ismā'īlites, for that of himself and the 'Abbāsid Caliph. But it was during the reign of Malik-Shāh (d.485/1092), Alp-Arslān's son who succeeded him after his death in 465/1072, that the power of the Seljūqs reached its peak. Malik-Shāh's Empire "stretched from Central Asia and the Indian frontier to the Mediterranean, and from the Caucasus and the Aral Sea to the Persian Gulf, with a slight measure of control over Mecca and Medina."[9] The life-span of al-Ghazzālī, who died in 505/1111 at the age of fifty-three, thus almost coincides with that brief but politically volatile period in the history of the Islamic world that witnessed the rise and expansion of the Seljūq dynasty. Al-Ghazzālī also lived to witness the swift decline of the dynasty following the assassination of Malik-Shāh in 485/1092.

The Seljūq rulers, like al-Ghazzālī, were Shāfi'ites in law and Ash'arites in their theological persuasion. Consequently, under their leadership, al-Ghazzālī gained respect. The most important political figure of Seljūq rule to be associated with al-Ghazzālī's scholarship was Niẓām al-Mulk. He was wazīr for about thirty years, first to Alp-Arslān[10] and then to Malik-Shāh. He brought stability to the Seljūq Empire and succeeded in reducing religious tensions and conflicts among the various schools of *fiqh* and *kalām*.

Ash'arism was officially condemned along with Shī'ism[11] during Ṭughrul-Beg's rule, apparently on the injunction of his wazīr 'Amīd al-Mulk al-Kundurī. Al-Juwaynī (d.478/1085), the leading

Ash'arite of Naishapur and later one of the principal teachers of al-Ghazzālī, was forced to live in exile in Mecca and Medina where he spent several years teaching.[12] Niẓām al-Mulk reversed this decision and adopted Ash'arism as the official theology[13] of the Seljūq Empire. He promoted Sunni learning in competition with the better established Shī'ite system of the Fāṭimid caliphate.[14] He established about a dozen colleges (sing. *madrasah*) modelled on earlier Shī'ite institutions.[15] In contrast to the latter, however, the Niẓāmīyah *madrasahs* deemphasized the philosophical sciences[16] and promoted religious sciences like *fiqh* and *kalām*.[17]

7.2 Al-Ghazzālī's Early Education and Intellectual Interest

Al-Ghazzālī received his early education in Ṭūs itself. Not long before he died, his father entrusted the education of al-Ghazzālī and his younger brother Aḥmad (d. 1126) to a pious Sufi friend. Al-Ghazzālī's education included learning the Qur'ān and *ḥadīths*, listening to stories about saints, and memorizing mystical love poems.[18] After his educational trust fund was exhausted, he was sent to a *madrasah* where he first learned jurisprudence from Aḥmad al-Rādhkānī.[19]

Later, before he was fifteen years old, al-Ghazzālī went to Jurjān in Mazardāran to continue his studies in jurisprudence under Abū Naṣr al-Ismā'īlī.[20] At seventeen, he returned home to Ṭūs. Before his twentieth birthday, he went to Naishapur to study *fiqh* and *kalām* under al-Juwaynī. At this time al-Ghazzālī composed his first work, entitled *al-Mankhūl min 'ilm al-uṣūl* (*A Resume of Science of Principles*),[21] on legal theory and methodology. He was appointed al-Juwaynī's teaching assistant and continued to teach at the Niẓāmīyah in Naishapur until the latter died in 478/1085.

It is important to note that al-Ghazzālī studied *kalām* with al-Juwaynī. The latter played a significant role in the philosophization of Ash'arite *kalām*.[22] This "philosophization" influenced al-Ghazzālī's own vision and treatment of *kalām* as a discipline.[23]

Al-Subkī claims that al-Juwaynī introduced al-Ghazzālī to the study of philosophy (*falsafah*) including logic and natural philosophy.[24] Since al-Juwaynī was a theologian (*mutakallim*), not a philosopher, he must have imparted his knowledge of philosophy through the discipline of *kalām*.[25] Al-Ghazzālī was not satisfied with what he had learned from his teacher. He later wrote in *al-munqidh* that not a single Muslim religious scholar before him had directed his attention and endeavour to a thorough study of philosophy.[26] The knowledge of *falsafah* that he gained through his study of al-Juwaynī's discourse on *kalām* and possibly through other writings as well was sufficient, however, to acquaint him with the methodological claim of the philosophers that they are the people of logic and demonstration (*ahl al-manṭiq wa'l-burhān*). This is because that claim had been current since al-Fārābī and could not have been unknown to al-Juwaynī, a leading intellectual opponent of the philosophers.

Another area of study which engaged al-Ghazzālī's mind during his stay in Naishapur was Sufism. He studied its theory and practice under the guidance of al-Fārmadhī.[27] Al-Ghazzālī at this time may also have been acquainted with the claim of the Ta'līmites or Ismā'īlīs that they were the unique possessors of authoritative instruction (*al-ta'līm*) and the privileged recipients of knowledge acquired from the Infallible Imām. However, the generally accepted view is that al-Ghazzālī did not begin to study the doctrines and teachings of the Ta'līmites until the ascendancy of al-Mustaẓhir to the caliphate in 1094.[28] Nevertheless, he informs us that he was already aware of some of the claims of the Ta'līmites before the caliphal order came[29] and, in fact, that he had long sought to know their position.[30]

I am inclined to believe that al-Ghazzālī began to study the Ta'līmite position at Naishapur. *Ta'līm* was the fundamental principle of Ismā'īlism for more than a century before al-Ghazzālī and the Ismā'īlīs were a sizeable segment of the population in al-Ghazzālī's native province.[31] Also al-Ṣabbaḥ's missionary activities in Persia already were prominent at the time that al-Ghazzālī resided in Naishapur.

7.3 Al-Ghazzālī's Intellectual Crisis

Al-Ghazzālī's initial encounter with the methodological claims of the *mutakallimūn*, philosophers, Ta'līmites, and Sufis contributed to his first personal crisis. The true nature of the crisis seems to have been epistemological since it was essentially a crisis of finding the rightful place for each of the human faculties of knowing within the total scheme of knowledge.[32] In particular, it was a crisis of establishing the right relationship between reason and intellectual intuition.[33] As a young student, al-Ghazzālī must have been troubled by the confrontation between reliance on reason on the one hand, as in the case of the *mutakallimūn* and the philosophers, and reliance on supra-rational experience on the other, as in the case of the Sufis and the Ta'līmites.[34] He, in fact, came to doubt the reliability of both sense-data (*ḥissīyāt*) and rational-data of the category of self-evident truths (*ḍarurīyāt*).

Al-Ghazzālī claims that he was delivered from the crisis not through rational arguments or rational proofs but as the result of a light (*nūr*) which God cast into his breast. He once again accepted the reliability of rational-data of the category of *darurīyāt*. But he now affirms that intellectual intuition is superior to reason.[35] This affirmation is of crucial importance to a proper understanding of his classification of the sciences. He also concluded that the four classes of knowers exhaust all the paths to truth.

7.4 Post-Crisis Intellectual Life and Works

Thus, al-Ghazzālī, upon the resolution of the crisis, proceeded to undertake a thorough study of them. He applied himself first to *kalām* while still at Naishapur. He claims in *al-Munqidh* that he wrote some works on the subject during this period. However, his extant works do not support the claim. Contrary to W.R.W. Gardner,[36] *The Golden Mean in Belief*, his earliest known work on *kalām*, was composed subsequent to his study of philosophy.[37]

According to Watt, when al-Ghazzālī claimed to have written works on *kalām* he meant books on the principles of jurisprudence.[38] Moreover, Watt thinks that *al-Munqidh* is arranged schematically rather than in a strict chronological order. Thus, the problem of discrepancy between al-Ghazzālī's claim and the extant works does not arise. However, it seems to me that there is no contradiction involved in taking seriously al-Ghazzālī's claim that he really studied *kalām* and wrote works on this subject prior to his indepth study of philosophy in Baghdad. Furthermore, many of his works have not survived, so it is not impossible that he did write on *kalām* during this period. For example, among his lost early works are a study on the science of disputation (*'ilm al-jadal*), which falls within the domain of *kalām*, and may be an early work from this period.[40]

During his stay in Baghdad al-Ghazzālī completed his promised in-depth study of the four classes of knowers. This was also his most prolific period of writing.[41] Having studied *kalām* and written several works on the discipline, al-Ghazzālī devoted himself to the study of philosophy. He tells us that he studied thoroughly the writings of the philosophers without the help of a master during his hours of free time when he was not writing and lecturing on the religious sciences.[42] He claimed that the best writings by Muslim philosophers were those of al-Fārābī and Ibn Sīnā.[43]

Al-Ghazzālī completed his first work on philosophy, *Maqāṣid al-falāsifah* (*The Purposes of the Philosophers*),[44] between 486/1093-1094 and 487/1094.[45] It is a summary of Peripatetic philosophy based on Ibn Sīnā's *Dānishnāma-yi 'alā'ī* (*The Book of Science Dedicated to 'Alā' al-Dawlah*). It was written as a preamble to his *Tahāfut al-falāsifah* (*The Incoherence of the Philosophers*), which was completed on Muharram 11, 488/January 21, 1095.[46]

The *Tahāfut* is a negative polemic against the philosophies of al-Fārābī and Ibn Sīnā. It does not expound al-Ghazzālī's own affirmative views on the questions in dispute. As for his positive views that were to take the place of the errors of the philosophers, al-Ghazzālī mentions in that work that he intends to expound them in a book to be entitled *Qawā'id al-'aqā'id* (*The Foundations of the Articles of Faith*).[47]

In the words of Nasr, the *Tahāfut* "broke the back of rationalistic philosophy and in fact brought the career of philosophy, as a discipline distinct from gnosis and theology, to an end in the Arabic part of the Islamic world."[48] In al-Ghazzālī's *Jawāhir al-Qur'ān* (*The Jewels of the Qur'ān*), he tells us that the *Tahāfut* is a work of *kalām* because it repels "errors and heresies", removes "doubts" and guards "the layman's religious belief against the confusion created by the heretics."[49]

Long before al-Ghazzālī wrote the *Tahāfut*, he embraced the Sufi doctrine that the "light of intuition" (*kashf*) is superior to reason. Now, after an in-depth study of philosophy, he accuses the philosophers of claiming the power to understand everything through reason alone,[50] in opposition not only to the theological claim that reason is subservient to revealed faith but also to the Sufi claim that *kashf* is the real key to certainty.

The author of the *Tahāfut* may also be regarded as a "philosopher", because he knows philosophy and his criticism is philosophic. However, as a "philosopher" he curtailed Muslim rationalism and paved the way for the spread of the Illuminationist (*ishrāqī*) doctrines of Suhrawardī (d. 587/1191) and the gnosis (*'irfān*) of the school of Ibn 'Arabī (d. 638/1240).[51] The best known response to al-Ghazzālī's attack on Muslim Aristotelianism was Ibn Rushd's *Tahāfut al-tahāfut* (*The Incoherence of the Incoherence*). However, Ibn Rushd failed to influence the subsequent course of Muslim intellectual history.

The character of the *Tahāfut* as a work of both *kalām* and philosophy can further be seen from the nature of the two works mentioned there which he intended to follow. One is the *Mi'yār al-'ilm* (*The Standard for Knowledge*),[52] a work on Aristotelian logic that explains methods of reasoning and technical terms of the philosophers used in the *Tahāfut*. The other is the *Qawā'id al-'aqā'id* mentioned above.

Other works on logic and philosophy during this period include the *Miḥakk al-naẓar fi'l-manṭiq* (*The Touchstone of Logical Thinking*)[53] and the *Mīzān al-'amal* (*The Criterion of Action*).[54] The latter work deals primarily with ethics in the tradition of al-Fārābī and Ibn Sīnā.[55] As we shall see below, it contains important references to al-Ghazzālī's classification of the sciences.

Contemporary bibliographical studies of al-Ghazzālī's works point to the fact that he studied the Ta'līmites and wrote the *al-Mustaẓhirī* against them after between writing the *Maqāṣid* and the *Tahāfut*.[56] This means that al-Ghazzālī paid more attention to the philosophers than the Ta'līmites. Of al-Ghazzālī's five books against Ta'līmism,[57] the *al-Mustaẓhirī* is the most important. It is also the only work that we can be certain he composed during his first stay in Baghdad. His second work in this series of refutations, entitled *Ḥujjat al-ḥaqq wa qawāṣim al-bāṭinīyah* (*The Proof of the Truth and Fragments of Bāṭinism*) which is now considered lost, most probably also was written during this period.[58]

The *al-Mustaẓhirī* provoked a debate between al-Ghazzālī and the Ta'līmites that lasted many years.[59] The *al-Qisṭās al-mustaqīm* (*The Correct Balance*), the fifth in the above series, is an account of a personal conversation and debate between al-Ghazzālī and "a companion who belonged to the group professing *al-ta'līm*" on a journey.[60] Unfortunately, the response of the Ta'līmites is lost. H. Corbin's study of the Ismā'īlī response to the *al-Mustaẓhirī*, contained in the work of 'Alī ibn Muḥammad ibn al-Walīd (d. 612/1215), the fifth dā'ī (missionary) of the post-Fāṭimid period in the Yemen, does provide some help in understanding of the debate. However, al-Walīd's response was made long after al-Ghazzālī died.[61]

The main significance of the *al-Mustaẓhirī* is that it was written for the reigning Caliph al-Mustaẓhir to attack a religio-political movement that threatened the 'Abbāsid caliphate. The assassination of al-Ghazzālī's patron, Niẓām al-Mulk, and the Great Seljūq Malik-Shāh, about two years before this work was written, was linked to members of this movement. Al-Ghazzālī considered it to be a work of *kalām*.[62] However, it is also a juridical-political work inasmuch as al-Ghazzālī dealt as a jurist with the question of religio-political status of the Ta'līmites before the *Sharī'ah* as interpreted in the Sunni legal tradition, as well as with the question of the legitimacy of the caliphate (*Imāmah*) of al-Mustaẓhir.[63]

Al-Ghazzālī informs us that the heart and essence of this work is "the establishment of the legal apodeitic demonstrations of the validity of the holy, prophetic, Mustaẓhirite positions on the basis of rational and juristic proofs."[64] It is with justification, therefore, that Henri Laoust judged the *al-Mustaẓhirī* to be an important contribution to Sunni political theory.[65]

7.5 Al-Ghazzālī's Spiritual Crisis

In his autobiography, al-Ghazzālī claims that in the month of Rajab 488/July 1095, about six months after the completion of the *Tahāfut*, he experienced a second personal crisis because of his study of Sufism. He claims to have mastered the doctrines and teachings of Sufism both through the writings of Sufis such as al-Muḥāsibī (d. 243/837), al-Junayd (d. 298/854), al-Shiblī (d. 334/945), and al-Basṭāmī (d. 261/875) and through oral teachings.[66] He did not identify these oral sources, but it is likely that one of them was his own brother Aḥmad.[67]

Al-Ghazzālī concluded that "the Sufis were masters of states (*arbāb al-aḥwāl*) and not purveyors of words (*aṣḥāb al-aqwāl*)."[68] He came to realize that there is a great difference between theoretical knowledge and "realized knowledge"[69] and that his only hope of attaining certitude and beatitude in the afterlife lay in following the way of the Sufis.[70]

This second crisis was far more serious than the first because it involved a decision to abandon one kind of life for another which is essentially opposed to the former. It affected his emotional and physical health. It caused an impediment of his speech which prevented him from teaching. His physical powers were so weakened that the physicians could not treat him. He tells us that when he completely lost his ability to make a choice, God delivered him.[71]

In Dhū'l-Qa'dah 488/November 1095 al-Ghazzālī left Baghdad on the pretext of making the pilgrimage to Mecca. In reality he abandoned his teaching career and other occupations to devote himself completely to the Sufi path. For eleven years, he led an ascetic and contemplative life, with occasional returns to his family and society. In Dhū'l-Qa'dah 499/July 1106, he again assumed public teaching at Naishapur.

Al-Ghazzālī's withdrawal from public life has been much discussed from his own time[72] until the present day. Different motives have been suggested by modern scholars, ranging from Father Jabre's proposal of al-Ghazzālī's personal fear of an assassination by the Bāṭinites to al-Baqarī's suggestion that al-Ghazzālī sought another kind of fame and glory as a religious reformer. McCarthy argues that al-Ghazzālī's own account of his motive should be accepted, namely his conversion to Sufism.[73]

7.6 Spiritual Retreat and Scholarly Output

Al-Ghazzālī spent his first spiritual retreat at the Umayyad Mosque in Damascus. Various traditional sources have linked his choice of Damascus with the presence in that city of a Sufi master by the name of Abu'l Fatḥ Naṣr ibn Ibrāhīm al-Maqdisī al-Nābulusī (d. 490/1097) who was also the leading scholar of the Shāfi'ī school in Syria.[74] In 489/1096, al-Ghazzālī moved to Jerusalem and stayed at a *zāwiyah* (Sufi convent) situated in the vicinity of the Dome of the Rock. In the same year, after visiting Abraham's tomb at Hebron, al-Ghazzālī set out for Mecca to perform the pilgrimage. He returned to Damascus in early 490/1097 to discover that Shaykh Naṣr had just passed away. He remained in that city but some time not later than the month of Jumādā II 490/June 1097[75] he returned to Baghdad because of "certain concerns and the appeals of my children."[76]

The period of al-Ghazzālī's stay in Syria,[77] including the time he spent on the pilgrimage, is less than two years.[78] During this period he composed certain parts of his *Iḥyā' 'ulūm al-dīn* (*The Revivification of the Religious Sciences*),[79] and completed *al-Risālat al-qudsīyah fī qawā'id al-'aqā'id* (*The Jerusalem Epistle on the Principles of the Faith*).[80] Abū Bakr ibn al-'Arabī says that he heard al-Ghazzālī expound the *Iḥyā'* in Baghdad following his return from Damascus.[81]

In Baghdad al-Ghazzālī could not fully continue his spiritual life because of family matters and other distractions. This dissatisfaction led him to leave Baghdad for his native city Ṭūs, possibly about 492/1099.[82] Apparently, on the way he spent some time at Hamadān.[83]

On the basis of the available evidence, modern scholars are not yet in a position to determine precisely when and where al-Ghazzālī completed the four-volume *Iḥyā'*. What is certainly known is that between the completion of the *Iḥyā'* and his return to public teaching at Naishapur in Dhū'l-Qa'dah 499/July 1106, he wrote at least five other works, including the previously mentioned *The Jewels of the Qur'ān* and the well-known *Kīmīyā-i sa'ādat (The Alchemy of Happiness)*.[84] The latter is an abridged popular version of the *Iḥyā'* in Persian. It is likely that the *Iḥyā'* was completed at Ṭūs some time before the year 499/1106.

Al-Ghazzālī's eleven-year period of spiritual retreat had convinced him that "the Sufis are those who uniquely follow the way to God, their mode of life is the best of all, their way the most direct of ways, and their ethic the purest."[85]

Having himself attained the highest level of spiritual realization, al-Ghazzālī reflected on the state of moral and religious decadence in the Muslim community of his time. He questioned whether he should remain in seclusion. Then Fakhr al-Mulk ordered him to teach at the Niẓāmīyah *Madrasah* at Naishapur. Fakhr al-Mulk was the *wazīr* of the Seljūqs. He was the son of Niẓām al-Mulk, who had his court in Khurāsān.

Al-Ghazzālī taught there for at least three years. Around 503-504/1110[86] he returned to his home in Ṭūs. In Naishapur he wrote his autobiography, *al-Munqidh min al-ḍalāl* and a work on legal theory entitled *al-Mustaṣfā min 'ilm al-uṣūl* (*Quintessence of the Science of Principles of Jurisprudence*).[87]

At Ṭūs, al-Ghazzālī set up a *madrasah* for students of the religious sciences and a *khānqāh* (Sufi convent) for the Sufi adepts. Here, he spent the rest of his life as a religious teacher and a Sufi master.[88] At the same time he applied himself to a deepening of the science of Traditions.[89] Every moment, says al-Fārisī, was filled with study, teaching, and spiritual devotion until his death on Monday the 14th of Jumādā II, 505/December 18, 1111 at the age of fifty-three.[90]

Some of the most important and well-known works have already been discussed. His greatest work, the *Iḥyā'*, written in his capacity as a Sufi, has been mentioned but its significance is yet to be discussed. Two other writings deserve mention because they are essential to understanding al-Ghazzālī's classification of the sciences. One is the *al-Risālat al-ladunīyah* (*Treatise Concerning Divine Knowledge*),[91] and the other is *Mishkāt al-anwār* (*The Niche for Lights*).[92] Both were written after al-Ghazzālī's retirement to Ṭūs.

7.7 The Authenticity of Some Works Attributed to al-Ghazzālī

Since there are sometimes inconsistencies between what al-Ghazzālī wrote in different works, some modern scholars have raised doubts about the authenticity of some of them.[93] It is beyond the scope of this study to deal with all of the disputed texts. However, I will examine three of them – *Mīzān al-'amal*, *Mishkāt al-anwār*, and *al-Risālat al-ladunīyah* – because of their relevance to al-Ghazzālī's classification of the sciences.

Watt's provisional list of al-Ghazzālī's "spurious works" is the longest. He argues that the veils-section of the *Mishkāt* is not by al-Ghazzālī but the rest of it is genuine.[94] The *Mīzān* likewise is judged partly genuine and partly not.[95] He rejects the whole of *al-Risālat al-ladunīyah*.[96] Watt bases his conclusions on three general criteria of authenticity. These criteria in turn are based on the assumption that the real views of al-Ghazzālī are sufficiently represented by the *Tahāfut*, *Iḥyā'*, and the *Munqidh*.[97] The three criteria are: (1) al-Ghazzālī, in the post-*Iḥyā'* period, affirms the superiority of prophetic revelation and "religious intuition" over reason; consequently, no work ascribing primacy to reason can belong to that period; (2) al-Ghazzālī arranges his works in an orderly and logical fashion; (3) allowing for the possibility of an "anti-orthodox phase" and an "early Neoplatonic period," al-Ghazzālī was orthodox throughout his life.

The authenticity of *Mīzān al-'amal*, *Mishkāt*, and *al-Risālat al-ladunīyah* is accepted by traditional Muslim sources and many modern scholars. Consequently, I need only to show that Watt's arguments are insufficient to disprove the traditional claim. M. A. Sherif has rebutted Watt's arguments against the authenticity of certain sections of *Mīzān al-'amal*.[98] There is no need to reproduce his arguments here. There remains only the question of the authenticity of *al-Risālat al-ladunīyah* and of the "veils-sections" of the *Mishkāt*.

The main objection that can be raised against Watt's methodology is that he defines the real al-Ghazzālī in terms of the views expounded in the *Tahāfut*, *Iḥyā'*, and *al-Munqidh*. In so doing he is begging the question of what al-Ghazzālī's authentic views are. Watt's stated criteria depict al-Ghazzālī's intellectual and religious position correctly.[99] But he is wrong in concluding that the *al-Risālat al-ladunīyah* and *Mishkāt* fail to satisfy the criteria. He arrives at that conclusion because he identifies Islamic orthodoxy with the teachings embodied in the *Tahāfut*, *Iḥyā'*, and *al-Munqidh*. However, traditional Muslim scholars like Ibn Rushd and Ibn Ṭufayl maintained that al-Ghazzālī's orthodox writings contain both exoteric and esoteric teachings.[100] The *Tahāfut*, *Iḥyā'*, and *al-Munqidh* comprise al-Ghazzālī's exoteric teachings. His esoteric views are contained in works like *al-Risālat al- ladunīyah* and *Mishkāt*. He himself often alludes to the latter views in a number of his works including the *Iḥyā'*.[101]

In traditional Islamic scholarship opposition between the exoteric and the esoteric is acknowledged. It was in the light of the traditional distinction between the two kinds of teachings that Ibn Rushd and Ibn Ṭufayl deal with discrepancies between al-Ghazzālī's exoteric works and the esoteric ones. Consequently, it is not sufficient to judge esoteric works attributed to al-Ghazzālī solely on the basis of exoteric orthodoxy.

Watt is aware of al-Ghazzālī's distinction between esoteric and exoteric teaching. But he claims that the distinction cannot amount to a real opposition or contradiction between them.[102] My reply is that if there is no opposition whatever then it does not make sense for al-Ghazzālī to insist so often in many of his works that esoteric teachings are not to be divulged except to the qualified few. From the point of view of the esoterist, insofar as he remains within the bounds of orthodoxy the problem of opposition between the esoteric and the exoteric does not arise. He accepts the truth and validity of the exoteric at its own level. But from the more limited point of view of exoterism, the opposition between the two is real. This fact was taken fully into consideration by al-Ghazzālī. One of his achievements was to restore the equilibrium between the exoteric and esoteric dimensions of Islam. As for the *Mishkāt*, many believed that it was not intended for public dissemination at all, because it presents al-Ghazzālī's most esoteric teaching.[103]

Concerning *al-Risālat al-ladunīyah,* Watt maintains that it cannot be authentic because two passages in this treatise are also found in Ibn 'Arabī*'s Risālat fi'l-nafs wa'l-rūḥ.*[104] Watt considers this textual evidence alone to be sufficient to establish his conclusion. However, the textual evidence cited above is not sufficient to prove Watt's point, since we do know that al-Ghazzālī's writings on Sufism exercised a great influence upon Ibn 'Arabī.[105] It is possible that the latter reproduced the passages in question from al-Ghazzālī's work.

Watt further claims that *al-Risālat al-ladunīyah* ascribes primacy to reason. He therefore applies his first criterion to show that it could not belong to al-Ghazzālī's latest period, as claimed by those who accept its authenticity. He then argues that neither could it belong to al-Ghazzālī's earlier period. His argument is that in that work a distinction is drawn between what prophets come to know by revelation (*waḥy*) and what religious persons come to know by inspiration (*ilhām*). If al-Ghazzālī had once been so interested in this distinction, says Watt, then it does not make sense why he should ignore it completely in the *Mundiqh* and *Mishkāt*. Watt concludes that the work cannot belong to any stage prior to that of the *Munqidh*.[106]

It is not the case that *al-Risālat al-ladunīyah* affirms the supremacy of reason. Watt's case is solely based on the following passages: ".... the knowledge of the Unseen produced by revelation is stronger and more perfect than acquired knowledge"[107] and "it is the overflowing of the Universal Reason (*al-'aql al-kullī*) which produces revelation."[108] In these statements, Watt says he fails to see "the contrast and opposition between prophetic and rational knowledge" so clearly discernable in the authentic works of al-Ghazzālī.

In response, neither the statements cited by Watt nor any other passage from the treatise supports his claim. On the contrary, the statements show that the author of *al-Risālat al-ladunīyah* believes in the superiority of prophetic revelation over reason. However, the superiority is described in esoteric terms. Revelation and reason are both seen as microcosmic manifestations of the Universal Intellect (*al-'aql al-kullī*) in man.[110] What is emphasized here is the "immanent reality" of revelation and its "essential continuity" with reason. Revelation, however, remains superior to reason, since it is a more perfect manifestation of the Universal Intellect than the latter. In contrast, what is emphasized in exoteric teaching is the "transcendent reality" of revelation with respect to reason. Revelation and reason are seen as two "discontinuous realities" having no common measure. Therefore, there appears, to use Watt's own words, this "contrast and opposition between prophetic and rational knowledge."

The fact is that al-Ghazzālī always believed in the superiority of prophetic revelation and intellectual intuition over reason. No part of *al-Risālat al-ladunīyah* is found to contradict this fact. Consequently, there is no contradiction in claiming that the work belongs to al-Ghazzālī's latest period. Neither is there contradiction in assuming it to be a pre- *Munqidh* work so long as it is not anterior to al-Ghazzālī's conversion to Sufism. Watt's conclusion that the work could not belong to the pre- *Munqidh* period is only based on a subjective impression. He finds it strange that al-Ghazzālī, having been interested in the distinction between revelation and inspiration in *al-Risālat al-ladunīyah*, chose to ignore that distinction in the *Munqidh* and *Mishkāt*.[111] In my view, there is no logical necessity for al-Ghazzālī to deal with the distinction in the latter two works. He had written the three works with different objectives in mind.

On the basis of the above discussion, I do not think Watt has succeeded in proving his case against the authenticity of *al-Risālat al-ladunīyah*.

Let us now turn to the question of authenticity of the "veils-section" of the *Mishkāt*. W.H.T. Gairdner was the first Western scholar to raise the problem of doctrinal consistency between the *Mishkāt*, especially the veils-section, and al-Ghazzālī's more well-known and popular writings. Gairdner, however, believes that the *Mishkāt* is authentic, as do Goldziher, Macdonald, Bouyges, and Jabre.[112] Watt based his judgement upon some of the problems raised by Gairdner.

Watt presents three main arguments to support his view: (1) the doctrine of the divine attributes in the veils-section contradicts what al-Ghazzālī says in the rest of the *Mishkāt* and his later works like the *Munqidh;* (2) there is no mention of prophethood or the prophetic spirit in that section, although in the rest of the *Mishkāt* and in the *Munqidh* these have a central place in the thought of al-Ghazzālī; and (3) while the rest of the *Mishkāt* is a closely argued whole, it appears to be unrelated to the veils-section.

It is true that several doctrines in the veils-section contradict al-Ghazzālī's position in, say, the *Munqidh*. This opposition, however, results from the fact that the *Mishkāt* is an esoteric work dealing with highly subtle metaphysical issues. These doctrinal problems are solved when a distinction is made between the exoteric and esoteric dimensions of al-Ghazzālī's writings.

Let us now consider the possibility of any contradiction between the veils-section and the rest of the *Mishkāt*. I begin with Watt's last claim that the two parts of the *Mishkāt* are totally unrelated. To support his claim Watt argues that the veils-tradition, the subject of commentary in the veils-section, is interpreted without any prior explanation of how light can be a veil. This is not true. The way that light can be a veil is clearly explained by al-Ghazzālī in the early part of the *Mishkāt*. It is explained not in a logical or discursive fashion, but by allusion to the symbolic meanings of the grades of luminousness of the stars, the moon, and the sun, as these are illustrated in the well-known story of Abraham in the Qur'ān.[113] Through his symbolic illustration of this story, al-Ghazzālī conveyed the message that light becomes a veil when man is deceived by its luminosity into thinking that there is no brighter light. Al-Ghazzālī takes up this message again in the veils-section where he presents his three-fold classification of those veiled by pure light.[114]

Concerning his second argument, Watt argues that in the main part of the *Mishkāt*, al-Ghazzālī divides men into three categories with respect to their "attitude" to the prophetic spirit: those at the levels of (1) faith (*īmān*), (2) ratiocination or discursive knowledge (*'ilm*), and (3) supra-rational experience (*dhawq*).[115] In the veils-section, the classification of the different religious groups is based not on the above "attitude" but on their views on divine unity. However, contrary to Watt, the two classifications are not incompatible.[116] The basis of the first classification is methodological. It pertains to the three fundamental modes of accepting revealed truth. In contrast, the basis of the second classification is doctrinal. The two are not unrelated. The second clarifies and explains the first. For example, there are various degrees of *dhawq* and to each degree corresponds some particular doctrinal formulation of *Tawḥīd* (divine unity). In the classification in the veils-section, those who attain (*al-wāṣilūn*) fall under the category of *dhawq* of the first division. However, al-Ghazzālī divides them into a number of sub-classes which are distinguished from one another by the different levels of their vision or realization of divine unity.[117]

Watt's first and most important argument is that the doctrine of divine attributes in the veils-section contradicts the rest of the *Mishkāt*. In the latter al-Ghazzālī often mentions "several of the Attributes of God in the course of his explanation of the phrase that Adam was created 'in the image of the Merciful'."[118] Yet the author of the veils-section, in the manner of the Neoplatonists, claims that ascription of attributes to God denies his unity.[119] There is an apparent contradiction here. However, the problem is resolved by synthesizing the two positions as the Sufis have done. Al-Ghazzālī, as I shall explain below, is presenting the two conceptions of Divine Attributes from different standpoints.

In the veils-section, al-Ghazzālī is primarily concerned with what veils man in various degrees from God as the absolutely transcendent or God in his state of absolute unity. From the point of view of absolute transcendance (*tanzīh*) or absolute unity (*fardānīyah*), God is above all qualities. To ascribe attributes or qualities to God is therefore to negate His absolute Transcendance and Unity. It is to veil oneself from his true reality. In his commentary on the Light-verse "God is the Light of the Heavens and the Earth," al-Ghazzālī's standpoint is that of *tashbīh* (analogy or comparison). His primary interest here is to explain the various grades of light as manifestations of God, the true Light.[120] It is in this context that he affirms Divine Attributes. The above doctrine of divine attributes is completely in conformity with the accepted teachings of Sufi metaphysics.

I have thus shown that Watt's arguments lack the necessary basis to lead one to reject the authenticity of the veils-section of the *Mishkāt*.

7.8 Significance of al-Ghazzālī's *Iḥyā'* and His Sufism

This chapter ends with a brief discussion of the significance of the *Iḥyā'* and of al-Ghazzālī as a Sufi. The whole of the *Iḥyā'* was translated into the various languages of the Muslim peoples. Numerous commentaries have been written.[121] It is one of the most extensive and influential works on Sufi ethics. It is the most visible fruit of al-Ghazzālī's attempt to restore equilibrium and harmony between the exoteric and esoteric dimensions of Islam. The developments of the two dimensions prior to and during his time had generated considerable tension in Muslim society. Highly critical of the legalism of many of the jurists, al-Ghazzālī sought through this work to reassert the supremacy of the spiritual life within the framework of the *Sharī'ah* and to revive the spiritual teachings embodied in the latter. Likewise, he criticized those esoterists who sought to belittle or negate the injunctions of the *Sharī'ah*.

Al-Ghazzali's personality and influence made it possible to teach Sufism in official, formal religious circles. He was also generally accepted as one of Islam's greatest *mujaddids* (revivers). As the title of the *Iḥyā'* indicates, al-Ghazzālī believed that true Islamic revival means the revival of Muslim communal ethics through individual moral transformation. He says in the *Munqidh*:

> I now earnestly desire to reform myself and others.....I ask Him (God) then to reform me first, then to use me as an instrument of reform; to guide me, then to use me as an instrument of guidance...[122]

ENDNOTES

Chapter 7

[1]His original name was simply Muhammad. The name Abū Ḥāmid was given later apparently because he had a son of that name who died in infancy. Although generally known as al-Ghazzālī, he was also sometimes referred to in traditional sources as al-Shāfiʿī and al-Naishapuri (Ar: al-Nīsābūrī).

There is a long and intense dispute going back to some of the earliest traditional biographers concerning whether his name should be spelled with one or two z's. Following J. Homāʾī (*Ghazzālī nāmah*, Tehran), I have adopted the spelling with two z's. See also D.B. MacDonald, "The Name al-Ghazzālī," *Journal of the Royal Asiatic Society*, 1902, 18-22. See also S.M. Zwemer, *A Moslem Seeker after God: Showing Islam at its Best in the Life and Teaching of al-Ghazzālī, Mystic and Theologian of the Eleventh Century*, New York, Chicago, London and Edinburgh, 1920, pp. 63-65 and 140-43.

[2]See al-Hujwīrī, *The Kashf al-mahjūb: The Oldest Persian Treatise on Sufism*, trans. R.A. Nicholson, Lahore, 1980, pp. 173-74.

[3]For an account of the history of Ṭūs, see A.V.W. Jackson, *From Constantinople to the Home of Omar Khayyam*, New York, 1911.

[4]On Niẓām al-Mulk, see for example M.R. Hassan, "Niẓām al-Mulk al-Ṭūsī," in M.M. Sharif (ed.), *A History of Muslim Philosophy*, Vol. I, 747-74.

[5]On al-Fārmadhī, see Jāmī, *Nafahāt al-uns*, ed. W.N. Lees, Calcutta, 1850, pp. 419-422.

[6]The most important and authentic source for al-Ghazzālī's life, especially concerning the development of his intellectual and spiritual life, is his semi-autobiographical work *al-Munqidh min al-ḍalāl* (*Deliverance From Error*). (Here after this work will be cited as *Munqidh*). R.J. McCarthy's annotated English translation, based on the earliest available manuscript, is the latest and perhaps the best in a European language. See R.J. McCarthy, *Freedom and Fulfillment: An Annotated Translation of al-Ghazzālī's 'al-Munqidh min al-ḍalāl' and Other Relevant Works of al-Ghazzālī*, Boston, 1980, pp. 61-143. For references to other translations of the *Munqidh*, see *ibid*, p.xxv.

The principal traditional biographies of al-Ghazzālī in Arabic have been compiled by ʿAbd al-Karīm al-ʿUthmān, under the title *Sīrat al-Ghazzālī wa aqwāl al-mutaqaddimīn fīhi* (*The Life of al-Ghazzālī and Its Accounts by the Early Biographers*), Damascus, 1960. Included in this collection of biographies are those of ʿAbd al-Ghāfir al-Fārisī (d.529/1129), Ibn ʿAsākir al-Dimashqī (d. 571/1175), Abū'l-Faraj ibn al-Jawzī (d. 597/1200), Tāqūt al-Hamawī (d. 681/1282), Ibn Khallikān (d. 681/1282), al-Subkī (d. 771/1369) and the above cited S. Murtaḍā, known also as al-Zabīdī (d. 1205/1790). Of these biographies, the most authentic appears to be that of al-Fārisī, who was a close friend of al-Ghazzālī. For an English translation of the major portion of al-Fārisī's biography, see R.J. McCarthy, *op.cit.*, pp. xiv-xx.

For modern accounts of his life, see in particular W.M. Watt, *Muslim Intellectual: A Study of al-Ghazālī*, Edinburgh University Press, 1963; also his "al-Ghazālī", in *Encyclopaedia of Islam*, 2nd ed., Leiden-London, 1960, pp. 2038-41; M. Smith, *al-Ghazāli the Mystic*, London, 1944; S.M.Zwemer, *op. cit.*; W.R.W. Gardner, *An Account of al-Ghazzālī's Life and Works*, Madras 1919; M.S. Sheikh, "Al-Ghazālī" in M.M. Sharif (ed.), *op. cit.*, pp. 581-87; and D.B. MacDonald, "The Life of al-Ghazzālī with Special Reference to His Religious Experiences and Opinions", *Journal of the American Oriental Society*, XX (1899), 71-132.

[7]There are numerous studies, traditional as well as modern, which deal with the political and religious conditions in the 'Abbāsid caliphate in the period just prior to and including the fifth/eleventh century. The earliest of such studies is that of Ibn 'Aqīl (d. 513/1119), a well-known Ḥanbalite jurist and contemporary of al-Ghazzālī who in fact attended the latter's inaugural lecture when he was appointed as Professor of Shāfi'ī law at the Niẓāmīyah *Madrasah* in Baghdad. See G. Makdisi, *Ibn 'Aqīl et la Resurgence de l'Islam Traditionalists au XIe Siecle (Ve Siecle de l'Hegire)*, Damascus, 1963. For studies by modern scholars, see for example; E.G. Brown; *Literary History of-Persia*. vol. II; and W.M. Watt, *The Majesty that was Islam*, pp. 193-255. There are also a few studies of al-Ghāzzālī that describe the religious and political conditions at this time. See for example M. Umaruddin, *The Ethical Philosophy of al-Ghazzālī*, pp. 14-49.

[8]Earlier, in 432/1041, al-Qā'im officially recognized Ṭughrul-Beg as the governor of provinces he had conquered up to that time. See. W.M. Watt, *op. cit.*, p. 200.

[9]W.M. Watt, *ibid.*, p. 241.

[10]In his *Ganj-i Dānish* (Teheran, 1305/1887, p. 350), Taqī Khān mentions that Niẓām al-Mulk's association with Alp-Arslān went back to the day he was appointed *mushīr* (counsellor) and *kātib* (secretary) of the latter, then only a governor of Khurāsān. (Quoted by M.R. Hassan, *op. cit.*, p.749). On his accession to the sultanship in 455/1065, Al-Arslān appointed Niẓām al-Mulk as one of his two *wazirs*, the other being al- Kundurī, Ṭughrul-Beg's *wazīr*. About a year later, when al- Kundurī was put to death, Niẓām al-Mulk became the sole *wazīr* for the rest of Alp-Arslān's ten-year rule.

[11]See W.M. Watt, *op.cit.*, p. 251. MacDonald mentions specifically the Rāfiḍites, not Shī'ites, apart from the Ash'arites. See his "The Life of al-Ghazzālī", *op. cit.*, p. 79. Al-Kundurī, a Hanafite in law and a Māturīdite in *kalām*, was said to be particularly opposed to the Shāfi'ites, Ash'arites and Shī'ites. As for the religious views of the Rāfiḍites, they were to be severely criticized later by al-Ghazzālī himself. See his *The Infamies of the Bātinites and the Virtues of the Mustaẓhirites*, trans. R.J. McCarthy, *op. cit.*, Appendix II, pp. 184, 194, 202 and 245; also his *The Criterion for Distinguishing between Islam and Godlessenss*, in *ibid*, Appendix I, p. 165.

[12]Consequently, al-Juwaynī came to be known as *Imām al-haramayn* (The Imām of the two sanctuaries).

[13]Ash'arism may not be regarded as the "official theology" during Niẓām al-Mulk's wazirship in the sense of being the sole theological doctrine that was adopted by the entire ruling establishment and then imposed on the whole Empire. Rather, it is official in the sense that it was adopted by Niẓām al-Mulk himself, with all the power and influence that he wielded, as the theology most suited to serve the unification of the Sunni Seljūq Empire and to counter Fāṭimid Ismā'īlism. Cf. I. Goldziher: *Le dogma et la loi de l'Islam: histoire du developmen̈t dogmatique et juridique de la religion musulmane*, Paris, 1920, p. 98.

[14]For an extensive acount of the Fāṭimid educational institutions, especially the *Dār al-'ilm*, see Y. Eche, *Les bibliothèques arabes publiques et semi-publiques en Mesopotamie, en Syrie et en Egypte au Moyen-Age*, Damas 1967.

[15]See S.H. Nasr, *IICD*, p. 18; and Y. Eche, *op.cit.*, pp. 253-54. On the Niẓāmīyah *madrasahs* and other educational institutions in the fifth/eleventh century 'Abbāsid caliphate, see G. Makdisi, "Muslim Institutions of Learning in Eleventh-Century Baghdad" in *Bulletin of the School of Oriental and African Studies*. XXIV (1961), 1-56; also his *The Rise of Colleges: Institutions of Learning in Islam and the West;* and A.L. Tībāwī, "Origin and Character of *al-Madrasah*", *Bulletin of the School of Oriental and African Studies*, XXV (1962), 225-38.

Concerning Niẓām al-Mulk's primary motive in establishing the Niẓāmīyah *madrasahs*, Ṭibāwi writes:

> A safe guide for an understanding of this question is to relate it internally to Niẓām's general administrative reforms, and externally to what had been going on in the rival Fāṭimid Caliphate. For the Fāṭimids, with their religious fervour and vigorous propaganda, were in more than one sense pioneers in initiating various centres for teaching and preaching as well as centers for study and research. The example set by Jauhar, al-'Azīz, and Ibn Killis could not have been lost on Niẓām. As chief minister to a conquering race, he too needed to educate his subjects, and to provide the state if not with outright propagandists at least with efficient religious and civil servants. That is not an unreasonable assumption to make, especially if we consider it later on in relation to what was actually done.

See *ibid.*, pp. 233-34.

[16]This is best reflected in the fact that there was far greater scientific activity in Fāṭimid Egypt than in Seljūq Baghdad.

[17]S.H. Nasr, *Science and Civilization in Islam*, p. 72.

[18]See M. Smith, *Al-Ghazzālī The Mystic*., p. 11.

[19]See al-Fārisī's *Life of al-Ghazzālī* in R.J. McCarthy, *op. cit.*, p. xv.

[20]Al-Sukbī, *Tabaqāt al-Shāfi'iah al-Kubrā*, Cairo, 1324/1906, III, 37 and IV, 103, 104.

[21]For a discussion of the authenticity of this work, which has been edited by Muḥammad Ḥassan Hitu and published (Damascus, 1390/1970), see M. Bouyges, *Essai de Chronologie des Oeuvres de al-Ghazali*, Imprimerie Catholique, Beyrouth (1959), pp. 8-9.

[22]By philosophization of *kalām* I mean the integration into that science of philosophical doctrines, conceptions, and arguments of the *falāsifah*. As far as philosophization of Ash'arite *kalām* is concerned, al-Juwaynī played an "intermediate" role between that of al-Bāqillānī and that of al- Ghazzālī. Al-Shahrastānī (d. 1153) in his *al-milal wa'l-nihal* (*The Nations and Their Beliefs*) seems to associate al- Juwaynī with such a role. For references by modern scholars to al-Juwaynī's philosophization of Ash'arite *kalām*, see S.H. Nasr, "Fakhr al-Dīn Rāzī," in M.M. Sharif (ed.), *A History of Muslim Philosophy*, I, 643; W.M. Watt, *Islamic Philosophy and Theology*, Edinburgh University Press, 1962, p. 112; H.A. Wolfson, *The Philosophy of the Kalām*, Harvard University Press, 1976, pp. 693-908; also H. al-Fakuri and al-Jarra, *Tārīkh al-falsafat al-'arabīyah*, Beirut, 1957, II, 267.

[23]Al-Ghazzālī's treatment of *kalām* marks a new turning point in the history of that discipline. He accepted the total application of syllogistic arguments of the philosophers. For this reason, Ibn Khaldūn (*The Muqaddimah*, trans. F. Rosenthal, III, 40) describes al-Ghazzālī as the religious scholar who introduced 'the method of the later *mutakallimūm* (*ṭarīqat al-muta'akhkhirīn*).'' And according to Wolfson, Maimonides makes a reference to "a skillful one among the later *mutakallimūn*." Wolfson identifies him with al- Ghazzālī. See H.A. Wolfson, *op. cit.*, p. 41, n. 167 and p. 6595.

[24]Al-Sukbī, *op. cit.*, IV, 103.

[25]For a discussion of some of the "philosophical" content of al-Juwaynī's theological works, see for example R.M. Frank, "Bodies and Atoms: The Ash'arite Analysis", in M.E. Marmura (ed.), *Islamic Theology and Philosophy: Studies in Honor of George F. Hourani*, SUNY Press, Albany, 1984, pp. 39-53. H . Wolfson, *op. cit.*, deals at numerous places with the extent of al-Juwaynī's familiarity with *falsafah* as reflected in several of his theological works.

[26]See R.J. McCarthy, *op. cit.*, sec. 26, p. 70.
[27]"Al-Fārmadhī guided him, and he followed his path (*ṭarīqah*) and imitated all the practices that were put before him. He took part in *dhikrs* (invocations) and passed through all the laborious and wearying life of the Sufi neophyte, but did not attain what he sought." D.B. MacDonald, *op. cit.*, p. 89.
[28]This view is based on al-Ghazzālī's claim that he wrote the *al-Mustaẓhirī*, his first polemical work against the Ta'līmites, following an order from the new Caliph. The *al-Mustaẓhirī* is *The Infamies of the Bāṭinites and the Virtues of the Mustaẓhirites* (see n. 11 of this chapter). It was so named because al-Ghazzālī dedicated it to al-Mustaẓhir. Subsequent citations of this work will be from McCarthy's translation.
[29]See *al-Munqidh*, sec. 61, p. 82.
[30]*al-Mustaẓhirī*, sec. 5, p. 178.
[31]See M.G.S. Hodgson, "The Ismā'īlī State", in J. Boyle (ed.), *Cambridge History of Iran*, Cambridge, 1968, 5, 422-82; and his *The Order of Assassins*, The Hague, 1955. See also A. Esmail and A. Nanji, "The Ismā'īlīs in History", in S.H. Nasr. (ed.), *Ismā'īlī Contribution to Islamic Culture*, p. 247.
[32]The epistemological nature of the crisis is affirmed by McCarthy, V.M. Poggi and G. Furlani. See R.J. McCarthy, *op. cit.*, p. xxix; V.M. Poggi, *Un Classico della Spiritualita Musulmana*, Liberia dell' Universita Gregoriana, Rome, 1967, p. 171; and G. Furlani, "Dr. J. Obermann, Der Philos. und Regligiose Subjektivismus Ghazalis," (Recensione) in *Revista trimestrale di studi filosoficie reliqiosi*, Perugia, 1922, III (3), 340-53.
[33]In the *Munqidh* intellectual intuition is symbolized by light which God casts into the breast. See *al-Munqidh*, p. 67.
[34]In the context of this "epistemological confrontation" the *mutakallimūn* and the philosophers, as my analysis of their methodologies will later demonstrate, may justifiably be grouped together on the side of reason although there exist significant differences between them. Similarly, in relation to reason, the *ta'līm* of the Ta'līmites and the *kashf* of the Sufis possess certain common characteristics which justify these two classes to be grouped together on the side of supra-rational experience.
[35]The key to the greater part of knowledge, says al-Ghazzālī, is that light which God casts into man's breast.
[36]W.R.W. Gardner, *An Account of al-Ghazzālī's Life and Works*. p. 38.
[37]This view is confirmed by the fact that it makes references to a number of his works on logic and philosophy, including the *Tahāfut*, and also to the *Mustaẓhirī*. Consequently, it could not have been composed before 488/1095. See M. Bouyges, *op. cit.*, pp. 33-34; also G.F. Hourani, "The Chronology of Ghazālī's Writings," in *Journal of the American Oriental Society*, 79(4), 1959, 228.
[38]W.M. Watt, *Muslim Intellectual*, p. 117.
[39]Traditional bio-bibliographers like Ibn Khallikān, al-Subkī, Ḥājjī Khalifah, and Murtaḍā had all attributed such a work to al-Ghazzālī although there are variations in their designation of its title. For a discussion of the identity of this work on the basis of these traditional sources, see M. Bouyges, *op. cit* pp. 16-17.
[40]The researches of M. Bouyges (*ibid*) and al-Kurdi support this view. Al-Kurdī's view is given in his *Tarjamat al-muṣannif* (pp. 5-6), his biographical notice of al-Ghazzālī placed at the beginning of his edition of the *Mi'yār al-'ilm* (Cairo, 1329/1911).
[41]For a comprehensive list of al-Ghazzālī's works said to have been composed during this period, and a discussion of them, see M. Bouyges, *op. cit.*, pp. 22-40.
[42]*al-Munqidh*, sec. 27, p.70.
[43]See his *The Incoherence of the Philosophers*, trans. S.A. Kamali, Pakistan Philosophical Congress, Lahore, 1963, p. 5; also *al-Munqidh*, sec. 34, p. 72. Subsequent

citations of *The Incoherence* will be made from Kamali's translation.

[44]This work was translated into Latin by Dominicus Gundissalinus, under the title *Logica et Philosophia Algazelis Arabis*, and into Hebrew in two versions, first by Issac Albalag in the 13th century under the title *De'ot ha-Pilusufim*, and second by Judah Nathan in the following century under the title *Kawwanot ha-Pilusufim*.

There is a German translation of the first two chapters of the work, by G. Beer, based on his own critical edition. See his *Maqāsid*, Leipzig, 1888.

[45]See M. Bouyges, *op. cit.*, p. 24; G.F. Hourani, *op. cit.*, p. 227.

[46]This date was recorded in a manuscript of the work discovered in Istanbul.

[47]The question of the real identity of this promised treatise has not been satisfactorily resolved by scholars until now. Ibn Rushd, in his *Tahāfut al-tahāfut*, says that he could not get hold of the work and suggests that perhaps it was never written. However, in my judgement the *al-Risālat al Qudsīyah fī qawā'id al-'aqā'id* (*The Jerusalem Epistle*), written as an independent treatise but incorporated later into the *Ihyā'* (first volume, second book, third section), is al-Ghazzālī's promised constructive work in theology.

G.F. Hourani claimed that the work in question is the *al-Iqtiṣād fi'l-i'tiqād* rather than "the actual *qawā'id al- 'aqā'id* which is later and is but a part of the *Ihyā'*" (Hourani, *op. cit.*, p. 228). However, it is not necessary, he says, to amend the textual reading *Qawā'id al-'aqā'id* in the *Tahāfut*, as S. Van den Burgh has done in his translation of Ibn Rushd's *Tahāfut al-tahāfut* (London, 1954), since al- Ghazzālī "may well have changed his mind about the title of the book."

However, contrary to Hourani's thesis, al-Ghazzālī describes the *Jerusalem Epistle* and the *al-Iqtiṣād* in the *Jewels of the Qur'ān* as two works of the same nature except that the latter is far more comprehensive than the former. This means that contentwise, regardless of the chronological positions of tho two works, the *Jerusalem Epistle* would be the first affirmative work that should be read after the *Tahāfut*.

[48]S.H. Nasr, *Islamic Life and Thought*, p. 72.

[49]*The Jewels of the Qur'ān*, trans. M.A. Quasem, p. 38.

[50]S.H. Nasr, *op. cit.*, p. 71; see also *al-Munqidh*, sec. 61, pp. 81-82.

[51]S.H. Nasr, *Three Muslim Sages*, p. 55.

[52]See the anticipation of this work in *The Incoherence of the Philosophers*, pp. 10, 12. M. Bouyges reads *Mi'yār al-'āql*, instead of *Mi'yār al-'ilm*, in his edition of the *Tahāfut* (pp. 17, 20).

[53]This work which mentions the *Tahāfut* was written after the *Mi'yār* but completed earlier than the latter.

[54]For a discussion of its chronological place among al-Ghazzālī's writings, see M. Bouyges, *op. cit.*, pp. 28-29.

[55]See M.A. Sherif, *Ghazālī's Theory of Virtues*.

[56]See G.F. Hourani, *op. cit.*, p. 227.

[57]See *al-Munqidh*, sec. 76, p. 88.

[58]Al-Ghazzālī described the work as "an answer to some of their arguments proposed to me in Baghdad". See *ibid* ; also *The Jewels of the Qur'ān*, p. 39.

[59]In *al-Munqidh* (sec. 76, p. 88), al-Ghazzālī tells us that he wrote *The Detailed Exposition of the Disagreement* as a reply to the criticisms by the Ta'līmites made against him in Hamadān and the *al-Durj al-marqūm bi'l-jawādil* as a response to their arguments made in Ṭūs. These took place after he left Baghdad for the last time.

[60]See *The Correct Balance*, trans. R.J. McCarthy in his *Freedom and Fulfillment*, p. 287. Subsequent citation of this work will be made from this translation.

[61]H. Corbin, 'The Ismā'īlī Response to the Polemic of Ghazālī,' in S.H. Nasr (ed.), *Ismā'īlī Contributions to Islamic Culture*, pp. 69-98.

[62]*The Jewels of the Qur'ān*, p. 39.

[63]Chapters 8 and 9 of the *al-Mustaẓhirī* deal specifically and at length with these two questions.
[64]*al-Mustaẓhirī*, sec. 15, p. 181.
[65]See H. Laoust, *La Politique de Gazali*, Librairie Orientaliste Paul Geuther, Paris, 1970, 82.
[66]*al-Munqidh*, sec. 81, p. 90.
[67]I have heard from Nasr that Aḥmad was very likely one of al-Ghazzālī's teachers in Sufism.
[68]*al-Munqidh*, sec. 83, p. 90.
[69]To realize spiritual knowledge is to transform one's soul in conformity with that knowledge so that knowledge and being are one.
[70]*al-Munqidh*, sec. 84, p. 91.
[71]*Ibid*, sec. 89.
[72]See *ibid*, secs. 89-90 where al-Ghazzālī describes the various explanations offered by his contemporaries concerning his abandonment of public teaching. In terms of modern scholarship, see R.J. McCarthy, *op. cit.*, pp. xxix-xlii.
[73]*Ibid.*, sec. 49, p. xxix.
[74]See A.L. Ṭībāwī, *Arabic and Islamic Themes*. London, 197O̱, pp. 203-208, where he refers to the various traditional sources in which al-Ghazzālī was linked to this Sufi figure.
[75]It cannot be later than this date, because Abū Bakr ibn al-'Arabī claimed in his *al-Qawāṣim wa'l-'awāṣim* that in that month he met al-Ghazzālī in Baghdad and heard him expound the *Iḥyā'*. See A. Badawī, *Mu'allafāt al-Ghazālī*, Cairo (1961), p. 546; also F. Jabre, "La Biographie et l'oeuvre de Ghaẓālī reconsiderées à la lumiere de *Tabaqat* de Sobki," in *Mélanges de l'institut Dominican d'Etudes Orientales du Caire*, i (1954) pp. 75, 92.
[76]*al-Munqidh*, sec. 93, p. 94.
[77]Ṭībāwī (*op. cit.*, p. 200) has drawn the attention of scholars to the necessity of translating the Arabic *al-Shām*, mentioned in the *al-Munqidh*, as Syria and not Damascus. For Damascus, al-Ghazzālī uses the Arabic *Dimashq*. That translation error makes al-Ghazzālī stay for "nearly two years" at Damascus (cf. Hourani, *op. cit.*, p. 229) and this second error has given rise to several discrepancies between al-Ghazzālī's account of this period and those of his biographers.
[78]Assuming that his return to Baghdad took place in Jumādā II, 490/June 1097, his period of stay in Syria came to only 18 months. Consequently, as suggested by Watt (*Muslim Intellectual*, pp. 145-146) it seems best to assume that al-Ghazzālī used the phrase "nearly two years somewhat loosely."
[79]On the basis of reports from Ibn al-Athīr (A.L. Ṭībāwī, *op. cit.*, p. 206) and Mujīr al-Dīn (*ibid.*, p. 207), we may infer that the first few parts of the *Iḥyā'* were first composed either in Jerusalem or in Damascus.
[80]In a manuscript of a portion of the *Iḥyā'* dated 1160 A.H. (Ṭībāwī, *op. cit.*, p. 209), a statement supposed to come from al-Ghazzālī himself, reads: "and I completed *al-Risālah al-Qudsīyah*, which I concluded in this section, in the Aqṣā Mosque, in answer to the request of its people."
[81]See n. 75.
[82]According to Ibn al-Athīr, al-Ghazzālī left Baghdad before the fall of Jerusalem in 492/1099. See M. Bouyges, *op. cit.*, p. 4, n. 1.
[83]This is based on al-Ghazzālī's own account (cf. our n. 59) that he received certain criticisms from the Ta'līmites while he was at Hamadān.
[84]There are two English versions of this work, one rendered from Turkish by H.A. Homes (Albany 1875) and the other from Urdu by C. Field (London, 1910).

[85]*al-Munqidh*, sec. 94, p. 94.
[86]See M. Bouyges, *op. cit.*, p. 6; W.M. Watt, *op. cit.*, pp. 147-48.
[87]According to Ibn Khallikān (*Biographical Dictionary*, II, 622), this work was completed on Muḥarram 6, 503/August 56, 1109.
[88]Some of the well-known disciples of al-Ghazzālī are mentioned by M. Bouyges, *op. cit.*, pp. 4-5.
[89]According to al-Fārisī (McCarthy, *op. cit.*, sec. 25, p. xix), al-Ghazzālī during this period frequented the company of those devoted to the science of Tradition.
[90]On the folklore surrounding his death, see M. Smith, *Al-Ghazzālī the Mystic*, pp. 35-36. Al-Ghazzālī was buried outside Ṭābarān in a grave near that of the poet Firdawsī.
[91]This work has been translated into English by M. Smith and published in *Journal of the Royal Asiatic Society*, Pt. II, April 1938, pp. 177-200 and Pt. III, July 1938, pp. 353-74. Citations of this work will be made from this translation.
[92]This work, whose aim is to give a commentary of the Light-verse in the Qur'ān (XXIV: 35) and the prophetic *ḥadīth* on the "seventy thousand veils of light and darkness", has been translated by W.H.T. Gairdner (Lahore) 1952) and first published as *Monograph vol. XIX of the Royal Asiatic Society* (London, 1924). Subsequent citation of this work will be made from this translation.
[93]See M. Asin Palacios' list of what he calls works of "apocryphal or doubtful authenticity" in his *La espiritualidad de Algazel y su sentide cristiano*, Madrid-Granada (1934-1941), iv, 385-90; and W.M. Watt, "The Authenticity of the Works Attributed to al-Ghazzālī," in *Journal of the Royal Asiatic Society*, 1952, 24-45.
[94]W.M. Watt, "A Forgery in al-Ghazzālī's Mishkāt?" in *Journal of Royal Asiatic Society*, 1949, 5-22.
[95]W.M. Watt, "The Authenticity of the Works Attributed to al-Ghazzālī," *op. cit.*, pp. 38-40.
[96]*Ibid.*, pp. 33-34.
[97]*Ibid.*, p. 26.
[98]See M.A. Sherif, *Ghazālī's Theory of Virtue* Appendix I, pp. 170-76.
[99]However, Watt stands to be corrected on the following points: first, al-Ghazzālī believed in the supremacy of prophetic revelation and intuition in the pre-*Iḥyā'* period as well; second, al-Ghazzālī never had an anti-orthodox phase in his life.
[100]For Ibn Rushd's view, see his *Al-Kashf 'an manāhij al- adillā'*, ed. Muller, p. 71. (The relevant passage has been translated by W.H.T. Gairdner in his "Al-Ghazālī's *Mishkāt al-anwār* and the Ghazālī-Problem," in *Der Islam*, V, 1914, 133). For Ibn Ṭufayl's view, see his introduction to his *Ḥayy ibn Yaqẓān*, ed. L. Gauthier (Beirut, 1936), pp. 15-16; see also M.A. Sherif, *op. cit.*, p. 173; and W.H.T. Gairdner, *op. cit.*, p. 147.
[101]In the first book of the first volume of the *Iḥyā'*, *The Book of Knowledge*, trans. N.A. Faris, Lahore, 1979, p. 6, al-Ghazzālī, in referring to the science of revelation (*'ilm al-mukāshafah*), says: "... one is not permitted to record in writing, although it is the aim of saints and the desire of the eyes of the sincere."

In *The Jewels of the Qur'ān*, p. 44, al-Ghazzālī claims in fact to have written esoteric books while in the *Mīzān* (see *Critere de L'action*, trans. H. Hachem, Librairie Orientale et Americaine, Paris, 1945, pp. 147-48), he speaks of three sets of doctrines possessed by every perfect man, the third being the esoteric.
[102]W.M. Watt, "A Forgery in al-Ghazālī's Mishkāt?," *op. cit.*, 21.
[103]See W.H.T. Gairdner, *op. cit.*, p. 121.
[104]This work was edited and translated by Asin Palacios in his "La Psicologia segun Mohidin Abenarabi" in *Congres XIVe International des Orientalistes*, vol. iii. This Spanish Orientalist first drew attention to the two identical passages in question. See his *La espiritualidad de Algazel*, iv, 388.

[105]See S.H. Nasr, *Three Muslim Sages*, pp. 88-89.
[106]W.M. Watt, "Authenticity of Works Attributed to al-Ghazzālī," *op. cit.*, p. 34.
[107]*Ibid.* (Cf. *al-Risālat al-laduniyah*, p. 364).
[108]*Ibid.* (Cf. *al-Risālat al-laduniyah*, p. 366).
[109]*Ibid.*
[110]The Intellect *(al-'aql al-kullī)* "which is at once the source of revelation and exists microcosmically within man, must not be mistaken for reason alone. The *'aql* is at once both *intellectus* or *nous* and *ratio* or reason. It is both the supernal sun that shines within man and the reflection of this sun on the plane of mind which we call reason." S.H. Nasr, *Sufi Essays*, SUNY Press, Albany, 1973, p. 54.
[111]The distinction is mentioned in the *Iḥyā'* which is pre-*Munqidh*. See R.J. McCarthy, *op. cit.*, 45, p. 378.
[112]According to Jabre, he discovered a photograph of a manuscript of the *Mishkāt* which its coypist claimed to have completed copying in the month of Ramaḍān 509, that is four years after the death of al-Ghazzālī. The manuscript contains the "veils-section" in full, which agrees with the other known editions of the *Mishkāt*, except for a few minor variations. Says Jabre: "Le fait est au moins hautement en faveur de l'authenticite de la "veils-section," s'il ne la prouve pas d'une facon apodictique." See F. Jabre, *La notion de la ma'rifa chez Ghazali*, Beyrouth, 1958, p. 106, n. 1.
[113]*Mishkāt al-anwār*, pp. 126-28; for the story of Abraham in question, see *the Qur'ān* (XV: 75-78).
[114]*Mishkāt*, pp. 169-71.
[115]Watt, "A Forgery in al-Ghazālī's Mishkāt?", *op., cit.*, p. 10. (cf. *Mishkāt*, pp. 148-9).
[116]According to Watt, all the groups in the first classification hold the same dogmas and differ only in their "attitude" towards them, while those in the second classification hold different dogmas and the question of their "attitude" is irrelevant (*op. cit.*, p. 10). This is a misinterpretation of the *Mishkāt*. Al-Ghazzālī does not say that all the groups in the first classification hold the same beliefs. He speaks only of their different "attitudes" toward revealed truth in general. A better interpretation would be "modes of accepting" than Watt's "attitudes"). Watt fails to see that different "modes of understanding" the same revealed truth do lead to different doctrines. For example, the revealed truth of the Oneness of God is accepted by all Muslims, but the Sufi doctrine of Unity is not the same as that of the *mutakallimūn*.

Similarly, it is not true that the question of "attitude" does not enter into the second classification. Associated with the idea of God of each group is a particular "mode of understanding." Watt's misinterpretation led him to attribute incompatibility to the two classifications.
[117]The highest degree of realization is that of whom al-Ghazzālī calls *khawāṣṣ al-khawāṣṣ* (Elite of the elite). See *Mishkāt*, p. 173.
[118]*Mishkāt*, pp. 134-6.
[119]Watt, *op. cit.*, p. 7. (Cf. *Mishkāt* pp. 170, 172).
[120]*Mishkāt*, pp. 101-3.
[121]Murtaḍā al-Zabīdī's ten-volume commentary in Arabic written in the 18th-century has been mentioned. Of special significance is the 17th-century attempt by Muḥsin-i Fayḍ, one of the most famous students of Mullā Ṣadrā, to rewrite the *Iḥyā'* in the light of religious and spiritual teachings embodied in Shi'ism. His work; entitled *Maḥajjat al-badya'fī iḥyā' al-iḥyā'*, reflects the influence of al-Ghazzāli even in Shi'ite philosophical circles. See S.H.Nasr, "The School of Isfahan," in Sharif (ed.), *A History of Muslim Philosophy*, II, 926-30; also, H. Corbin, "The Ismā'īlī Response to the Polemic of Ghazāli," *op. cit.*, p. 71.
[122]R.J. McCarthy, *op. cit.*, sec. 139, p. 107.

CHAPTER 8

AL-GHAZZĀLĪ'S CLASSIFICATION OF SEEKERS AFTER KNOWLEDGE

8.1 Basis of Classification

In the *Munqidh*, al-Ghazzālī divides the seekers after knowledge of his time into four classes, namely the *mutakallimūn*, the philosophers (*al-falāsifah*), Ta'līmites (*al-bāṭiniyah*, and the Sufis (*al-ṣufiyah*). The primary basis of the classification is methodological for he describes each intellectual school essentially in terms of its methodological claim to truth. He first conceived the idea of this methodological basis immediately following the resolution of his epistemological crisis. However, he did not put down that classification into writing until after he had completed his verification of the different methodological claims. In his view, the classification not only embraced the whole spectrum of Islamic epistemological thought, but exhausted all the possible avenues of knowledge open to man.[1]

Al-Ghazzālī claimed that his interest in the various classes of knowers was generated by his inner quest for the knowledge of the true reality of things. The highest of this knowledge, he says, is the knowledge of God.[2] Al-Ghazzālī excluded the jurists from the stream of seekers after knowledge of God. He recognized the jurists as the most important religious group from the point of view of the general welfare and goodness of the individual and the community in the life of the world.[3] The school of jurists as a whole could not be qualified as a class of seekers because they were not concerned with the knowledge of the true nature of things. There were individual jurists who sought after this kind of knowledge, but they did not do so as jurists. In relation to the path to God, the jurists are analogous to "those who build and maintain houses of refuge and provide facilities along the way to Mecca to the pilgrimage."[4]

There is a similar fourfold classification of knowers by 'Umar Khayyām (d. 526/1131), al-Ghazzālī's older contemporary.[5] Regardless of whether or not there was any influence of the one upon the other in relation to the classification,[6] it is significant that the two thinkers share the idea of the methodological basis of the classification and the view that the classification embraced the total hierarchy of modes of knowing. For the two figures were of different intellectual perspectives and philosophical persuasions. Khayyām regarded himself as a Phythagorean[7] and a follower of the philosophical school of Ibn Sīnā. He also described himself as a Sufi. But Pythagoreanism and Ibn Sīnā's Peripatetic philosophy came under severe criticism from al-Ghazzālī. It seems that al-Ghazzālī and 'Umar Khayyām had adopted the same classification of knowers because of their common Sufi perspective. The Sufi perspective is thus presented as the most universal point of view that transcends and comprehends all other points of view.[8]

8.2 Al-Ghazzālī's Views Concerning the Four Classes

I will now discuss al-Ghazzālī's treatment of each of the four classes of knowers. In this discussion, the emphasis is not on the doctrines of each school, but rather on its epistemology and methodology.

8.2.1 The Mutakallimūn

Al-Ghazzālī's criticism of *kalām* is significant. For his authority in that discipline was widely recognized. He had studied under the greatest Ash'arite theologian of the day and mastered the works of the most meticulous *mutakallimūn*. He himself became the leading theologian of his time with several excellent works on the subject to his credit. In his description and criticism of the *mutakallimūn*, al-Ghazzālī did not deal with the different schools of *kalām*.[9] What mainly attracted al-Ghazzālī was not the differences among these schools in matters of doctrine, but the common "methodological stand" that they had adopted.

Al-Ghazzālī describes the *mutakallimūn* as those who claim themselves as men of independent reasoning and intellectual speculation (*ahl al-ra'y wa'l-naẓar*).[10] This characterization of the *mutakallimūn* as *ahl al-ra'y* was principally related to their generally positive stand on the use of reason in understanding articles of faith. As a legal term in Islamic jurisprudence, *ra'y* connotes the use of reason to produce a well-considered opinion or a sound judgement in legal matters on the basis of explicit textual evidence. More precisely, the term refers to the application of that method of reasoning called *qiyās* (analogy) in connection with problems of law. According to Ibn Khaldūn, *kalām* borrowed this method of reasoning from *fiqh* and applied it to the domain of faith.[11]

There is some validity in the claim that the *mutakallimūn* were men of sound reasoning if they are contrasted with those who had been collectively called the *ahl al-taqlīd*.[12] These latter groups were opposed to the use of reason in explaining religious beliefs. They condemned rational discussion in matters of faith as innovation (*bid'ah*) and sin.[13]

Al-Ghazzālī approves of the aim of *kalām* and its role in society, namely the defense of the common religious beliefs of the community by "repelling errors and heresies and removing doubts and confusion relating to those beliefs." He even praised the *mutakallimūn*, describing them as people who had been inspired by God "to champion orthodoxy by a systematic discussion (*kalām*) designed to disclose the deceptions introduced by the contriving innovators contrary to traditional orthodoxy."[14] In *The Jewels of the Qur'ān*, al-Ghazzālī claims that the science of *kalām* has its roots in the Qur'ān.[15]

However, al-Ghazzālī was highly critical of certain aspects of the methodology of *kalām*. He considers the methods of *kalām* to be defective both to satisfy his thirst for the knowledge of the reality of things and to inflict intellectual defeat on the opponents of *kalām*. He writes:

> ....they [i.e., the *mutakallimūn*] relied on premises which they took over from their adversaries, being compelled to admit them either by uncritical accepance (*taqlīd*), or because of the Community's consensus (*ijma'*), or by simple acceptance deriving from the Qur'ān and the Traditions. Most of their polemic was devoted to bringing out the inconsistencies of their adversaries and criticizing them for the logically absurd consequences of what they conceded. This, however, is of little use in the case of one who admits nothing at all except the

> primary and self-evident truths. So *kalām* was not sufficient in my case, nor was it a remedy for the malady of which I was complaining.[16]

By the adversaries of *kalām*, al-Ghazzālī means the philosophers. His portrayal of *kalām* as a discipline that had become influenced by *falsafah* was confirmed by later authorities like Shahrastānī, Maimonides, and Ibn Khaldūn.[17] Concerning premises borrowed form the philosophers – on which the *mutakallimūn* had placed so much reliance in their demonstration of religious beliefs – al-Ghazzālī must have been referring to ideas and concepts related to atomism. The doctrinal status of atomism within Ash'arite *kalām* was transformed by al-Bāqillānī from being a mere premise in support of certain religious beliefs[18] to being an essential part of the creed.[19]

Al-Ghazzālī criticizes the methodology of *kalām* from two points of view. First, he criticizes *kalām* in his capacity as a *mutakallim*. In this capacity, we saw, he accepts the religious and intellectual perspective of that school,[20] and even affirms its necessity (*farḍ kifāyah*).[21] But he criticizes the inadequacy of its methodological tools to confront its intellectual opponents. Second, he criticizes *kalām* from the point of view of a seeker after a direct spiritual experience of God and the inner reality of things.

According to al-Ghazzālī, the *mutakallimūn* were orthodox because they subordinated reason to revelation. But the use of reason in *kalām* had not been exercized to the fullest extent possible. He saw much scope for improvement in *kalām's* methodology. He was not against the borrowing of premises from the opponents of *kalām*. What he was against was the uncritical acceptance of premises by the *mutakallimūn*. For example, the premises they used to demonstrate fundamental religious beliefs like the divine creation of the world were not at all primary and self-evident truths. Moreover, their proofs involve tedious arguments.[22] Al-Ghazzālī criticized his contemporary *mutakallimūn* for being too closely bound in their views to previous authorities in *kalām*, like al-Ash'arī and al-Bāqillānī, even in matters relating to premises and proofs.[23]

Al-Ghazzālī found defects in the *mutakallimūn's* use of syllogism.[24] To remedy this defect, he wrote several works on Aristotelian logic in a manner that had not been attempted before by any jurist or theologian.[25] Al-Ghazzālī also considered undersirable the *mutakallimūn's* emphasis on the logical inconsistencies in the arguments of their opponents. Instead, says al-Ghazzālī, they should concentrate on the refutation of the fundamental doctrines of their opponents, insofar as these doctrines were viewed as heretical. The *Tahāfut* was seen as advocating this new approach of *kalām*.[26]

From the point of view of his quest for a direct experience of God and for knowledge of the true nature of things, al-Ghazzālī sees inherent limitations in the methodology of *kalām*. As a science, *kalām* "does not concentrate on the intuitive knowledge (*kashf*) of realities."[27] He acknowledged the fact that some of his predecessors among the *mutakallimūn* had sought to defend orthodoxy "by the study of the true natures of things" like the study of substances and accidents and their principles. However, their discussion of the subject was not thoroughgoing, because "that was not the aim of their own science."[28]

In al-Ghazzālī's view, the methodology of *kalām* was comprised of faith (*īmān*) and ratiocination tainted by false syllogism,[29] The premises of *kalām* were mainly accepted on the basis of faith. For al-Ghazzālī, faith implies a particular level of knowledge and certainty. It is of a lower grade than scientific knowledge (*'ilm*) based on apodeictic demonstration (*al-burhān*), and much more so than mystical experience (*dhawq*).[30] At the level of faith, and even of ratiocination, one does not know the truth directly or with immediacy. In the *Mishkāt* al-Ghazzālī speaks of faith as a degree of light and false syllogism as a veil of darkness. There, he identifies the *mutakallimūn* with the best group "among those veiled by mixed light and darkness."[31]

Al-Ghazzālī accepts the fact that some people might find their thirst for knowledge and certitude quenched by the science of *kalām*. But as far as he is concerned, *kalām* could not deliver the certitude that he sought. Al-Ghazzālī seems to be asserting that the spiritual and the intellectual needs of man are not the same for all individuals.

8.2.2 **The Philosophers**

To establish himself as a respected critic of *falsafah*, al-Ghazzālī realized that it was necessary and sufficient for him to claim that he had read and understood al-Fārābī and Ibn Sīnā, for these two thinkers were the most well-known Muslim philosophers.[23] The extent of his acquaintance with the philosophic works of the two thinkers could not yet be established. What is certain is that Ibn Sīnā exercized a far greater influence upon him than did al-Fārābī.[33] It appears that al-Ghazzālī was also familiar with some of the writings or doctrines of other schools of Muslim philosophy. The Peripatetics were not the only disciples of Greek learning among Muslims. The *Ikhwān al-Ṣafā'* (The Brethren of Purity) claimed themselves as followers of the tradition of Pythagoras and Nicomachus.[34] Ismā'īlī philosophy was closely related to Hermetic-Pythagorean school of Greek philosophy.[35] Al-Ghazzālī knew well the *Rasā'il* (Epistle) of the *Ikhwān*.[34] His cosmological formulations in *al-Risālat al-laduniyah* betray a strong influence of the *Ikhwān'* cosmology.There are a few indications[37] that al-Ghazzālī had some knowledge of Ismā'īlī philosophy.

Al-Ghazzālī describes the *falāsifah* as those who claim that they are "men of logic and apodeictic demonstration (*ahl al-manṭiq wa'l-burhān*)."[38]This description fits well the Peripatetic school, but excludes the Hermetic-Pythagorean schools, since their methodological approach is primarily based upon a metaphysical and symbolic interpretation of things. Yet al-Ghazzālī includes the ethical writings of the *Ikhwān* in his discussion of the ethical sciences of the philosophers.[39] By implication, he identifies the *Ikhwān* with the above methodological claim as well.

Al-Ghazzālī appears not to be concerned with the division of Muslim philosophers into the different schools, having different methodological approaches to the sciences. Similarly, he did not make a careful distinction between the different schools of Greek philosophy.[40] His general characterization of Muslim philosophers, whether they are followers of Aristotle or of other Greeks, is that they rely on reason to know all things and that, consequently, *falsafah* ought to be identified with rational truths or human wisdom rather than with revealed *ḥikmah*. It is on the basis of this presupposition concerning *falsafah* that al-Ghazzālī carried out his criticism of the philosophers.

Al-Ghazzālī extended his criticism of Muslim philosophers to their Greek "masters." He recognized only one stream of theistic Greek philosophy, and Aristotle as the peak of its achievement.[41] He identifies the fundamental method of Greek philosophy with the logical and rational methods systematized by Aristotle. He seems to accept Aristotle's refutation of Plato, Pythagoras, and other Greek philosophers by means of this "philosophic" method as decisive and consequently to entertain the view that to demonstrate the incoherence of Aristotelianism is to descredit the methodological claims of philosophy as such.[42]

Al-Ghazzālī denies the "philosophic" method of its competence to comprehend metaphysical truths. A significant part of the philosophers' knowledge concerning such things as prophecy and spiritual psychology is simply borrowed truths taken over from prophets and saints.[43] Al-Ghazzālī attempts to prove this limitation of the "philosophic" method by means of that method itself. He maintains that the metaphysical sciences of the philosophers are plagued with errors and inconsistencies[44] precisely because in this domain "they could not carry out apodeictic demonstration according to the conditions they had postulated in logic."[45] These errors and inconsistencies, says al-Ghazzālī, show that it is not possible to arrive at certainty of metaphysical truths through the "philosophic" method.

Al-Ghazzālī's portrayal of the philosophers' methodology needs closer examination. For there seems to be a discrepancy between his portrayal and the actual methodology of the philosophers. In *al-Risālat al-laduniyah* al-Ghazzālī indirectly acknowledges that the philosophers may have a direct experience of metaphysical truths when he refers to their rational soul as being of the same substance as the "spirit" or "heart" of the Sufis.[46] But nowhere did he ever come to regard supra-rational experience as a major element in the methodology of the philosophers. We do know, however, that the philosophic or demonstrative method as understood, for example, by al-Fārābī and Ibn Sīnā is inseparable from metaphysical or intellectual intuition.[47]

Al-Ghazzālī's critique of the philosophers' methodology appears to be motivated by "certain theological and perspectival interests." First, he "wishes to reserve for the Sufis the monopoly of spiritual knowledge."[48] To minimize the importance of supra-rational experience in the philosophers' methodology is to help guarantee a greater glory for the method of the Sufis. At the same time, it helps to diminish an image of superiority of the philosophers in the eyes of certain segments of the community. One of al-Ghazzālī's declared aims in the *Tahāfut* is in fact to "disillusion those who think too highly of the philosophers and consider them to be infallible."[49] Second, he wishes to defend the theological perspective that makes reason subservient to revelation. Since that perspective assigns metaphysical truths to the realm of revealed faith, the claim of reason to comprehend those truths independently of revelation must be denied. The denial of the claim is sought to be affirmed by emphasizing the negative aspects of reason.

In fact, al-Ghazzālī belongs to that category of Muslim thinkers who "have emphasized the negative aspect of purely human reason as veil and limitation and its inability to reach the divine verities."[50] In contrast, philosophers like al-Fārābī and Ibn Sīnā have sought to reach transcendent truths through reason itself, and to make use of logic and the rational faculties of man to lead man above and beyond these faculties and planes. The philosophers do not deny the fact that there is a negative aspect to reason when the latter is dimmed by the passions of the animal soul. The philosophers' reason (*'aql*) that seeks to reach the divine truths found in revelation is not one that is obscured by the passions, but rather one that is wholesome and balanced (*salīm*). It is in this connection that al-Fārābī makes the assertion that the power of the rational faculty becomes sharpened when man purifies his soul and directs his desire toward the Truth instead of the sensual pleasures.[51]

Third, al-Ghazzālī wishes to draw the "legitimate" boundaries of *falsafah*, that would be acceptable to orthodoxy defined by *kalām*. The believers, he says, should not entertain any negative attitude or prejudice toward the philosophical sciences except with regards to the errors and "heresies" that he has enumerated in the *Tahāfut*. Thus, he reproached those Muslims who opposed the legitimate philosophical sciences – that is, those which are not in conflict with any religious principle – just because these sciences have been ascribed to the philosophers.

Al-Ghazzālī's conception of "legitimate philosophy" is elaborated in his account of the six philosophical sciences in the *Munqidh*.[52] The mathematical sciences, he says, are purely quantitative or exact sciences which "do not entail denial or affirmation of religious matters" because "they concern vigorously demonstrated facts."[53] Al-Ghazzālī's philosophy of mathematics does not embrace the symbolic and metaphysical meanings of numbers as emphasized, for example, by the *Ikhwān*. Seen as qualities and symbols, numbers and figures are not neutral with respect to spiritual truths, but rather lend support to them. The *Ikhwān* affirmed the view of Pythagoras that "the knowledge of numbers and of their origin from unity is the knowledge of the Unity of God". Further, "the knowledge of the properties of numbers, their classification and order is the knowledge of the beings created by the Exalted Creator, and of His handiwork, its order and classification."[54]

Al-Ghazzālī considers the logical sciences to be philosophically and religiously neutral as well.[55] These sciences are merely methodological tools that may be used by philosophers and the *mutakallimūn* alike. In his view, there is no necessary connection between the theory of causality of the philosophers and their method of demonstration. Al-Ghazzālī rejected their theory of causality, but accepted their demonstrative method as an important tool for the attainment of rational certainty in many of the sciences.

The natural sciences too generally fall within the domain of "legitimate philosophy."[56] It is in metaphysics, political philosophy, and ethics that much of the redrawing of "boundaries" needs to be done. This redrawing of the "boundaries of philosophy" is to a certain extent reflected in al-Ghazzālī's classification of the sciences treated in the next chapter. Politics and ethics are incorporated into the religious sciences. "Legitimate philosophy" is enumerated as the intellectual sciences. One important consequence of al-Ghazzālī's treatment of the confrontation between *kalām* and *falsafah* is that the pursuit of philosophy in the Sunni world became inseparable from *kalām*. *Kalām* became more philosophized.

8.2.3 **The Ta'līmites**

The question of al-Ghazzālī's acquaintance with the teachings of the Ta'līmites is problematic. On the one hand, he never mentions by name the sources with which he claims to be acquainted, whether written or oral. On the otheı hand, his claim of a sound knowledge of Ismā'ī lism is not truly reflected in his exposition and critique of their doctrines. In the *Muniqdh*, he mentions, in addition to "their writings,"[57] two oral sources. One of these oral sources is an associate "who frequented my company after he had affiliated himself with them and professed their doctrines."[58] The other source is referred to simply as "one who claimed to know some of their lore."[59] In Islamic tradition authentic oral sources are certainly important especially in the case of an esoteric movement like Ismā'īlism or Sufism. However, in the case of the sources cited by al-Ghazzālī, it is not possible to determine their authenticity or to estimate their worth, since he neither identifies them nor specifies their peculiar doctrines.

All that al-Ghazzālī relates to us is that the substance of what he learned from the second source is Pythagoreanism, which he describes as "the feeblest of all philosophical doctrines."[60] The second source is likely authentic. For there was an important link between Ismā'īlism and Pythagoreanism. By al-Gazzālī's time, Ismā'īlism had absorbed into its esoteric teachings much of what has been called "Oriental neo-Pythagoreanism."[61]

In *al-Mustaẓhirī* al-Ghazzālī is likewise content to tell us that he had read books about Bāṭinism, without mentioning their authors. Only in the *Iḥyā'* al-Ghazzālī mentions explicitly that one of the major sources of his knowledge of Ismā'īlism is al-Bāqillānī's work on the refutation of the Bāṭinīs.[62] According to al-Walīd, al-Ghazzālī had written *al-Mustaẓhirī* by "simply gathering together all the accusations in the different heresiographers, without ever referring to an authentic Ismā'īlī source."[63] Al-Walīd's criticism is to a large extent valid. Badawī has shown that *al-Mustaẓhirī* relies quite substantially on al-Baghdādī's *al-Farq bain al-firaq* (Schisms and Sects).[64] But Badawī also thinks that al-Ghazzālī was acquainted with Ḥasan al-Ṣabbāḥ's *The Four Points*.[65] It appears that *The Four Points* is the only Ta'līmite writing known by name, with which al-Ghazzālī was likely acquainted.

Al-Ghazzālī describes the Ta'līmites as those "who claim to be the unique possessors of *al-ta'līm* and the privileged recipients of knowledge acquired from the Infallible *Imām*."[66] To understand the Ta'līmites as a distinct class of seekers after the Truth within Islam, it is necessary to know their fundamental doctrine of esotericism which has given rise to the appellation *al-bāṭinīyah* itself. Al-Walīd summarized the doctrine as follows:

> If one should call us that (i.e., *Bāṭinīyah*), it is because we believe that for every exoteric meaning (*ẓāhir*) of the Holy Book, there is an esoteric sense (*bāṭin*) which is its true meaning. That is our firm belief, and the very form of our divine service. And it is also that to which the Book of God refers, that to which the Messenger of God called men, and that for the sake of whose transmission (*ta'līm*) he has established his *waṣī* (spiritual successor) and the Imāms who are his successors. We shall set forth the proofs verifying all that when .....it will be necessary to demonstrate the bases of the esoteric (*ta'wīl*).[67]

The Ta'līmites, thus, believe that the Qur'ān contains knowledge of the true reality of things, which constitutes the essence of prophetic knowledge. The key to that knowledge is the application of the esoteric method of *ta'wīl* to the Qur'ānic verses. Moreover, the Ta'līmites believe that that knowledge can only be acquired through the spiritual and divinely guided teaching (*ta'līm*) of the Imāms who are the inheritors of the Prophet's esoteric function of interpreting the inner meaning of the Qur'ān. In Ismā'īlism, and indeed in Shi'ism in general, this esoteric function is called *wilāyah*.[68] Belief in the superiority of the esoteric over the exoteric constitutes the necessary ideational basis of the religious community.

Al-Ghazzālī's account of the Ta'līmite doctrine of esotericism is in general agreement with al-Walīd's.[69] Al-Ghazzālī, however, also understood the appellation *al-bāṭinīyah* in a pejorative sense.[70] The Ismā'īlīs were called Bāṭinīs, he says, because they wanted to do away with the prescriptions of the *Sharī'ah* and to base all their actions on the esoteric meanings. Al-Walīd rejected this accusation and went at great length to defend the Ismā'īlīs against charges of infidelity to the *Sharī'ah*. Like his critique of the philosophers, al-Ghazzālī 's critique of the Ta'līmites is coloured with several sectarian considerations. His defense of theological and perspectival interests, namely Sunnism and Sufism, may be best illustrated by referring to his generalized judgment on the Ismā'īlīs concerning the *Sharī'ah* and his critique of *ta'līm*.

In *al-Mustaẓhirī* al-Ghazzālī associates with Ismā'īlism various religious and spiritual movements like the *Qarāmiṭah* (Carmathians), *Khurramīyah* (or *Kurramidīnīyah*) and Bābakīyah.[72] He uses the term Bāṭinīs to refer collectively to the Ismā'īlīs and these different movements. Among the Bāṭinīs associated with Ismā'īlism in one way or another, there were definitely those, such as the Qarāmites, who in the name of esotericism considered the abandonment of adherence to the *Sharī'ah* admissible.[73] Al-Ghazzālī, however, generalizes this tendency among certain Bāṭinīs to the whole of Ismā'īlism.

In my view al-Ghazzālī's generalization has to do with the purpose for which *al-Mustaẓhirī* was composed.[74] The work was written under the order of the Sunni Caliph with the aim of refuting Ismā'īlī doctrines, whose socio-political order is represented by the Fāṭimid Caliphate, and with the aim of belittling their religious significance before Sunni orthodoxy. In this polemic it matters little to al-Ghazzālī that the Fāṭimid Caliphate succeeded in maintaining an equilibrium between the esoteric and the exoteric. What is emphasized is the threat perceived to be inherent in Bāṭinism as a whole to the supremacy of the *Sharī'ah* as the ideational basis of the Islamic community.[75]

In the *Munqidh* al-Ghazzālī does not discuss at all the significance of the term *bāṭin* for the Ismā'īlī methodology. The doctrine of *ta'wīl* is not even mentioned. Instead, he deals exclusively with the idea of *ta'līm* and infallible teacher (*al-Imām al-ma'ṣūm*) and its implications for the independent use of reason. In *al-Mustaẓhirī* however, he offers a critical discussion of the Bāṭinīs' method of *ta'wīl* as applied to the Qur'anic verses and their use of numerical and alphabetical symbolism.

In Ismā'īlism, *ta'wīl* and *ta'līm* or *Imāmah* are two doctrinal principles with which the idea of *bāṭin* is inseparably linked. In the perspective of Ismā'īlism, and of Shi'ism generally, the Imām is, next to the Prophet, the supreme authority on the *ta'wīl*. The Arabic term *ta'wīl* means literally to take something back to its beginning or origin. In the technical sense used in both Shi'ism and Sufism, *ta'wīl* means the symbolic and esoteric interpretation of the Qur'ān.[76] It is the process of penetrating into the inner (*bāṭin*) meaning of the sacred text. The two meanings are not unrelated, since to penetrate into the inner mysteries of the Qur'ān is precisely to reach back to its Origin, which is the most inward.[77]

In several of his works,[78] al-Ghazzālī defends the notion of *ta'wīl* as the process of penetrating into the deep and hidden meanings of Qur'ānic verses, in a manner that implies a significant change in his attitude toward *ta'wīl* from the position he adopted in *al-Mustaẓhirī*.

The idea of the inner meanings of the Qur'ān finds strong support, he says, in numerous *ḥadīths*, the sayings of the Prophet's companions, and the sayings of early Muslims.[79] In support of *ta'wīl*, al-Ghazzālī refers, among others, to the well-known prayer of the Prophet for Ibn 'Abbās, in which he asked God to "bestow upon the latter the understanding of the religion and teach him the *ta'wīl* of the Qurān."[80]

Al-Ghazzālī argues further that the secret meanings of the Qur'ān are unveiled only to those established in the esoteric sciences (*'ulūm al-mukāshafāt*) and those possessed of purified souls.[81] He identifies these "men of understanding" with the Sufis whose *ta'wīl*, he says, must not be equated with the explanation of the Qur'ān by personal opinion *(ra'y)* prohibited by the Prophet. Al-Ghazzālī possibly realized that in accepting the Sufi idea of *ẓāhir-bāṭin* distinction and the related doctrine of *ta'wīl* he had in fact embraced the Ismā'īlī understanding of these doctrines as well. It is significant that in the *Iḥyā'* his criticism of the Bāṭinīs concerning *ta'wīl* no longer pertains to the method of *ta'wīl* as such, as in *al-Mustaẓhirī*, but is directed toward certain of their interpretations and at their use of *ta'wīl* as proof of the idea of the infallible Imām. Al-Ghazzālī, in conformity with his Sunni perspective, rejects that there is a necessary link between *ta'wīl* and the *Imāmah*.

In the *Munqidh* al-Ghazzālī excludes *ta'wīl* from the methodologies of both the Ta'līmites and the Sufis, because I believe he later realized that *ta'wīl* was not unique to either group. However, to exclude *ta'wīl* and the idea of the *bāṭin* from a discussion of Ismā'īlī methodology means that the initiatic and supra-rational character of *ta'līm* could hardly be appreciated. This character of *ta'līm* is further eclipsed in al-Ghazzālī's discussion, since he opposed the Shi'ite doctrine of extending the prophetic quality of *iṣmah* (purity and infallibility) to the Imāms. *Ta'līm*, as understood by the Ismā'īlīs themselves, derives its significance as a method of knowing the Truth from the belief that the Imām carries within himself the "Muhammadan light" which is identified with the source of all prophetic knowledge. It is by virtue of this light that the Imām gains his authority as the interpreter of the Divine Law and the religious sciences.[82]

In his exposition of *ta'līm* al-Ghazzālī's approach is not that of an independent scholar who stands above the Sunni-Shi'ite division, but rather that of a defender of the Sunni perspective. He presents the Ismā'īlī claim concerning the Imām's quality of inerrancy in spiritual and religious matters but remains silent on the question of the "Muhammadan light" as the spiritual basis of that quality. Al-Ghazzālī wishes to reserve this prophetic light for the Sufi saints. By virtue of carrying the prophetic light within their own beings, the Sufi saints possess the knowledge of the inner meaning of revelation.

Al-Ghazzālī presents *ta'līm* as something essentially opposed to *ijtihād* (the exercise of personal judgment) and the use of logic and reason. He admits the necessity of an infallible *Imām*, but this Imām is none other than the Prophet and his complete teaching is to be found in the Qur'ān and the *Sunnah* (prophetic traditions).[83] While individual *ijtihād* may not be free from error, the Qur'ān and the *Sunnah* as interpreted collectively by the *'ulamā'* constitute sufficient guides for the Islamic community. From the point of view of Shi'ism, however, *itjihād* is not incompatible with the idea of guide from the Imām. On the contrary, for the Shī'ites, a true *mujtahid* (he who can exercise his religious opinion) is one who is in inner contact with the Imām.

Al-Ghazzālī's omission of the initiatic and supra-rational character of *ta'līm* has greatly diminished the significance of the Ismā'īlī methodology as a means of attaining knowledge of the true reality of things. What he had emphasized instead is the relation of *ta'līm* to exoteric knowledge, which helps to bring into focus the Sunni-Shi'ite theological division, but which is of little relevance to his quest for the inner knowledge of things.

8.4 The Sufis

Al-Ghazzālī describes the Sufis as those who claim that they alone enter into the Divine Presence, and as men of mystic vision (*mushāhadah*) and illumination (*mukāshafah*).[84] The Sufi is thus presented as an intimate witness of God. The Sufi way of witness is described as authentic vision and the unveiling of the mysteries. According to al-Ghazzālī, to witness the Divine Presence is to attain the highest possible state of spiritual experience.[85]

The Sufi's mystic vision refers to sapiential knowledge, that is, realized knowledge that is inseparable from the transformation of the knower's being. Certitude derived from realized knowledge is the highest kind of certitude, which in Qur'anic terminology is called *ḥaqq al-yaqīn* (the truth of certainty). To perceive a truth inwardly through contemplation (*mushāhadah*) is to see through the eye of the truth of certainty, since "contemplation is more real and clearer than seeing with the (physical) eye."[86] Realized knowledge is free from error and doubt, because it is not based on conjecture or mental concepts but on the heart's direct vision of spiritual truths. In the Sufi perspective, the heart (*qalb*) is the real key to sapiential knowledge.[87]

The Sufis, says al-Ghazzālī, maintain that "the heart possesses an organ of sight like the body, and outward things are seen with the outward eye, and inward realities with the eye of the heart."[88] The knowledge gained through the vision of the eye of the heart (*'ayn al-qalb*) has the immediacy and directness of sensual knowledge but concerns the spiritual world. This spiritual knowledge identified with the heart is what the Sufis call presential knowledge (*'ilm huḍūrī*). The other kind of knowledge gained by the mind through the help of intermediary concepts is called attained knowledge (*'ilm ḥuṣulī*).[89]

According to al-Ghazzālī , although the heart of every man is created to know the "invisible divine world," what we find generally is that "man has veiled it by his lusts and worldly preoccupations and he has ceased to see with it."[90] In other words, the eye of the heart is blinded by passion so that there is a veil between the heart and the spiritual world. What the Sufis seek to do is to remove the veil from the heart. In Sufism there is a whole science associated with this goal, comprising theory as well as practice.[91]

The purification of the heart with the view of being totally absorbed in the remembrance (*dhikr*) of God and of attaining annihilation (*fanā'*) in God is what characterizes the Sufi methodology. The science to which this methodology belongs al-Ghazzālī calls *'ilm al-mukāshafah* (the science of the unveiling of the divine mysteries).[92] The purification of the heart, when it is efficacious, results in a transmutation of the very substance of the soul of the adept so that he becomes an accomplished Sufi.[93] In his newly realized mode of being he becomes a witness (*shāhid*) to the Divine Truth.

Al-Ghazzālī calls the Sufis masters of states (*arbāb al-aḥwāl*).[94] The Sufis are concerned with the different spiritual states and stations which the adept must experience before he can reach the Divine Presence. According to al-Ghazzālī, when the Sufis experience *mushāhadah, mukāshafah,* and *dhawq* they see even when awake "the angels and the spirits of the prophets and hear voices from them and learn useful things from them."[95] Then their state ascends from the vision of forms to "stages beyond the narrow range of words." These experiences, says al-Ghazzālī, are in reality "the first stages passed through by the prophets."[96]

Al-Ghazzālī maintains that pure Intellect, which is immanent in the heart of every man, is actualized at different degrees and in different modes only in those who follow the way of Sufism. Since the Sufi carries within his own being the prophetic light, he experiences directly the reality of prophecy (*ḥaqīqat al-nubuwwah*). Says al-Ghazzālī, "What became clear to me of necessity from practicing the Sufi way was the true nature and special character of prophecy."[97]

Although sapiential or presential knowledge is gained independently of reason, it is by no means irrational. The role of reason with respect to supra-rational reality is to acquaint us with this reality, to give assent to its truth, to understand it, and to certify "its own blindness to perceiving what the 'eye' of prophecy perceives."[98] Al-Ghazzālī is here affirming the traditional view that the heart or intellect is the principle of reason.

Since al-Ghazzālī was himself a distinguished Sufi, he was able to give an authentic and authoritative account of the Sufi methodology. As far as the way to the knowledge of the true reality of things is concerened, al-Ghazzālī considers the Sufi method the most excellent of methods, and the Sufis the most excellent of the knowers of the Truth. For he himself found the light of certainty in the spiritual path of the Sufis.

8.3 Significance of the Classification

The empiricists, that is, those who maintain that the sourse of all knowledge is sensual experience, are not included in the classification. They were almost non-existent in Islamic Society. Movever they could not be qualified as seekers after the Truth and knowledge of the inner reality of things.

Al-Ghazzālī's study of the methodological claims of the four classes of knowers enables us to establish his epistemological perspective. While affirming the superiority of the Sufi mode of knowing to all other modes, he remained attached to the *kalām* perspective. His internal criticism of *kalām* was aimed at refining the rational methods of that science. As a defender of *kalām*, he subordinates reason to revealed faith. As a Sufi, he subordinates reason to mystical intuition (*kashf*) and *dhawq*. The consequence in each case is the superiority of the religious sciences to the intellectual sciences.

ENDNOTES

Chapter 8

[1]In speaking of the four classes, al-Ghazzālī says: "The truth cannot be outside these four classes. These are the people who tread the paths of the quest for truth. If the truth is not with them, no point remains in trying to apprehend the truth." See McCarthy, *Freedom and Fulfillment*, p. 67

[2]In *The Jewels of the Qur'ān* (p. 43), for example, al-Ghazzālī writes: "The highest and noblest knowledge is the knowledge of God because all other forms of knowledge are sought for the sake of it and it is not sought for anything else. The manner of progression in regard to it is to advance from divine works to divine attributes, and then from divine attributes to divine essence; thus there are three stages. The highest of these stages is knowledge of divine essence and it is not possible for most people to understand this."

In *The Book of Knowledge* (p. 47), al-Ghazzālī enumerates the content of the knowledge of things "as they really are" attained through *kashf*. He mentions the knowledge of God as the highest.

[3]*The Jewels of the Qur'ān*, p. 40

[4]*Ibid.* pp. 41-2.

[5]Khayyām's classification has been studied by Nasr in his *Science and Civilization in Islam*, pp. 33-8. Although al-Ghazzālī and Khayyām have the same fourfold classification, their detailed exposition of each class is different. A comparative study of the two classifications has not yet been undertaken.

[6]According to al-Bayhaqī, who knew Khayyām personally and whose father was a close friend of the latter, al-Ghazzālī and Khayyām knew each other. See M.Meyerhof, "Ali al-Bayhaqī's *Tatimmat Ṣiwān al-ḥikma:* Biographical Work on Learned Men of Islam," *Osiris*, 6(1948), p. 173.

Khayyām, who was ten years older than al-Ghazzālī but who survived the latter's death by twenty years, may have already composed his *Risālat-i wujūd*, which contains his fourfold classification, by the time al-Ghazzālī pursued his studies at Naishapur. For by the time Khayyām was twenty-six (467/1074), he was already an accomplished mathematician. It is possible that al-Ghazzāli 's theoretical acceptance of a similar fourfold classification was inspired by Khayyām. But since we do not know exactly when Khayyām's above treatise was composed, it is also possible that al-Ghazzālī 's classification was his original idea.

[7]Khayyām, *Risālat-i wujūd*, quoted by Nasr, *op. cit.*, p. 34

[8]A 'perspective' or 'point of view' that sees all things "as they really are" is in reality not one perspective among several perspectives. It comprehends all perspectives. The use of a similar term is only meant as a comparison.

[9]His description and criticism of the *mutakallimūn* applies to all schools. However, he praises one group, without mentioning its name, for having "ably protected orthodoxy and defended the creed (*al-'aqīdah*) which had been readily accepted from the prophetic preaching." See McCarthy, *op. cit.*, sec.23, p. 68.

There is no doubt that al-Ghazzālī has the Ash'arites in mind since in several of his works such as *The Golden Mean in Belief* and the *Tahāfut* he expresses his disapproval of many of the theological views of the Mu'tazilites, the other major school of Sunni *kalām*.

[10]McCarthy, *op. cit.*, p. 67.

[11]The main characteristic of this method, as it was used in *fiqh*. is that it is based upon a mere likeness between things, unlike the philosophic use of analogy which is based upon an equality of relations.

[12]These include the literalists (*al-ẓāhirīyah*), the traditionists (*al-muḥaddithūn*) and the anthropomorphists (*al-mujassimiyah*).

[13]Al-Ash'arī wrote a special work against the *ahl al-taqlīd*, entitled *Istiḥsān al-Khauḍ fī 'ilm al-kalām* (*Justification of Engaging in the Science of Kalām*) explaining the necessity of justifying religious beliefs on rational grounds in the light of newly arisen problems relating to those beliefs. See M.A.Hye, "Ash'arism," *A History of Muslim Philosophy*, I, pp. 224-5

[14]McCarthy, *op. cit.*, sec.22, p. 68

[15]*The Jewels of the Qur'ān*, pp. 37-42. *Kalām* is there described as one of three inferior sciences of the pith of the Qur'ān, the other two being jurisprudence and the science concering "the knowledge of the stories narrated in the Qur'ān and of what is related to the prophets, to the deniers of God and to His enemies." *Kalām* concerns "God's arguments with the infidels and His dispute with them." (p. 38).

[16]McCarthy, *op. cit.*, sec.23, pp. 68-9

[17]For a detailed analysis of the accounts of *Kalām* given by these three figures, see. H.A.Wolfson, *The Philosophy of the Kalām*, pp. 3-58.

[18]Al-Ash'arī, al-Bāqillānī's teacher, used the theory of atoms as the basis of demonstration of the creation of the world and hence also of the existence of God. *Ibid.* pp. 40, 386.

[19]According to al-Bāqillānī, the demonstrations of the articles of faith or the premises upon which the demonstrations are based "hold the same position as the articles of faith themselves" and are "next to the articles of faith in the necessity of believing them." *Ibid.* p. 40

[20]More precisely, we are speaking here of the Ash'arite theological perspective. This perspective depicts the unlimitedness of divine omnipotence to which all other Divine Qualities are subordinated. The overwhelming motive, if not the only one, for God's actions, according to the Ash'arites, is "what He wills" and "because He wills." Applied to God's activity in nature, this activity gave rise to that important idea known as *occasionalism* which has been defined as the belief in the exclusive efficacy of God, of whose direct intervention the events of nature are regarded as the overt manifestation or occasion. Occasionalism implies a substantial discontinuity of things and hence a denial of causality, a view shared by al-Ghazzālī.

For a metaphysical critique of the religious and intellectual perspective of Ash'arism, see F.Schuon, *Islam and the Perennial Philosophy*, Suhayl Academy, Lahore, 1985, chapter entitled *Dilemmas within Ash'arite Theology*. pp. 118-51; see

also M. Fakhry, *Islamic Occasionalism*, London, 1958; and S.H.Nasr, *Islamic Life and Thought*, pp. 61-3.
[21]On the *farḍ kifāyah* status of *kalām*, see *The Book of Knowledge*, pp. 53-4. The concept of *farḍ kifāyah* as applied to knowledge is treated in the next chapter.
[22]*Tahāfut*, p. 202.
[23]In his *Fayṣal al-tafriqat bayn al-islām wa'l-zandaqah* (*The Clear Criterion for Distinguishing between Islam and Godlessness*), al-Ghazzālī criticizes those who claimed that "deviating from the doctrine of al-Ash'arī by even so much as a palm's width is unbelief (*kufr*) and that differing from him in even a trivial matter is error and perdition." McCarthy, *op. cit.*, p. 146.

In the same work (p. 149) al-Ghazzālī criticizes al-Bāqillānī for having included premises and proofs used in the demonstration of religious beliefs as an integral part of the creed.
[24]McCarthy, *op. cit.*, sec.26, p. 70
[25]This was claimed by al-Ghazzālī himself. See *The Jewels of the Qur'ān*, p. 39; and *Tahāfut*, p. 10.
[26]Ibn Khaldūn refers to this approach as the method of the later *mutakallimūn*. See *The Muqaddimah*, III, p. 41
[27]*The Jewels of the Qur'ān*. p. 38
[28]McCarthy, *op. cit.*, sec.24, p. 69
[29]*Ibid*, sec.26, p. 70
[30]*Ibid*, secs.98-9, pp. 95-6; also *Mishkāt*, pp. 148-9
[31]*Ibid*, pp. 168-9; W.H.T.Gairdner, "al-Ghazzālī's *Mishkāt al-anwar* and the Ghazali-Problem," *op. cit.*, p. 125.
[32]Says al-Ghazzālī, "....one cannot recognize what is unsound in any of the sciences unless he has such a grasp of the farthest reaches of that science that he is the equal of the most learned of those versed in the principles of that science; then he must excel him and attain even greater eminence so that he becomes cognizant of the intricate profundities which have remained beyond the ken of the acknowledged master of the science. Then and then only, will it be possible that the unsoundness he alleges will be seen as really such....." McCarthy, *op. cit.*, sec.25, pp. 69-70.
[33]M.E.Marmura has shown that the *Mi'yār al-'ilm* is heavily indebted to Ibn Sīnā's exposition of logic in his *al-Shifā'* (or its summary in *al-Ishārāt wa'l-tanbīhāt*). See Marmura, "Ghazali and Demonstrative Science," in *Journal of the History of Philosophy*. 3:2 (Oct.1965), pp. 183-204; also his "Ghazali on Ethical Premises," in *The Philosophical Forum*, new series, 1:3 (1969), pp. 393-403; and "Ghazali's Attitude to the Secular Sciences and Logic," in G.F.Hourani, ed., *Essays on Islamic Philosophy and Science*. pp. 100-11.

Similarly, M.A.Sherif has shown that al-Ghazzālī 's *Mizān al-'amal* draws much of its materials from various works of Ibn Sīnā. See Sherif, *Ghazali's Theory of Virtue*.

Al-Ghazzālī's commentary on the Light-Verse of the Qur'ān was very likely inspired by Ibn Sīnā's commentary, since the *Kitāb al-ishārāt wa'l-tanbihāt* (*The Book of Directives and Remarks*) in which his commentary is found was known to al-Ghazzālī. Similarly, al-Ghazzālī's *Risālat al-ṭair* (*Treatise of the Bird*) was very likely inspired by Ibn Sīnā's work of the same title.
[34]S.H.Nasr, *IICD*, p. 37
[35]Among the outstanding figures of Ismā'īlī school of philosophy were Abū Ya'qūb al-Sijistānī, Abū Ḥātim al-Rāzī and Nāṣir-i Khusraw.
[36]Nasr, *IICD*, p. 36
[37]For these "indications" see my discussion of al-Ghazzālī's acquaintance with Ismā'īlism in the next section (8.2.3).

[38]McCarthy, *op. cit.*, p. 67
[39]*Ibid*, secs. 56, 58, p. 80
[40]In the *Tahāfut* he states clearly that he is not concerned with the fine distinctions that exist among the philosophers. In the *Munqidh* he gives a threefold 'theological' division of the Greek philosophers into the Materialists (*al-dahrīyūn*), the Naturalists (*al-ṭabī'īyūn*) and the Theists (*al-ilahīyun*). Socrates, Plato and Aristotle were theistic philosophers who "never denied the validity of the religious laws." On the contrary, "they believed in God and had faith in His messengers" as well as believed in the Last Day. See McCarthy, *op. cit.*, secs. 29-33, pp. 71-2; and *Tahāfut*, p. 3
[41]Says al-Ghazzālī: "It was Aristotle who systematized logic for the philosophers and refined the philosophical sciences, accurately formulating previously imprecise statements and bringing to maturity the crudities of their sciences....Then, Aristotle refuted Plato and Socrates and the Theists who preceded him in such thorough fashion that he disassociated himself from them all." McCarthy, *op cit.*, sec.33, p. 72
[42]*Ibid*, sec.78, p. 89; *Tahāfut*, p. 4
[43]McCarthy, *op. cit.*, sec.58, p. 80. In making these assertions, al-Ghazzālī is implying two things. One is his denial of the possibility of these truths being known outside the prophetic experience of which the Sufi *dhawq* is a stage. The other is that these "borrowed" truths cannot be characterized as philosophical doctrines.
[44]In the *Tahāfut* al-Ghazzālī enumerates twenty problems, drawn mainly from the metaphysical sciences, in which he seeks to show that the philosophers' views are in error, self-contradictory and incoherent (pp. 11-2). He brands the philosophers with infidelity on three questions: (1) their view of the eternity of the world, (2) their assertion that "Divine Knowledge does not encompass individual objects," and (3) their denial of the resurrection of bodies. As regards the rest of the problems, al-Ghazzālī considers the views of the philosophers as innovations (pp. 249-50).
[45]McCarthy, *op. cit.*, sec.46, p. 76
[46]*al-Risālat al-ladunīyah*, p. 194.
[47]This fact was amply demonstrated in my discussion of al-Fārābī.
[48]F.Schuon, *Sufism: Veil and Quintessence*. p. 117
[49]*Tahāfut*. p. 8
[50]Nasr, *Sufi Essays*, p. 55
[51]I have referred to this view of al-Fārābī in chapter three.
[52]The six philosophical sciences are the mathematical, logical, natural, metaphysical, political, and the ethical. See McCarthy, *op. cit.*, secs. 37-60, pp. 73-81
[53]*Ibid*, sec. 37, p. 73
[54]Nasr, *Science and Civilization in Islam*, p. 155
[55]McCarthy, *op. cit.*, sec. 43, p. 74.
[56]Al-Ghazzālī accepts the natural sciences of the philosophers except with regards to a number of "errors" he identifies in the *Tahāfut*.
[57]McCarthy, *op. cit.*, sec. 61, p. 82
[58]*Ibid*, sec. 63, pp. 82-3
[59]*Ibid*, sec. 78, p. 89
[60]"Aristotle had already refuted him (Pythagoras) and had even regarded his teaching as weak and contemptible." *Ibid.*
[61]Nasr, *Science and Civilization in Islam*. p. 37. Al-Ghazzālī 's reference to the link between Ismā'īlism and Pythagoreanism raises the question of the possibility of his familiarity with the writings of the Ismā'īlī philosophers. The integration of Pythagoreanism into Ismā'īlism was achieved at the hands of philosophers like al-Sijistānī, Abū Ḥātim al-Rāzī, and Nāṣir-i Khusraw.
[62]Al-Ghazzālī mentions the title as *Kashf al-asrār wa hatk al-astār (The Unveiling of*

the Secrets).

[63]H.Corbin, "The Ismā'īlī Response to the Polemic of al-Ghazālī," *op. cit.*, p. 70

[64]See the introduction to his edition of the *al-Mustaẓhirī*, Cairo, 1964, p. (d).

[65]In a debate that took place in Transoxiana in the sixth/twelfth century between a certain Sharaf al-Dīn al-Mas'ūdī and Fakhr al-Dīn Rāzī, the former brought up the question of al-Ghazzālī's refutation of *The Four Points* for the latter's comments. Rāzī, who apparently was well-versed with that work, did not dispute al-Ghazzālī's acquaintance with it but accused al-Ghazzālī of having misunderstood al-Ṣabbāh's arguments. See P.Kraus, "The *Controversies* of Fakhr al-Dīn Rāzī," *Islamic Culture*, 12:2(April, 1938), pp. 147-8.

[66]McCarthy, *op. cit.*, p. 67

[67]H.Corbin, *op. cit.*, p. 79

[68]For a detailed discussion of this concept, see Nasr, *Ideals and Realities of Islam*, pp. 161-2

[69]For al-Ghazzālī 's account, see McCarthy, *op. cit.*, sec. 17, p. 181.

[70]During the early history of Islam, the word *bāṭin*, which occurs many times in the Qur'ān, simply means inwardness. Al-Hasan al-Baṣrī (d.110/728) is known to refer to the Sufis as *ahl al-bātin* (people of inwardness). Later, however, when the Shi'ites came to emphasize the fact that everything has an outward (*ẓāhir*) and an inward (*bāṭin*) aspect, the appellation *al-baṭinīyah* was given to the Ismāīlīs in a pejorative sense. See *ibid*, p. 182

[71]Al-Walīd cites letters of various Fātimid Imāms, including al-Ḥākim, the sixth Fātimid Caliph, which command the *dā'īs* and the believers "to remain attached to all the obligations of religious practice" and which warn them that "whoever abandons the practices imposed by the *Sharī'ah* is at the same time abandoning Rel igion and thus becomes an unbeliever." See H.Corbin, *op. cit.*, p. 89

[72]McCarthy, *op. cit.*, p. 182. For al-Walīd's reply to al-Ghazzālī's attempt to connect Ismā'īlism to these various movements, see Corbin, *op. cit.*, pp. 80-7.

[73]M.H. Ṭabāṭabā'ī, *Shi'ite Islam*, trans. S.H.Nasr, SUNY Press, Albany, 1975,. p. 80.

[74]The highly polemical and political nature of this work was referred to in the previous chapter.

[75]McCarthy, *op. cit.*, p. 182

[76]For a more detailed treatment of *ta'wīl*, see Nasr, *Ideals and Realities of Islam*, pp. 58-61; and S.M.N.al-Attas, *The Concept of Education in Islam*, ABIM. Kuala Lumpur. 1980.

[77]"*Ta'wīl* for Sufism, or Shi'ism, does not possess the same meaning as it does in Mu'tazilite theology and in jurisprudence. It has nothing to do with the debate between the Ash'arites and Mu'tazilites over the literal meaning of the Qur'ān versus rational interpretation of it. *Ta'wīl* in the sense used by the Sufis and Shi'ite sages is the penetration into the symbolic–and not allegorical–meaning of the text which is not a human interpretation but reaching a divinely pre-disposed sense placed within the Sacred Text through which man himself becomes transformed. The symbol has an ontological reality that lies above any mental constructions. Man does not make symbols. He is transformed by them. And it is as such that the Qur'ān with the worlds of meaning that lie hidden in its every phrase transforms and remakes the soul of man." Nasr, *op. cit.*, p. 61

[78]See, for example, *al-Risālat al-ladunīyah*. p. 190; and M.A.Quasem, trans., *The Recitation and Interpretation of the Qur'ān: al-Ghazzālī's Theory*. 1982, chapter 4.

[79]For example, he quotes the prophetic *hadīth:* "The Qur'ān has an outward aspect, an inward aspect, a limit and a prelude." He also quotes the saying of a certain religious scholar that "for every Qur'ānic verse there are sixty thousand understandings

(comprehensible to man)." See Quasem, *op. cit.*, p. 87; (cf. *al-Risālat al-ladunīyah*. p. 354, where al-Ghazzālī quotes the following *ḥadīth*: "There is not a verse of the Qur'ān but has a literal sense and an esoteric sense, and its esoteric sense includes another esoteric meaning up to seven esoteric meanings.")

[80] Quasem, *op. cit.*, p. 91

[81] *Ibid*, p. 102

[82] Nasr, *op. cit.*, pp. 162-3

[83] McCarthy, *op. cit.*, sec.64, p. 83 and sec.68, p. 85

[84] *Ibid*. p. 67

[85] This state of experience, says al-Ghazzālī, is beyond the ability of words to express. He thinks that such terms as *hulūl* (indwelling), *ittiḥād* (union) and *wuṣūl* (arrival or connection) should not be used to describe this state of spiritual experience, because they convey misleading ideas about God-man relationship in that state. For him, the scriptural symbolism of the servant's "nearness *(qurb)* to God" is the most apt to be used. See *ibid*, sec.96, p. 95, and pp. 355-60.

[86] *The Book of Knowledge*. p. 141

[87] In both the Qur'ān and *ḥadīths* there are numerous references to the heart as being the seat of intelligence and knowledge. The Sufi doctrine of the knowledge of the heart is in fact developed on the basis of this teaching contained in the Qur'ān and *ḥadīths*. Concerning the heart as the instrument for the attainment of supra-rational knowledge, al-Ghazzālī writes:

> ....then know that that which seeks to press toward God in order to attain a place in His neighbourhood is the heart and not the body. And by the heart I do not mean the palpable matter of flesh but one of the mysteries of God which the bodily senses fail to perceive; a spiritual substance (*laṭīfah*) from God, sometimes indicated by the word spirit (*rūḥ*) and at times by the calm soul (*al-nafs al-muṭma'innah*).

See *The Book of Knowledge*, pp. 141-2.

[88] *al-Risālat al-ladunīyah*, p. 198

[89] Al-Ghazzālī's distinction between the two kinds of knowledge (*ibid*, pp. 360-8) will be treated in the next chapter. He identifies presential knowledge with *al-'ilm al-ladunī* of the Qur'ān (XVIII:64).

[90] *Iḥyā'*, IV, 430, quoted by M. Smith, *al-Ghazzālī the Mystic*, p. 198.

[91] McCarthy, *op. cit.*, sec. 80, p. 90

[92] *The Book of Knowledge*, p. 142

[93] McCarthy, *op. cit.*, sec. 82, p. 90

[94] *Ibid*, sec. 83, p. 90

[95] *Ibid*, sec. 96, pp. 94-5.

[96] *Ibid*, sec. 97, p 95.

[97] *Ibid*, sec. 101, p. 96

[98] *Ibid*, sec. 124, p. 102

CHAPTER 9

AL-GHAZZĀLĪ'S CLASSIFICATION OF THE SCIENCES

My study of al-Ghazzālī's classifications of the sciences is based on two main sources: *The Book of Knowledge* of the *Iḥyā'* and *al-Risālat al-laduniyah.* I have also made significant use of his two other works, namely, *The Jewels of the Qur'ān* and the *Mīzān al-'amal.* Al-Ghazzālī refers in these works to four different systems of classification:

(1) division into theoretical and practical sciences[1].
(2) division into "presential" (*ḥuḍūrī*) and attained (*ḥuṣūlī*) knowledge[2]
(3) division into religious (*shar'īyah*) and intellectual (*'aqlīyah*) sciences[3]
(4) division into *farḍ 'ayn* (obligatory on every individual) and *farḍ kifāyah* (obligatory on all) sciences.[4]

Of the four systems, the one that receives the most extensive treatment from al-Ghazzālī is the division into the religious and the intellectual. His discussion of it incorporates the first and the fourth systems as well. The division into presential and attained knowledge is treated separately. According to al-Ghazzālī, all four systems of classifications are valid even though they do not have the same degree of validity. Each classification is based on a particular aspect of the relationship between man and knowledge as well as on a particular perspective of looking at that relationship. The more fundamental and universal the aspect or perspective in view, the greater will be the validity of the resulting division. Before going into a detailed discussion of al-Ghazzālī's treatment of these classifications, the basis of each as he has defined it should be explained.

9.1 Basis of the Division into Theoretical and Practical Parts

In the *Maqāṣid* al-Ghazzālī divides philosophy or the science of wisdom (*al-'ilm al-ḥikmī*) into its theoretical and practical parts. The theoretical part makes known the states of beings as they are. The practical deals with man's actions. It aims at finding out the human activities conducive to man's well-being in this life as well as in the next. Al-Ghazzālī is reproducing here the popular distinction made by philosophers between theoretical and practical knowledge. But he confirms this division as his own in both *Mīzān al-'amal* and *al-Risālat al-ladunīyah*[5].

Like al-Fārābī and Ibn Sīnā, al-Ghazzālī maintains that the above division is based on the corresponding distinction between theoretical and practical intellect.[6] In the case of al-Ghazzālī, however, this Aristotelian division has been applied mainly to the science of religion. His discussion of the philosophical sciences, in contrast to that of al-Fārābī, contains few references to the theoretical-practical division of each of those sciences.

9.2 Basis of the Division into "Presential" and Attained Knowledge

The division into presential and attained knowledge has been discussed briefly in the last chapter. This division is based on the most fundamental distinction pertaining to modes of knowing. Presential knowledge is direct, immediate, supra-rational, intuitive, and contemplative. Al-Ghazzālī refers to this knowledge under several names. Among them are *'ilm ladunī* (knowledge from on high) and *'ilm al-mukāshafah* (knowledge of unveiling of the divine mysteries). Attained or acquired knowledge is indirect, rational, logical, and discursive.[7]

Presential knowledge is superior to attained knowledge because it is free of error and doubt and it confers the highest certitude concerning spiritual truths. Sensual knowledge is also direct and immediate but it pertains only to the physical world. The division of knowledge into the presential and the attained is therefore based on the distinction between immediate and indirect knowledge concerning the intelligible or spiritual world.

9.3 Basis of the Division into Religious and Intellectual Sciences

In the *Book of Knowledge*, al-Ghazzālī defines the religious sciences (*al-'ulūm al-shar'īyah*) as "those which have been acquired from the prophets and are not arrived at either by reason, like arithmetic, or by experimentation, like medicine, or by hearing, like language."[8] Al-Ghazzālī's definition makes the religious sciences more specific than the transmitted sciences (*al-'ulūm al-naqlīyah*). The latter, as they appear in the classifications of many Muslim scholars including Ibn Khaldūn include linguistic science. But in the same work as well as in *al-Risālat al-ladunīyah,* al-Ghazzālī is also using the term religious sciences as synonymous with transmitted sciences. His fourfold classification of praiseworthy[9] (*maḥmūd*) religious sciences include not only linguistic sciences, but all the sciences that were traditionally identified with the category of transmitted knowledge.[10] He makes clear, however, that in itself linguistic science is not a religious science; but inasmuch as it is one of the preludes (*muqaddimāt*) of the religious sciences proper, it may, for the purpose of classification, be included under the category of the latter.[11]

By the intellectual sciences (*al-'ulūm al-'aqlīyah*) al-Ghazzālī means those sciences which are attained by the human intellect alone.[12] His enumeration of them shows that they are almost identical with the philosophical sciences contained in al-Fārābī's classification. The only difference is that in al-Ghazzālī's classification the sciences embraced by al-Fārābī's practical or political philosophy are placed under the science of religion instead of under the philosophical sciences. In viewing political and ethical sciences as religious rather than intellectual or philosophical, al-Ghazzālī is consistent with his definitions of "religious" and "intellectual" sciences. He makes quite clear in the *Munqidh* that the Muslim philosophers' teachings in the political and ethical sciences were drawn mainly from scriptures revealed to the prophets.[13] The philosophers did not arrive at their knowledge in these two sciences through an independent use of reason. Consequently, in conformity with the above definitions, political and ethical sciences must fall under the category of religious knowledge.

How then should we characterize the basis of al-Ghazzālī's distinction between religious and intellectual knowledge? It appears to me that the division of knowledge into the religious and the intellectual in the sense defined above is a logical consequence of the *mutakallimūn's* conception of the relationship between revelation and reason. This shows that this system of classification is closely related to the intellectual perspective of *kalām*. More specifically, the classification reflects al-Ghazzālī's exoteric theological attitude toward *falsafah* which I have discussed rather extensively in the last two chapters.

Revelation and reason are conceived as mutually exclusive sources of knowledge. Not that the theologians view the two sources as being opposed to each other. Al-Ghazzālī maintains that the religious and intellectual sciences complement each other and are never contradictory.[14] But the limitations of reason as a mode of knowing, and hence the subordination of reason to revelation, are greatly emphasized. The source of religious knowledge is revelation. Reason alone could not attain that knowledge. The source of intellectual knowledge is intellect. We may say, therefore, that al-Ghazzālī had drawn the distinction between religious and intellectual knowledge on the basis of their sources, that is to say whether they are based on revelation or reason.

9.4 Basis of the Division into "farḍ 'ayn" and "farḍ kifāyah" Sciences

Al-Ghazzālī's division of the sciences into the *farḍ 'ayn* and the *farḍ kifāyah* categories has been briefly noted in the last chapter.[15] The term *farḍ 'ayn* refers to a religious obligation that is binding on every Muslim. As for the term *farḍ kifāyah*, it refers to what is divinely ordained and binding for the Muslim community as a whole but is not necessarily binding for each member of the community. Al-Shāfi'ī, who was apparently the first to introduce this latter term, defines it as "the obligation which if performed by a sufficient number of Muslims, then the remaining Muslims who did not perform it would not be sinful."[16] In other words, the fulfillment of the obligation by a segment of the community would absolve the rest of it of that obligation.

The division of knowledge into the *farḍ 'ayn* and the *farḍ kifāyah* is therefore based on the distinction between two types of religious obligation pertaining to its acquisition. The idea of religious obligation in the acquisition of knowledge has its basis in the following prophetic *ḥadīth* quoted by al-Ghazzālī: "Seeking knowledge is an ordinance obligatory on every Muslim."[17] But, as reminded by al-Ghazzālī, Muslim scholars "disagreed as to what branch of knowledge man is obliged to acquire, and as a result split up into about twenty groups."[18] Al-Ghazzālī's own views on this question will be discussed later.

There is a related classification based on the same legal principles. This is the tripartite division of knowledge into (a) the praiseworthy (*maḥmūd*), (b) the blameworthy (*madhmūm*), and (c) the permissible (*mubāḥ*).[19] This division is at once more general and more detailed than the former, since the praiseworthy sciences comprise both the *farḍ 'ayn* and the *farḍ kifāyah*. It is as a jurist that al-Ghazzālī deals with these two closely related classifications. The perspective of jurisprudence is essentially based on ethical considerations. The legal status of the acquisition of each branch of knowledge is determined on the basis of the degree of its usefulness to the individual as well as the society in the light of the ultimate goals of the *Sharī'ah*. The basis of the division into the *farḍ 'ayn* and the *farḍ kifāyah* is therefore ethical in nature. This recalls my earlier assertion that al-Ghazzālī gives greater priority to the ethical basis of the classification of the sciences than does al-Fārābī.

9.5 Classification of the Religious and Intellectual Sciences

Having discussed and identified the basis of each of the four systems of classification mentioned by al-Ghazzālī, I will now present a summary of his most prominent classifications, namely the division into the religious and the intellectual, and the division into the *farḍ 'ayn* and the *farḍ kifāyah*.

9.5.1 The Religious Sciences

A. The science of fundamental principles (*al-uṣūl*)

1. The science of divine unity (*'ilm al-tawḥīd*)
2. The science of prophethood. This science is also concerned with the states of the Companions and their religious and spiritual successors.
3. The science of the hereafter or eschatology.

4. The science of the sources of religious knowledge. There are two primary or fundamental sources, namely the Qu'rān and the *Sunnah* (prophetic traditions). The other two are secondary sources: the consensus (*ijmā'*) of the Community and the traditions of the Companions (*āthār al-ṣaḥābah*). The science of the sources falls under two categories:
   (a) preludes or auxiliary (*muqaddimāt*) sciences. These include the science of writing and the various branches of linguistic science.
   (b) supplementary sciences (*mutammimāt*). These comprise:

      (1) the Qur'anic sciences including the science of interpretation.
      (2) the sciences of the prophetic traditions such as the science of transmission of *ḥadīths*.
      (3) the science of principles of jurisprudence (*uṣūl al-fiqh*).
      (4) biography – dealing with the lives of the prophets, the companions, and illustrious men.

B. The science of branches (*furū'*) or derived principles

1. The science of man's obligation to God. This is the science of religious rites and worship.
2. The science of man's obligations to society. This science comprises:
   (a) the science of transactions
   It deals mainly with business and financial transactions. Other kinds of transactions include *qiṣāṣ* (just retaliation).
   (b) the science of contractual obligations
   It deals mainly with family law.

3. The science of man's obligation to his own soul. This is the science of moral qualities (*'ilm al-akhlāq*)

9.5.2 The Intellectual Sciences

A. Mathematics

(1) arithmetic
(2) geometry
(3) astronomy and astrology
(4) music

B. Logic
C. Physics or the natural sciences
(1) medicine
(2) meteorology
(3) mineralogy
(4) alchemy

D. The sciences of beings beyond nature, or metaphysics
(1) ontology
(2) knowledge of the divine essence, attributes, and activities. knowledge of God's relation to the universe.
(3) knowledge of simple substances, that is, the intelligences and angelic substances
(4) knowledge of the subtle world
(5) the science of prophecy and of the phenomenon of sainthood. the science of dreams
(6) theurgy (*nairanjīyāt*). This science employs terrestrial forces to produce effects which appear as supernatural.

The above classification is based on al-Ghazzālī's enumeration of the sciences in *al-Risālat al-ladunīyah* and *The Book of Knowledge*. The division into the religious and the intellectual given in the former work is not the same as the one found in the latter. I have synthesized the two to produce a more detailed classification. Al-Ghazzālī has treated the same system of division differently in the two treatises because in each of them he has a different aim in mind. He states clearly in *al-Risālat al-ladunīyah* that one of the aims of the treatise is "to enumerate the sciences and their different classes."[20] The two classes of knowledge that come out prominently in this treatise are the religious and the intellectual. A proper treatment of these two classes is necessary because al-Ghazzālī wishes to demonstrate the extent of validity of the religious-intellectual dichotomy when seen in the light of Sufi epistemology. The question of this validity will be examined later.

In *The Book of Knowledge*, al-Ghazzālī's aim is not to study the religious-intellectual dichotomy as such. His reference to that dichotomy is made with a view of discussing the distinction between the *farḍ 'ayn* and the *farḍ kifāyah* sciences as well as the distinction between the praiseworthy and the blameworthy sciences. Consequently, the enumeration of the branches of both the religious and the intellectual classes in the former treatise is more comprehensive than in the latter.

9.6 Nature and Characteristics of the Religious Sciences

According to al-Ghazzālī, all religious sciences, as enumerated in the above classification, are praiseworthy.[21] These sciences fall under two categories. The first comprises those sciences whose acquisition is deemed *farḍ 'ayn* .The second comprises those sciences whose acquisition is deemed *farḍ kifāyah*. In al-Ghazzālī's view, the knowledge meant by the Prophet as incumbent upon every Muslim (i.e. a *farḍ 'ayn*) refers to the science of the path to the hereafter (*'ilm ṭarīq al-ākhirah*). Al-Ghazzālī divides this science into its exoteric and esoteric dimensions. He calls the former dimension the science of devotional practice (*'ilm al-mu'amalah*) and the latter the science of unveiling (*'ilm al-mukāshafah*).

The science of unveiling pertains to knowledge alone.[22] Being esoteric, it is not incumbent on every individual. It is meant only for the minority who are adepts on the spiritual path. This science is not discussed in the *Iḥyā'*. The science of devotional practice embraces both doctrines and practices. It deals with the fundamental articles of the Islamic faith, namely the doctrine of Divine Unity and other fundamental doctrines derived from it. The knowledge of these doctrines corresponds to the first three parts of the science of fundamental principles contained in al-Ghazzālī's classification.[23] According to al-Ghazzālī, this knowledge is obligatory because it is what saves the soul. The knowledge of God is not sought for the sake of any other knowledge but for itself and for the bliss of the hereafter.[24]

The science of devotional practice also deals with religious and spiritual practices in accordance with the knowledge of fundamental principles. The first and third parts of the science of the branches (*furū'*) in the classification corresponds to this "practical" dimension of the science of devotional practice. A part of the science of man's obligation to society is also viewed by al-Ghazzālī as *farḍ 'ayn*.

The science of devotional practice, which constitutes the whole concern of the *Iḥyā'*, is divided by al-Ghazzālī into two parts. One is the science of the outward (*ẓāhir*), which deals with the acts and functions of the external senses. The other is the science of the inward (*bāṭin*), which deals with the acts and functions of the heart.[25] The outward science is, in turn, divided into the science of acts of worship (*'ibādāh*) and the science of customs (*'ādāh*).[26] Similarly, the inward science is subdivided into the science of the destructive qualities of the soul (*muhlikāt*) and the science of its qualities that lead to salvation (*munjiyāt*). These four parts of the science of devotional practice determine the division of the *Iḥyā'* into its four "quarters" or "volumes."[27]

According to al-Ghazzālī, the category of *farḍ kifāyah* religious sciences includes (1) the science of the sources of religious knowledge (enumerated in the classification under the science of fundamental principles) and (2) the science of jurisprudence which forms a part of the science of the branches.[28] Therefore, sciences like the science of Qur'anic interpretation, the science of prophetic traditions, jurisprudence, and principles of jurisprudence belong to this category. Another important example of a *farḍ kifāyah* religious science is *kalām*.

Farḍ 'ayn knowledge is praiseworthy not partially but totally.[29] The more one acquires of it the better one becomes. In other words, there is no limit to its praiseworthiness or excellence. In contrast, the category of *farḍ kifāyah* knowledge is praiseworthy only within a certain limit. Al-Ghazzālī calls this limit the "limit of sufficiency."[30] Since this limit varies with individuals, disciplines, and changing needs of society, al-Ghazzālī formulated his conception of sufficiency only in general terms. He laid down three general principles governing the pursuit of *farḍ kifāyah* sciences. First, one should always maintain the supremacy and the priority of the *farḍ 'ayn* over the *farḍ kifāyah*.[31] Second, one should observe gradual progress in the study of the *farḍ kifāyah* sciences since they themselves are of various degrees of excellence.[32] Third, one ought to refrain from studying those sciences which have already been taken up by a sufficient number of people.[33]

Al-Ghazzālī maintains that there are three degrees of acquisition of knowledge, namely limitation (*iqtiṣār*), moderation (*iqtiṣād*), and thoroughness (*istiqṣā'*).[34] The science of knowing God is praiseworthy even to the limit of thoroughness. In the case of *farḍ kifāyah* sciences, however, the general rule is that they should not be pursued beyond the first two degrees. Al-Ghazzālī justifies this rule by the following argument:

> Do not spend all your life in one of these sciences seeking to exhaust the subject thoroughly, because knowledge is of varied and numerous (branches) and life is short. Furthermore, these sciences are only introductory means sought not for themselves but for the sake of something else. But in everything which is sought as a means for the attainment of another thing, one should not lose sight of the end.[35]

Al-Ghazzālī discusses the three degrees of acquisition of knowledge in relation to a number of *farḍ kifāyah* sciences of the religious as well as the intellectual categories. According to him, the science of interpretation of the Qur'ān and the science of *kalām* should not be pursued beyond the first two degrees.[36] But in the sciences of jurisprudence and of the prophetic traditions, it is praiseworthy even to attain to the degree of thoroughness. Al-Ghazzālī does not explain what he means precisely by each degree of acquisition. For each of the four religious sciences mentioned above he simply refers to a certain work as being equivalent to the degree of acquisition in question.[37]

Al-Ghazzālī's "degree of thoroughness" may be termed a kind of specialization. In the light of the various conditions imposed by him on the pursuit of *farḍ kifāyah* sciences, it is clear, however, that it cannot be equated with the modern idea of specialization. For al-Ghazzālī, specialization is subject to the general principles he had laid down concerning the "limit of sufficiency." Within this framework, specialization is praiseworthy because it is sought either for the sake of the *farḍ 'ayn* knowledge related to one's needs or for the sake of the legitimate religious and spiritual needs of the Community as a whole. Moreover, traditional specialization as envisaged by al-Ghazzālī is never divorced from its ultimate goal, namely the knowledge of God and of the hereafter.

9.7 The Ethico-Legal Status of the Intellectual Sciences

It has already been mentioned that al-Ghazzālī divides the intellectual sciences into three categories: praiseworthy, blameworthy, and permissible. Only the first two are discussed by him. Concerning praiseworthy intellectual sciences, he writes:

> They are those on whose knowledge the activities of this life depend. Examples are medicine and arithmetic. They are divided into sciences the acquisition of the knowledge of which is *farḍ kifāyah* and into sciences the acquisition of the knowledge of which is meritorious though not obligatory. Sciences whose knowledge is deemed *farḍ kifāyah* comprise every science which is indispensable for the welfare of this world: for example, medicine which is necessary for the life of the body, arithmetic for daily transactions and the division of legacies and inheritances, and others besides. These are the sciences the absence of which could reduce a community to serious straits.[38]

It appears from the above passage in the *Iḥyā'* that none of the praiseworthy intellectual sciences is viewed by al-Ghazzālī as a *farḍ 'ayn*. Yet some of the metaphysical sciences enumerated in the above classification necessarily belong to the category of the *farḍ 'ayn* he had defined. This is especially true of the science of divine unity. It is not that al-Ghazzālī has fallen into a contradiction. It is rather that in the *Iḥyā'* he does not view the praiseworthy metaphysical sciences as distinctive intellectual sciences. He has included these metaphysical sciences under the category of praiseworthy religious sciences.[39] According to al-Ghazzālī, this is admissible:

> Most of the branches of religious knowledge are intellectual in the opinion of him who knows them, and most of the branches of intellectual knowledge belong to the religious code, in the opinion of him who understands them.[40]

Concerning *farḍ kifāyah*, intellectual sciences, arithmetic and medicine are the only ones explicitly mentioned by al-Ghazzālī. It appears, however, that he also includes logic among them. This may be inferred from his assertion that logic is an indispensable tool of *kalām* and he regards the latter as a *farḍ kifāyah*.[41] Al-Ghazzālī also mentions a number of intellectual sciences the acquisition of which he considers permissible. These include geometry,[42] astronomy,[43] music[44] and the physical sciences.[45] He extends his positive attitude toward 'specialization' to the intellectual sciences as well. At least this is so in the case of arithmetic and medicine for he says:

> To go deep into the details of arithmetic and the nature of medicine – as well as such details which, while not indispensable, are helpful in reinforcing the efficacy of whatever is necessary – is considered meritorious, not obligatory.[46]

As for the blameworthy intellectual sciences, they are defined by al-Ghazzālī as those "whose knowledge is blameworthy whether it be in part or *in toto*."[47] He poses the following question: could a thing be knowledge and at the same time be blameworthy? He answers it by saying that since knowledge is seeing things "as they really are" and is one of the attributes of God, it is not held to be blameworthy in itself.[48] Knowledge is only regarded as blameworthy in the eyes of men for one of three reasons. First, when it leads to any harm, whether the harm should befall its practitioner or someone else.[49] Al-Ghazzālī cites the science of magic and talismans as an example. This science is considered blameworthy because "it is of no use except for harming people" and "the instrument of evil is in itself evil."[50] Moreover, says al-Ghazzālī, many practical aspects of the science are clearly contrary to Islamic law.

Second, a knowledge is blameworthy "when it is mostly harmful."[51] Such is the case with judicial astrology. Al-Ghazzālī makes a distinction between this science and mathematical astronomy. The study of the latter is permissible. The Qur'ān itself, says al-Ghazzālī, lends support to its study. But astrology has been declared blameworthy by Islamic law. He mentions several reasons why it has been forbidden.

First of all, astrology insists that the stars have influence on the course of events on earth. To attribute good and evil to the influence of the stars is harmful to the religious belief of the common people, because they might turn the stars into deities and objects of worship. According to al-Ghazzālī, no such harm could come to the learned, who are well grounded in knowledge because they would understand that "the sun, moon, and stars are themselves subject to the will of God."[52] But "most people", says al-Ghazzālī, "do not look beyond the immediate and earthly causes and therefore never arrive at the Cause of all causes." Since the *Sharī'ah* seeks to protect the religious and spiritual welfare of the Muslim community as a whole it declares belief in the stars as influencing good and evil blameworthy.

Secondly, astrology is concerned with the prediction of future events on the basis of present causes. However, according to al-Ghazzālī, this aspect of astrology is "purely guess-work," and is not determined "either with certainty or even with probability."[53] It must be pronounced blameworthy because of this ignorance, not because it is knowledge. The rare cases in which the astrologer happens to be correct are mere coincidences. Al-Ghazzālī maintains, however, that true knowledge of astrology was originally possessed by the prophet Idrīs (Enoch) but that knowledge "has now vanished and is no more."[54]

The third reason why astrology is considered blameworthy is because it is of no use at all. "The most which could be said on its behalf is that it is, at its best, an intrusion into useless things and a waste of time and life which is man's most precious belonging."[55] And to waste anything precious is blameworthy.

Al-Ghazzālī was not the first Muslim religious scholar to criticize astrology from the point of view of the *Sharī'ah*. The Andalusian jurist and theologian, Ibn Ḥazm (d. 456/1064) was in fact more categorical in his rejection of astrology.[56] Unlike al-Ghazzālī, he denies altogether the validity of astrology as a science. It constitutes no more than a set of claims which are impossible to verify. He rejects, for example, the claim concerning the influence of the stars on terrestrial phenomena save those that are verifiable by the senses. Although both scholars criticize astrology from the religious point of view, their positions on the subject differ in several respects. And as far as is known, al-Ghazzālī's discussion of the legal status of astrology within the context of a clearly formulated and detailed conception of blameworthy knowledge as opposed to praiseworthy knowledge is his original contribution to this subject.

The third reason mentioned by al-Ghazzālī, for which a certain knowledge may be pronounced blameworthy, is that at times "the pursuit of that kind of knowledge does not give the practitioner any real increase in knowledge."[57] This third criterion of blameworthiness is illustrated by al-Ghazzālī by means of the following examples: studying a trivial science before an important one; delving into the science of divine mysteries when one is not qualified to do so, with the result that one's secure and certain faith is replaced by doubt and confusion. Emphasizing the significance of the second example, al-Ghazzālī remarks further that there are indeed persons for whom ignorance of certain sciences is beneficial while exposure to them is harmful.

Al-Ghazzālī's three types of blameworthy knowledge may be described as corresponding to three different degrees of blameworthiness. These degrees are understood not in a qualitative but quantitative sense. What I mean by "quantity" refers to the number of people who are subject to the harmful effects of each of the above three types. The first type, exemplified by the science of magic, produces the greatest degree of harm in the sense that no one, not even a prophet or a saint, is immune from the evil of magic.[58] The second type, exemplified by astrology, corresponds to a lower degree of blameworthiness because there is a group of people, namely those "who are well grounded in knowledge" for whom the science is harmless although useless. As for the third type, exemplified on the one hand by trivial sciences and on the other by the science of divine mysteries, it is associated with the least degree of blameworthiness in the sense that it is harmful to the least number of people.

It is clear from my discussion of the *farḍ 'ayn-farḍ kifāyah* division – as well as the praiseworthy-blameworthy division – that al-Ghazzālī had given great attention to the question of the significance of the various sciences from the ethical point of view. He maintains, however, that, principally speaking, the division into the religious and the intellectual is prior to the two ethical divisions.[59] One must first enumerate the different sciences before one can speak of the ethico-legal status of the acquisition of each science. For al-Ghazzālī, that enumeration is best carried out in the context of the division of the sciences into the religious and the intellectual. Evidently, by choosing to discuss the ethico-legal status of the sciences in relation to the religious-intellectual division, al-Ghazzālī intends to show that, not only methodologically but also ethically, the religious sciences constitute a superior class of knowledge than the intellectual sciences.

9.8 Nature of the Theoretical-Practical Division

Al-Ghazzālī's discussion of the religious-intellectual division incorporates the theoretical-practical division. In *al-Risālat al-ladunīyah,* he describes the first class of religious knowledge, namely the knowledge of fundamental principles, as theoretical knowledge. He calls the second class of religious knowledge, that is to say the knowledge of the branches, practical knowledge. He does not explain in this work why the two divisions of religious knowledge should be described respectively as theoretical and practical science. But a careful examination of the content of each division will show that the above description is consistent with al-Ghazzālī's definitions of theoretical and practical science previously given. In the *Mīzān al-'amal,* al-Ghazzālī discusses the theoretical-practical division independently of any other division. But what constitutes theoretical and practical knowledge is the same in both works.

Al-Ghazzālī does not make any concerted attempt to apply the theoretical-practical division to the intellectual sciences in either work. In his view, the distinction between religious and intellectual knowledge is more fundamental than the distinction between theoretical and practical knowledge. This is evident from what I have just said regarding the way in which he has treated the theoretical-practical division in the two works. As far as the intellectual sciences are concerned, al-Ghazzālī appears to be more interested in another kind of division. In *The Marvels of the Heart,* in the seventh section entitled "The State of the Heart in Its Relationship to the Divisions of the Sciences", al-Ghazzālī divides the intellectual sciences into the worldly *(al-dunyawīyah)* and the other-wordly *(al-ukhrawīyah).*[60] The worldly rational sciences consist of medicine, mathematics and the like. The other-worldly rational sciences include the metaphysical science of divine unity and the science of the states of the heart (*'ilm ahwāl al-qalb*).

9.9 The Worldly and Other-Worldly Rational Sciences

How does he conceive of the distinction between the worldly and the other-worldly rational sciences? The perspective that he adopts in the above division is one in which the two kinds of sciences are viewed as mutually exclusive in their usefulness.The former sciences are regarded as being exclusively concerned with the affairs of this world while the latter are seen as being exclusively concerned with the affairs of the hereafter. For example, medicine is said to pertain only to the welfare of the body.[61] In contrast, the science of the states of the heart is directed solely toward the welfare of the soul or the spirit. It is true that al-Ghazzālī often portrays a number of the important intellectual sciences like mathematics and medicine as worldly sciences that are devoid of any religious and spiritual significance. This aspect of his views concerning the intellectual sciences has drawn severe criticism from a number of contemporary Muslim scholars.[62] But in fairness to al-Ghazzālī, there is another aspect of his views that ought to be mentioned. He is known to have often expressed the view that some of the intellectual sciences can contribute to the knowledge of God or bring man closer to God.[63]

It is characteristic of al-Ghazzālī to pass different judgments on the same science. This often gives rise to contradictory positions. The reason is that he adopts a different perspective in each judgment. In his division of the intellectual sciences into the worldly and other-worldly he is looking at the worldly sciences in their aspect of opposition or irrelevance to the science of the hereafter. Al-Ghazzālī would insist that the worldly sciences, at best, concern the hereafter only in an indirect way. In contrast, the other-worldly sciences pertain to the hereafter directly.

9.10 Significance of the Division into "Presential" and "Attained" Knowledge

Although the distinction between religious and intellectual knowledge has come out most prominenty in al-Ghazzālī's discussion of the different systems of classification, he himself does not consider that distinction the most fundamental. For him the division into presential and attained knowledge is more fundamental and universal than the religious-intellectual division. The former comprehends the latter but the converse is not true. Presential knowledge comprises both prophetic knowledge, which is derived from revelation (*waḥy*), and the knowledge of the saints, which is derived from inspiration (*ilhām*) and which is called knowledge from on high (*al-'ilm al-ladunī*).[64] It thus embraces religious knowledge irrespective of whether it is defined in terms of its original source, namely prophetic revelation, or in terms of the general mode in which its verities are accepted, namely faith. Similarly, insofar as it pertains to the intelligible or spiritual world, presential knowledge embraces intellectual knowledge.

Presential knowledge transcends the faith-reason dichotomy that al-Ghazzālī has also associated with his religious-intellectual division. When understood in relation to that dichotomy, the latter division is only valid up to a certain point. At the level of gnosis or *ma'rifah* of the Sufis, the distinction between the religious and the intellectual ceases to exist. Al-Ghazzālī maintains that the knowledge of the Sufis is at once religious and intellectual.[65] He refers specifically to their science of unveiling which constitutes the esoteric part of the "science of the path to the hereafter." Since this science transcends the religious-intellectual distinction, it is excluded from al-Ghazzālī's classification of the sciences.

9.11 **Conclusion**

I conclude this chapter with a brief discussion of two questions related to al-Ghazzālī's division of knowledge into the religious and the intellectual. First, in what fundamental respect does al-Ghazzālī's enumeration and classification of the sciences differ from that of al-Fārābī? Second, how valid is the criticism leveled by some contemporary Muslim scholars at the religious-intellectual division of the sciences. I will deal first with the second question.

According to Fazlur Rahman, the division of the sciences into the religious and the intellectual must be regarded as "the most fateful distinction that came to be made" in the intellectual history of Islam.[66] In his view, the seeds of the decline of science and philosophy in Islam can be partly traced to that distinction. He points out that Muslim religious scholars do not reject the "intellectual sciences" as such but their attitude of "discounting them as not conducive to one's spiritual welfare" could only be negative.[67] Rahman sees this attitude as one that is deeply rooted in the above division of the sciences.

It is not part of this study to examine the question of the decline of science in Islam. However, it is necessary to remark here that it is not the religious-intellectual division as such which constitutes the root cause of that decline.[68] My view is based on the following arguments. First, the basis of the division itself is epistemologically valid within a certain limit; insofar as it is true it is of a positive value.

Second, opposition to the intellectual sciences from among the religious scholars, especially the jurists, had been going on in Islamic society long before the explicit formulation of the religious-intellectual division. Third, many Muslim scientists and philosophers, before as well as after al-Ghazzālī,[69] accepted the validity of the division. They would not have done so if the division is inherently anti-philosophic or anti-scientific. It is important to note the fact that the division is not of the religious scholars alone but of the philosopher-scientists as well.[70] Fourth, al-Ghazzālī, in reproaching his contemporaries for neglecting the *farḍ-kifāyah*, intellectual sciences, particularly medicine, and for putting undue emphasis on jurisprudence,[71] finds no contradiction in defending at the same time his religious-intellectual division of the sciences. And, fifth, the decline of Islamic science did not take place until several centuries after the popularization of the classification in question.

I will now deal with the other question: what are the major features of al-Ghazzālī's religious-intellectual classification that distinguish it from that of al-Fārābī? Al-Ghazzālī incorporated into his classification all the major divisions of al-Fārābī's classification. However, he reorganized them in conformity with his theological and intellectual perspective and omitted some of their branches. Al-Ghazzālī includes the linguistic sciences under the category of religious knowledge, because he sees them mainly as the instruments of the latter. His science of language, therefore, refers to Arabic. In al-Fārābī's view linguistic science is a tool of all the sciences. He enumerated it as an independent category whose subdivisions are based on the anatomy of the human language itself rather than of Arabic alone.

Al-Ghazzālī's four subdivisions of the intellectual sciences differ slightly from al-Fārābī's philosophical sciences. Al-Ghazzālī consistently maintained the view that logic is one of the philosophical sciences. In al-Fārābī's classification, logic is not. Al-Ghazzālī's natural philosophy, contrary to al-Fārābī's, includes medicine and alchemy. Al-Fārābī's classification necessitates the exclusion of the two sciences. In the case of al-Ghazzālī's classification, the basis of which is the distinction between religious and intellectual knowledge, their inclusion is possible. Al-Ghazzālī also differs from al-Fārābī in the enumeration of the metaphysical sciences. Al-Ghazzālī adds a number of sciences to al-Fārābī's list, which the latter has either omitted or included under different categories. One of these is the science of theurgy. Al-Fārābī has omitted it because it belongs to the category of the occult sciences. Ibn Sīnā has placed it among the branches of natural philosophy.

Besides theurgy, al-Ghazzālī mentions the science of dream and the science of prophecy as among the metaphysical sciences. It has been shown in the case of al-Fārābī that the science of dreams is treated either under natural philosophy or political philosophy. As for the science of prophecy, he views it as a part of political philosophy although it is partly metaphysical in nature. It is understandable why al-Ghazzālī enumerated some of the occult sciences, the science of dreams and the science of prophecy under the category of intellectual sciences although the last of these is also viewed by him as a religious science. This is because all these sciences have been cultivated by the philosophers. Moreover, they satisfy his definition of the "intellectual".

It is necessary, however, to explain why al-Ghazzālī has placed them under the category of metaphysical sciences. The only plausible explanation is to be found in his two-fold division of reality. This division occurs in many of his works.[72] According to him, the whole of reality comprises two worlds. One is the visible world or the world of dominance (*'ālam al-mulk*) – also called *'ālam al-shahadah* (the world perceived by the senses). The other is the world of the unseen and the realm of the supernal (*'ālam al-malakūt*), also called *'ālam al-ghayb*. The latter is perceived by the light of inner vision (*nūr al-baṣīrah*). It is invisible to the majority while the former is perceived by all.[73]

The above division of reality provides the basis of his categorization of the intellectual sciences. The first category comprises the sciences of the material or the visible world. The second comprises the sciences of the world of the unseen. In one of his discussions of the knowledge of divine works,[74] al-Ghazzālī implies that the first category consists of two major divisions, namely the mathematical and the physical sciences. His "sciences of beings beyond nature" or metaphysical sciences clearly fit into the second category. The occult sciences, the science of dreams, the science of prophecy and of sainthood, and science of the subtle world all pertain to the world of the unseen. Therefore, al-Ghazzālī's classification of the intellectual sciences is closely related to his twofold division of the cosmos.

Al-Ghazzālī accepts the reality of the world that is intermediate between the material and the absolutely incorporeal or spiritual worlds.[75] However, in his popular division of the cosmos – and this is reflected in his division of the intellectual sciences – the intermediate world, that is the subtle domain, is not taken as a separate category of beings. Rather, it is included as a part of the spiritual world. Al-Ghazzālī's absorption of the intermediate world into the purely spiritual world clearly distinguishes him from al-Fārābī. The latter, as we have seen, conceives of mathematics and politics as intermediate sciences which provide the links between the physical world and metaphysical realities. As for al-Ghazzālī's mathematics, it appears to be concerned only with numbers and figures as they exist in the physical world. It does not make any reference to the reality of number and figures in the intermediate world of the mind. In my view, this is a consequence of al-Ghazzālī's attitude toward the intermediate world as reflected in his twofold division of reality.

Another significant difference between al-Ghazzālī's and al-Fārābī's classifications concerns the place of politics and ethics. On the whole, the former's religious sciences correspond to the latter's political philosophy, *fiqh*, and *kalām* and their branches. The reason why al-Ghazzālī places politics and ethics among the religious sciences has been explained. He does not conceive of political science as a discipline that embraces ethics. He views them as two independent sciences and regards ethics as the superior science. In contrast, al-Fārābī's conception of political science embraces both the politics and ethics of al-Ghazzālī.

END NOTES

Chapter 9

[1]See, for example, *Mīzān al-'amal.* pp. 36-37, 112-13; also *Maqāṣid al-falāsifah*, ed. Sulayman Dunya, Cairo, 1961, p. 134.

[2]*Ihyā'*, Lajnat nashr al-thaqafah al-Islamīyah, Cairo, 1356/1937 – 1357/1938, vol. III, Bk, I. p. 1376. See also *al-Risālat al-laduniyah*, pp. 360-368 (cf. my chap. 8, n. 116).

[3]See *The Book of Knowledge*, p. 36: here, al-Ghazzālī uses the term *ghayr shar'īyah* (non-religious) instead of *'aqlīyah*. See also *Al-Risālat al-ladunīyah*, p. 3.3; and M.A. Sherif, *op. cit.*, p. 8

[4]*The Book of Knowledge*, sec. II, pp. 30-72. See my explanation of the two terms in Chap. 8, n. 25.
[5]*al-Risālat al-ladunīyah*, p. 357.
[6]*Mīzān al-'amal*, p. 18.
[7]Al-Ghazzālī calls it "inferential knowledge". See *al-Risālat al-ladunīyah*, p. 362.
[8]*The Book of-Knowledge*, pp. 36-37.
[9]By praiseworthy knowledge al-Ghazzālī means both the knowledge necessary for the activities of this life and the knowledge necessary for the salvation of the soul in the afterlife. What constitutes necessary knowledge in each case is established on the basis of the fundamental doctrines and practices of the religion of Islam.
[10]In *The Book of Knowledge* (pp. 38-40), al-Ghazzālī divides the praiseworthy religious sciences into four parts: (1) science of the sourses (*uṣūl*) which are four in number, namely the Qur'ān, the Prophetic traditions (*Sunnah*), the consensus (*ijmā'*) of the community, and the Traditions of the Companions (*āthār al-ṣaḥābah*), (2) science of the branches (*furū'*), that is, jurisprudence and science of the hereafter, (3) preludes (*muqaddimāt*) for example, linguistic science, and (4) supplementary (*mutammimāt*) sciences such as the science of Qur'anic interpretation. All these sciences are enumerated by Ibn Khaldūn under the category of transmitted (*naqlīyah*) sciences. See S.H. Nasr, *Science and Civilization in Islam*, p. 63-64.

In *al-Risālat al-ladunīyah*, al-Ghazzālī gives a twofold classification of the religious sciences. He combines the science of the sources, the preludes and the supplementary sciences to form a single category called the science of the fundamentals. See below my synthesis of the two classifications.
[11]For al-Ghazzālī, linguistic science is the science of the Arabic language. According to him, it is appropriate to include linguistic science under the category of the religious sciences, because Arabic is the language of the sacred law of Islam. See *the Book of Knowledge*, p. 39.
[12]*Iḥyā'*, vol. III, Bk. I, pp. 1372-74. This should be understood in a qualified sense since al-Ghazzālī attributes certain knowledge comprised in some of the intellectual sciences like medicine and astronomy to the prophets.
[13]R.J. McCarthy, *op. cit.*, secs. 49-52, p. 77-78.
[14]See M.A. Sharif, *op. cit.*, p. 8.
[15]See our chap. 8, n. 25.
[16]A. Hasan, *The Early Development of Islamic Jurisprudence*. Islamabad, 1970, p. 39.
[17]*The Book of Knowledge*, p. 30.
[18]According to al-Ghazzālī, "each group insisted on the necessity of acquiring that branch of knowledge which happened to be its specialty." *Ibid.*
[19]*Ibid.*, p. 37.
[20]*al-Risālat al-ladunīyah, p. 360.*
[21]*The Book of Knowledge*, p. 37.
[22]This knowledge refers to the Sufi interpretation of the doctrine of divine unity and other fundamental doctrines of Islam.
[23]See classification above.
[24]*The Book of Knowledge*, p. 99.
[25]*Ibid.*, p. 6.
[26]By *'ibādāh* al-Ghazzālī means acts of devotion directed to God alone, and by *'ādāh* types of action directed to one's fellow men.
[27]*The Book of Knowledge*, p. 7.
[28]*Ibid*, p.d 38 (cf. classification above).
[29]*Ibid.*, p. 98.
[30]*Ibid*, p. 98, 100.

[31]According to al-Ghazzālī, one must first fulfill the task of acquiring the *farḍ 'ayn* sciences in accordance with one's own needs and ability. The *farḍ 'ayn* sciences pertain to beliefs, prohibitions and works. The obligation to acquire the knowledge of these three aspects of life is conditioned by the rise of new developments and changed circumstances in the life of the individual. Concerning beliefs, the basic doctrines one has to learn are the same throughout one's life. One's knowledge of them may vary in two respects. First, it becomes deeper with time. Second, feelings of doubt about the truth of the doctrines may arise in the mind at some stage of life. Then, it would be obligatory also upon the person to acquire the knowledge of whatever would remove that doubt.

Regarding obligatory works such as the canonical prayer, fasting and alms-giving the obligation to acquire the knowledge of them is partly governed by the factor of time. For example, one is not obliged to learn about fasting until the eve of Ramadan, the month of fasting.

As to prohibitions, al-Ghazzālī explains what he means by "according to one's needs" by giving the following example: the mute is not obliged to know what is unlawful in speech nor the blind to know what things are unlawful to see. See *The Book of Knowledge*, p. 31-35.

[32]*Ibid*, p. 101-02. The pursuit of the *farḍ kifāyah* sciences should begin with the study of the Qur'anic sciences, followed by that of the Prophetic traditions, of jurisprudence, the principles of jurisprudence (*uṣūl al-fiqh*), and the other sciences in that order.

[33]*Ibid*, p. 101.

[34]*Ibid*, p. 100.

[35]*Ibid*, p. 102.

[36]*Ibid*, p. 104.

[37]In the science of Qur'anic interpretation, the limited degree corresponds to *al-Wajīz* (Qur'anic commentary by 'Alī al-Wāhidi al-Naisābūrī, Cairo, A.H. 1305) and the moderate degree to *al-Wasīṭ* (by the same author). In *kalām*, the limited degree is equivalent to al-Ghazzālī's *Qawā'id al-'aqā'id* (*Iḥyā'* I, 2) and the moderate degree to his *The Golden Mean in Belief.*

In the science of Prophetic traditions, the limited degree is represented by the mastery of the *Saḥīhs* (*ḥadīth* collections) of al-Bukhārī and Muslim; the moderate degree would have to include, beside the two *Saḥīhs*, the other authoritative corpuses; beyond this is the degree of thoroughness which includes the study of *ḥadīth* transmissions. In the field of jurisprudence, the three degrees are represented respectively by al-Ghazzālī's *Khulāṣat al-mukhtaṣar, al-Wasīṭ min al-madhhab*, and *al-Basīṭ*. See *The Book of Knowledge*, p. 102-4.

[38]*Ibid*, p. 37.

[39]Al-Ghazzālī mentions specifically that the metaphysical science of divine unity is included under the religious science of *Tawḥīd.* See *ibid*, p. 53.

[40]*al-Risālāt al-ladunīyah*, p. 353.

[41]*The Book of Knowledge*, p. 54.

[42]*Ibid*, p. 53 (cf. *Tahāfut*. p. 9).

[43]To be more precise, al-Ghazzālī refers to mathematical astronomy as it has been defined by al-Fārābī. See *The Book of Knowledge*, p. 74. (Cf. R.J. McCarthy, *op. cit.*, sec. 37, p. 73).

[44]In the *Ihyā'* al-Ghazzālī defends the legitimacy of music as an aid to religious life and mystical devotion, as a means of pleasurable aesthetic enjoyment and as an effective moral agent. See his *The Books of the Laws of Listening to Music and Singing and of Ecstasy* (*Iḥyā'* II . 8), trans, D.B. Macdonald as " Emotional Religion in Islam as affected by Music and Singing," *JRAS*, 1901, Part I, 200-44; *The Alchemy of*

Happiness, trans. C. Field, Lahore, 1964, pp. 73-85. See also L.L. al-Faruqi, "Al-Ghazzālī on Samā'," in *Essays in Islamic and Comparative Studies* ed. I.R. al-Faruqi, International Institute of Islamic Thought, Herndon, 1982, pp. 43-50.
[45]See R.J. McCarthy, *op. cit.*, sec. 45, p. 76. See also *The Book of Knowledge*, p. 54.
[46]*Ibid*, p. 37.
[47]*Ibid*, p. 98.
[48]*Ibid*, p. 73.
[49]*Ibid*.
[50]*Ibid*, p. 74.
[51]*Ibid*.
[52]*Ibid*, p. 75.
[53]*Ibid*.
[54]*Ibid*.
[55]*Ibid*, p. 76.
[56]For Ibn Ḥazm's criticism of astrology, see A.G. Chejne, *Ibn Ḥazm*, pp. 180-84.
[57]*The Book of Knowledge*, p. 77.
[58]*Ibid*, p. 73.
[59]*Ibid*, p. 36.
[60]*Ihyā'*, III, 1, p. 1375.
[61]*The Book of Knowledge*, p. 45-46.
[62]See, for example, F. Rahman, *Islam and Modernity: Transformation of an Intellectual Tradition*, University of Chicago Press, Chicago-London, 1982, p. 34-45.
[63]See, for example, *The Alchemy of Happiness*. p.p. 29-30; also *The Book of Knowledge*, p. 57.
[64]See *al-Risālat al-laduniyah*, p. 365; also *The Marvels of the Heart* in R.J. McCarthy, *op. cit.*, secs. 44-46, p. 378.
[65]*al-Risālat-al-laduniyah*, pp. 359-60.
[66]F. Rahman, *op. cit.*, p. 33.
[67]*Ibid*, p. 34.
[68]Historically, the religious-intellectual division appeared in Islamic thought some time before the "golden age" of Islamic science in the tenth and eleventh centuries.
[69]Prior to al-Ghazzālī, Ibn Sīnā, for example, is known to have accepted the distinction between religious and intellectual sciences. He wrote a short epistle entitled *Fī aqsām al-'ulūm al-'aqlīyah* (On the Divisions of the Intellectual Sciences) in answer to someone who asked him to present a summary account of the intellectual sciences. See R. Lerner and M. Mahdi (eds.), *Medieval Political Philosoohy: A Sourcebook*. p. 96. Al-Fārābī himself had used the term religious sciences (*al-'ulūm al-shar'iyah*) to distinguish them from the philosophical sciences.

Among the post-Ghazzalian philosopher-scientists, we may mention Quṭb al-Dīn al-Shīrāzī whose classifications will be treated in Chapter 11. Al-Shīrāzī, however, adopts a different terminology. Religious sciences are designated by the term *'ulūm dīniyah* and intellectual sciences by *'ulūm ghayr dīniyah*.
[70]The philosopher-scientists who accepted the religious-intellectual division generally differed, however, from the religious scholars in that the former attached greater importance to the intellectual sciences than to the religious sciences.
[71]*The Book of Knowledge*, p. 51.
[72]See, for example, *Mishkāt al-anwār*, p. 122-23; *The Foundations of the Articles of Faith (Qawā'id al-'aqā'id)*, trans. N.A. Faris, Lahore, 1974, p. 119.
[73]*Mishkāt al-anwār*. p. 123.
[74]See *The Jewels of the Qur'ān*, p. 46-47.
[75]Al-Ghazzāli's acceptance of the 'intermediate world' pertains to both microcosmic

and macrocosmic realities. Microcosmically, he maintains that the imaginative spirit occupies an intermediate position between the sensory spirit and the rational spirit (see *Mishkāt al-anwār*. p. 144-145). Macrocosmically, he regards the world of the *jinns* (demons) as intermediate between the angelic world and the material world. He calls this intermediate world *'ālam al-jabarūt*. See F. Schuon, *Dimensions of Islam*, p. 150. (cf. *al-Risālat al-ladunīyah*, p. 359; *The Book of Knowledge*, p. 47).

He (Allāh) giveth wisdom (*ḥikmah*) unto whom He will, and he unto whom wisdom is given, he truly hath received abundant good.

The Qur'ān: The Cow, Verse 269.

PART III

Quṭb al-Dīn al-Shīrāzī's Classification

CHAPTER 10

THE LIFE, WORKS AND SIGNIFICANCE OF QUṬB AL-DĪN AL-SHĪRĀZĪ

10.1 Quṭb al-Dīn's Education and Intellectual Life

Quṭb al-Dīn Maḥmud ibn Ḍia al-Dīn Mas'ūd al-Shīrāzī was born in the Persian city of Shīrāz in 634/1236.[1] He belonged to a family of distinguished physicians and Sufis. His father, who hailed from Kāzirūn, was both a Sufi master attached to Shihāb al-Dīn Abū Ḥafṣ 'Umar al-Suhrawardī (d.632/1234)[2] and a famous physician. In Shīrāz Quṭb al-Dīn received his early education and training in both medicine and Sufism under his father's guidance. As a child he also learned numerous handicrafts with the notable exception of pottery.

Traditional biographers like Ibn Shuhbah and al-Subkī described Quṭb al-Dīn as a person of brilliant intelligence.[3] He was said to have begun his medical practice as early as the age of ten. At the age of twelve or so, as an aspirant of Sufism, he was considered worthy of receiving the "cloak of blessing" (*khirqah tabarruk*) from a Sufi master.[4] When he was fourteen, on his father's death, he was entrusted with his father's duties as physician and opthalmologist at the Muẓaffarī hospital in Shīrāz. He remained in that capacity for about ten years.

During this period he studied under three different teachers, all of them his uncles. First he became a pupil of Kamāl al-Dīn Abi'l-Khayr al-Kāzirūnī with whom he studied the *Kullīyāt* (*General Principles*) of Ibn Sīnās' celebrated medical work, *al-Qānūn fi'l-ṭibb* (*Canon of Medicine*).[5] His next teacher was Shams al-Dīn Muḥammad ibn Aḥmad al-Kabshī who later moved to Baghdad to become a professor there in 665/1266.[6] It is not known exactly in which discipline Quṭb al-Dīn received instruction from al-Kabshī. Very likely it was in the religious sciences.[7] Quṭb al-Dīn then joined the academic circle of Sharaf al-Dīn Zaki al-Bushkānī.[8] His association with this circle is significant in that it exposed him for the first time to a number of commentaries on Ibn Sīnā's *Canon*, including the one composed by Fakhr al-Dīn Rāzī. In addition, Quṭb al-Dīn had access to some of Rāzī's commentaries on Ibn Sīnā's other works.[9]

It was during this time that Quṭb al-Dīn became greatly attracted to the philosophy and medicine of Ibn Sīnā. But there was no one in Shīrāz who could give instruction on the subject to his satisfaction. Specifically, the *Canon* posed certain problems for him. He found that Rāzī's commentary as well as the other commentaries then avaibable to him were not sufficiently clear to help him solve these problems. He therefore entertained the idea of, some day, writing his own commentary upon the *Canon*. As a preparatory step, he had first to seek a master who could instruct him in the medicine and philosophy of Ibn Sīnā. With that aim in mind, Quṭb al-Dīn left his medical practice at the Muẓaffarī hospital and his native city around 658/1260.

Quṭb al-Dīn's quest for a master in Ibn Sīnā's philosophy took him to Marāghah where, around 660/1262, he became a student of Naṣīr al-Dīn al-Ṭūsī (d. 710/1311). Al-Ṭūsī, one of the leading philosophers and scientists of Islam, was then the director of the newly established Marāghah observatory[10] (begun in 657/1259 by the order of the Mongol ruler Hülagü,[11] grandson of Genghis Khān). In studying with al-Ṭūsī at the Marāghah observatory, Quṭb al-Dīn was blessed with the best possible opportunity of the time to develop his scientific talent and master the philosophical sciences. As far as philosophy was concerned, al-Ṭūsī was certainly the greatest authority in Ibn Sīnā's philosophy of that day.[12] He was responsible more than any other figure for reviving that philosophy after it had been severely criticized by prominent theologians like al-Ghazzālī and Fakhr al-Dīn Rāzī. He was also responsible for the revival of the study of astronomy and mathematics in thirteenth century Islam.

The famous Christian encyclopedist and philosopher Barhebraeus, who lectured at the observatory, described al-Ṭūsī as "a man of vast learning in all the branches of philosophy."[13] As far as the place of study was concerned, the Marāghah observatory was more than just a place for astronomical observation. "It was a complex scientific institution, in which nearly every branch of science was taught, and where some of the most famous scientists of the medieval period were assembled."[14] It was equipped with the best astronomical instruments, "some of them constructed for the first time."[15] There was also a huge library annexed to it. According to Ibn Shākir, the library contained more than 400,000 books.[16]

Quṭb al-Dīn spent many years at Marāghah as a distinguished member of al-Ṭūsī's scientific circle. There, al-Ṭūsī advised him to study mathematics and astronomy. Quṭb al-Dīn took up the study of the two sciences and later wrote numerous treatises on them, particularly the latter. Both teacher and student made important contributions to the field of astronomy. Together they proposed the first new medieval model for planetary motion, which was later employed by Copernicus. Copernicus most likely learned it through Byzantine Greek sources.[17] Quṭb al-Dīn worked out to completion a standard model of planetary motion for all the planets except Mercury, and a variation of the model for that planet. As a close associate of al-Ṭūsī at the observatory, Quṭb al-Dīn played a prominent role in bringing about a major renaissance of Islamic astronomy in the thirteenth century.

Under Al-Ṭūsī's guidance, Quṭb al-Dīn also studied the philosophy of Ibn Sīnā. The most important work of the latter which he read was *Kitāb al-ishārāt wa'l-tanbīhāt* (*Book of Directives and Remarks*). Quṭb al-Dīn's study of this philosophical masterpiece under al-Ṭūsī is not without significance. Fakhr al-Dīn Rāzī had earlier written a detailed and acutely critical commentary upon this work, taking the meaning of almost every sentence of it into consideration. The aim of this criticism was, without doubt, to destroy what was left of the philosophical influence of the Peripatetics after al-Ghazzālī's powerful attack on them, about a century earlier. It was al-Ṭūsī's defense of the school of Ibn Sīnā against the attack by Rāzī (as well as by al-Ghazzālī) that helped to revive Peripatetic philosophy.

It has been noted earlier that Quṭb al-Dīn probably had access to Rāzī's commentary upon the *Ishārāt* while still in Shīrāz. There, he had certain difficulties with the commentary, which no one could explain to his satisfaction. At Marāghah, he had direct access to the sources of the most powerful reply to the commentary – that is, al-Ṭūsī. It appears that al-Tūsī had helped him to resolve his difficulties with Rāzī's commentary. It is necessary to mention, however, that what Quṭb al-Dīn revived of Ibn Sīnā's philosophy was not in its original Peripatetic form. Rather, as pointed out by Nasr, it was given a new interpretation based mainly upon Suhrawardī's metaphysics of light.[18]

Quṭb al-Dīn was not a reviver of the Peripatetic school of Ibn Sīnā in the sense that al-Ṭūsī was. The latter was the real successor of Ibn Sīnā although on certain questions, such as that of God's knowledge of the world, he openly criticized the views of Ibn Sīnā in favour of those of Suhrawardī. However, because of his great interest in Ibn Sīnā and his major commentaries upon the latter's philosophical works, Quṭb al-Dīn made some contribution to the revival of Ibn Sīnā's philosophy. As will be discussed later, Quṭb al-Dīn's greater significance in Islamic intellectual history lies in his role as a key figure who brought the rapproachment of the four intellectual schools of theology (*kalām*), Peripatetic philosophy (*mashshā'ī*), illuminationist theosophy (*ishrāq*), and gnosis (*'irfān*).

During his stay at Marāghah, Quṭb al-Dīn continued to pursue his medical interest, especially in the study of the *Canon*. Some of the difficulties he had in understanding that work while in Shīrāz were resolved at Marāghah with the help of al-Ṭūsī.[19] It is not known who else at Marāghah might have become his teacher in medicine. As far as commentaries upon the *Canon* are concerned, the excellent library of the city's scientific complex probably furnished Quṭb al-Dīn with others that he had not known before. It appears that his decision to leave the city was partly motivated by the realization that the Marāghah institution could not fulfil all his intellectual aspirations and needs. In the field of medicine, he was still in search of further instruction or other commentaries on the *Canon* that would explain all the difficulties relating to general principles of the discipline as embodied in that treatise.

According to some historians, Quṭb al-Dīn decided to leave Marāghah because of his quarrel with al-Ṭūsī.[20] They claim that Quṭb al-Dīn was disappointed with the latter for not having assigned him the task of completing the *Ilkhānid Zīj* (Ilkhānid Astronomical Table) on which they originally had worked closely together.[21] Al-Ṭūsī had instead asked his son, Aṣīl al-Dīn, to take charge of its completion. In support of this claim, these historians point out that al-Ṭūsī, in his introduction to the *Ilkhānid Zīj,* failed to include Quṭb al-Dīn among his colleagues who participated in the above astronomical project.

In my view, Quṭb al-Dīn's disappointment over the *Ilkhānid Zīj* project could not have been the sole reason for his decision to leave the Marāghah institution. His wide intellectual and spiritual interests could not be served by his continuing to remain at Marāghah. As he tells us in his commentary on the *Kullīyat,* his unabating interest is to seek the acquaintance of scholars from the various parts of the Islamic world.

From Marāghah Quṭb al-Dīn went to a number of cities in Khurāsān and other parts of Persia, Iraq, and Syria. In Khurāsān, he studied with Najm al-Dīn Dabīrān Kātibī al-Qazwīnī (d.675/1276), a philosopher, astronomer and mathematician.[22] From Khurāsān, he set out for Qazvīn and then Isfahan. In Isfahan he became acquainted with its governor, Bahā' al-Dīn Muḥammad al-Juwaynī. Quṭb al-Dīn's major astromical work, *Nihāyat al-idrāk fī dirāyat al-aflāk* (*The Limit of Understanding of the Knowledge of the Heavens*), was dedicated to al-Juwaynī's son.[23]

Quṭb al-Dīn's next major destination was Baghdad. There he met a number of leading Sufi masters one of whom was Muḥammad ibn al-Sukrān al-Baghdādī. The visit could not have taken place later than 667/1269 since the latter died during that year. Most probably it was after his uncle, Shams al-Dīn al-Kabshī, became a professor at the Niẓamīyah school in 665/1266 since he was known to have stayed there. From Baghdad he set out for Anatolia where he met Jalāl al-Dīn Rūmī (d.672/1273), accounted by many the greatest mystical poet of Islam.[24] In Konya, he became a student of Ṣadr al-Dīn al-Qunawī (d.673/1274), the disciple of the Andalusian Sufi, Ibn 'Arabī, and himself a Sufi master. He studied with al-Qunawī both the science of the Sufi path and the religious science like Qur'anic commentary and the science of traditions. He stayed in Konya until al-Qunawī's death in 673/1274.

Quṭb al-Dīn's association with al-Qunawī is significant from the point of view of the understanding of the different dimensions of his intellectual life. It is also important for the understanding of the intellectual background of his role in helping to bring about the rapproachment of the major intellectual schools of Islam. This is because of al-Qunawī's own personal or intellectual contact at second hand with some of the most influential figures of Islamic scholarship. According to Jāmī (d.898/1492), al-Qunawī was connected with Rūmī "by special friendship and acquaintance."[25] Al-Qunawī was also known to have corresponded with Quṭb al-Dīn's teacher, al-Ṭūsī, regarding certain questions of metaphysics.[26] But above all, al-Qunawī was the most important expositor of the *'irfān* (gnosis)of the school of Ibn 'Arabī. Through his immediate teachers, Quṭb al-Dīn was, therefore, well exposed to the ideas of the different schools.

Quṭb al-Dīn's stay in Konya provided him with an excellent opportunity to seek the acquaintance of scholars. In Rūmī's time many scholars, arists, and mystics from all over the eastern Islamic world took refuge in Konya when the former was conquered and devastated by the Mongols. The migration of scholars to the city helped to stimulate its intellectual and religious life.[27]

After the death of al-Qunawī, Quṭb al-Dīn left Konya to become judge (*qāḍī*) in Sīwās and Molaṭya, part of Anatolia which was then ruled by the Ilkhānids, the Mongol dynasty founded by Hülagü Khān. Quṭb al-Dīn was appointed a judge either by the vizier Shams al-Dīn Juwaynī, the chief Mongol official in Anatolia and who was well-known as a patron of scholars, or by Mu'īn al-Dīn Parwānah, the Seljūq governor of Anatolia whom Quṭb al-Dīn came to know in Konya. Since most of his official work was done by his deputies, Quṭb al-Dīn spent his time teaching and writing. It was during his stay in Sīwās that he began to compose some of his major works including the *Nihāyah*. He made several visits to Tabrīz, the Ilkhānid capital of Persia. There, he became a friend of Aḥmad Takūdār, the son of Hülagü Khān. Aḥmad Takūdār was then the ruler of Persia.

According to certain traditional sources, Quṭb al-Dīn played a role in influencing the conversion of Aḥmad Takūdār to Islam.[28] His association with the Ilkhānids brought him a political role which also proved to be of major scientific significance for him. In 681/1282 Aḥmad Takūdār sent him with Kamāl al-Dīn Rāfi'ī, the Shaykh al-Islām, and Bahā' al-Dīn Pahlawān as ambassador to the court of the Mamlūk Sulṭān of Egypt, al-Manṣūr Sayf al-Dīn Qalā'ūn (678-89/1279-90). Quṭb al-Dīn's mission was to report Aḥmad Takūdār's conversion to Islam as well as to conclude a peace between the Mongols and the Mamlūks.[29]

Quṭb al-Dīn did not succeed in his peace mission. From the point of view of scientific research, however, his stay in Egypt was extremely fruitful. There he gained access to some of the most important commentaries upon the *Canon*. Among them were the *Mūjiz al-Qānūn (The Summary of the Canon*) of Ibn al-Nafīs (d.687/1288), the *Sharḥ al-kullīyāt min kitāb al-qānūn* (*Commentary on the General Principles of the Canon*) of Muwaffaq al-Dīn Ya'qūb al-Sāmarrī, and the *Kitāb al-shāfī fi'l-ṭibb* (*Book of Medical Healing*) of Ibn al-Quff. It is not known whether Quṭb al-Dīn met Ibn al-Nafīs or his Christian student Ibn al-Quff.[30] Ibn al-Nafīs, who was entitled 'the second Ibn Sīnā' wrote several commentaries on the *Canon*.[31] It is also not known whether all these commentaries of Ibn al-Nafīs came into the hands of Quṭb al-Dīn.

However, Quṭb al-Dīn's discovery of the above-mentioned works, regarded by later physicians as among the best commentaries upon the *Canon*, appeared to have fulfilled his long cherished goal of gaining access to commentaries that would help illuminate all the obscurities of the *Canon*.

From Egypt, Quṭb al-Dīn set out for Syria where he stayed for a while. Among his reported activities there was his teaching of the *Kitāb al-shifā'* of Ibn Sīnā.[32] He then returned to Anatolia. There, with the help of all commentaries known to him and the knowledge he had gathered from various medical authorities of the day, Quṭb al-Dīn finally wrote his celebrated commentary upon the *Canon*. Begun in 682/1283 the work was not completed until many years later. The treatise is known under several names. One of its titles is *al-Tuhfat al-sa'dīyah* (*The Presentation to Sa'd*) since Quṭb al-Dīn had dedicated it to Muḥammad Sa'd al-Dīn, the *wazīr* of Arghūn. It was also called *Kitāb muzḥat al-ḥukamā' wa rawḍat al-aṭibbā'* (*Delight of the Wise and Garden of the Physicians*) and *Sharḥ kullīyāt al-qānūn* (*Commentary upon the Principles of the Canon*)[33]

Sometime around 690/1290 Quṭb al-Dīn settled in Tabrīz. The Ilkhānid ruler at that time was Arghūn, Aḥmad Takūdār's succesor. In 694/1295 Ghāzān Khān (d. 705/1305) came to the throne. It was during his reign that Tabrīz became one of the major intellectual centers of the Islamic world. The person largely responsible for this was his *wazīr*, Rashīd al-Dīn Faḍlallāh (d. 718/1318). The latter, a physician, historian and philosopher, was a great patron of the arts and the sciences.[34] Apparently, Ghāzān Khān himself, who was an architect and a city planner,[35] took much interest in the intellectual and artistic activities of Tabrīz.[36] Quṭb al-Dīn met important scholarly figures who gathered in this city. He was a close friend of Rashīd al-Dīn. It appears that they had developed their friendship while Rashīd al-Dīn was still the court physician of Arghūn, Ghāzān Khān's predecessor. We infer this from the fact that Rashīd al-Dīn wrote to Quṭb al-Dīn from Sind informing the latter of the success of his political and medical mission to India ordered by Arghūn.[37]

Quṭb al-Dīn remained in Tabrīz for the rest of his life until his death on 17 Ramaḍān 710/February 1311. He spent the last fourteen years mostly in seclusion and devoted to writing. Toward the end of his life he ardently studied the *Ḥadīth* and wrote a few works on the subject.[38]

10.2 Quṭb al-Dīn's Works

Like al-Fārābī and al-Ghazzālī, Quṭb al-Dīn had a universal interest in nearly every branch of the sciences and the arts as well as in philosophy and theology. He wrote numerous works on medicine, geometry, optics, astronomy, geography, the sciences of language, philosophy, and religious sciences including commentaries upon the Qur'ān. On the basis of the traditional bibliographies, it is established that about fifty treatises have been attributed to Quṭb al-Dīn including a few poems which he wrote occasionally in both Arabic and Persian.[39] However, in the field of science and philosophy he has several encyclopedic works to his credit. In the centuries that followed these became among the most widely read works in Islam. They are also indispensable sources for the understanding of the later development of Islamic science and philosophy.

I will attempt to categorize Quṭb al-Dīn's writings according to the classification of the sciences which he has given in the *Durrat al-tāj* and which will be discussed in the next chapter. The above manner of classifying his works enables us to see whether the prominent position accorded to certain categories of the sciences in his classification is reflected in the number of works written on them. In the *Durrat al-tāj*, Quṭb al-Dīn divides knowledge (*'ilm*) into two kinds, the philosophical (*al-ḥikmī*) and the non-philosophical (*ghayr al-ḥikmī*).[40] The second category is in turn divided into the religious (*al-dīnīy*) and the non-religious (*ghayr al-dīnīy*). As will be shown in the next chapter, the philosophical sciences are viewed by him as superior to the non-philosophical sciences.

10.2.1 **Philosophical Works**

Let us first consider Quṭb al-Dīn's writings on the philosophical sciences by which he means metaphysics, mathematics, natural philosophy, and logic and their respective branches, which comprise theoretical philosophy (*ḥikmat-i naẓarīy*); to these he adds politics, economics, and ethics, which constitute practical philosophy (*ḥikmat-i 'amalīy*). Nearly two-thirds of the treatises attributed to Quṭb al-Dīn are devoted to this category of knowledge.[41] The greatest share of his philosophical works goes to mathematics, especially astronomy. He wrote nearly a dozen treatises on astronomy, most of which belong to the category of independent works rather than to commentaries. Historians of Islamic science generally agree that the two greatest astronomical works by Quṭb al-Dīn are the *Nihāyat al-idrāk fī dirāyat al-aflāk* (*The Limit of Understanding of the Knowledge of the Heavens*) and *al-Tuḥfat al-shāhīyah fi'l-hay'ah* (*The Royal Gift on Astronomy*). Referring to them, Wiedemann says:

> Ḳuṭb al-Dīn has given what is conceivably the best Arabic account of astronomy (cosmography) with mathematical aids. It closely follows the *al-Tadhkirat al-naṣiriyah*, the memorandum of Naṣīr al-Dīn al-Ṭūsī, his teacher. But Ḳuṭb al-Dīn's works are very much fuller and deal with many questions which Nasīr al-Dīn did not touch; they are therefore much more than commentaries.[42]

The *Nihāyah* consists of four books (introduction, the heavens, the earth, and the "quantity" of the heavens). It treats such subjects as cosmography, geography, geodesy, meteorology, mechanics, and optics. It contains detailed discussions of astronomical theories of Quṭb al-Dīn's well-known predecessors like Ibn al-Haytham (d.430/1309) and al-Bīrūnī (d. 442/1051). It also contains new scientific theories in optics and planetary motion. In fact, the *Nihāyah's* major contribution to the history of astronomy is in its new treatment of the problem of planetary motion.[43] In this work as well as in *al-Tuḥfat al-Shāhīyah*[44] Quṭb al-Dīn makes successive attempts to arrive at a more satisfactory planetary model for all the planets.[45] He succeeded in constructing a special model for the planet Mercury, which is the most irregular of the five planets visible to the naked eye, as well as a standard model for the other planets.

According to E.S. Kennedy, Quṭb al-Dīn's model for Mercury probably marks the apex of the techniques developed by the Marāghah school.[46] His attempt to apply these techniques to the construction of a new lunar model ended, however, in failure. It was left to the Damascene astronomer, Ibn al-Shāṭir (fl.1350), to produce a lunar model that was free of the very serious defects in the Ptolemaic one. That model is identical with that of Copernicus (fl. 1520).

The two above-mentioned works by Quṭb al-Dīn earn him an important place in the history of astronomy. They influenced the later development of astronomy in the West.[47]

Among important astronomical works by Quṭb al-Dīn we may also mention the *Ikhtiyārāt-i muẓaffarī* (*Muẓaffarī Selections*), two treatises on astronomical tables, and the *Sharḥ al-tadhkirat al-naṣīrīyah* (Commentary Upon the Memorial of Naṣ*īr al-Dīn*).[48] The *Muẓaffarī Selections*, which was written sometime before 703/1304, is actually a synopsis of the *Nihāyah* but composed in Persian. Quṭb al-Dīn's two works on astronomical tables are known as *Taḥrīr al-zīj al-jadīd al-riḍwānī* (*Recension of the New Riḍwānī Astronomical Tables*) and *al-Zīj al-sulṭānī* (*The Sulṭānī Astronomical Tables*).

There are no independent and separate treatises by Quṭb al-Dīn on other branches of mathematics except geometry. Four geometrical works are attributed to him. One is a Persian translation of al-Ṭūsī's *Taḥrir uṣūl uqlīdus* (*Recension of the Elements of Euclid*). We may also mention *Risālat fī ḥarakat al-daḥrajah wa'l-nisbah bayn al-mustawī wa'l-munḥanī* (*Treatise on the Motion of Rolling and the Relation Between the Straight and the Curved*). This second treatise is one of the few works of Quṭb al-Dīn to have been analyzed thoroughly in a European language.[50]

Quṭb al-Dīn also wrote on other branches of mathematics but not in the form of separate treatises as already indicated. His treatment of music, for example, is contained in the *Durrat al-tāj* – in Book Four dealing with mathematics. The material on the subject is drawn mainly from the writings of al-Fārābī, Ibn Sīnā, and 'Abd al-Mu'min. Quṭb al-Dīn's interest in music is not merely theoretical. He was known as an excellent player of the viol (*rabābah*). This was one of his two favourite pastimes, the other being the game of chess at which he was claimed to be brilliant.[51]

Next to astronomy, the significance of Quṭb al-Dīn as a mathematician lies in optics. His discussion of optics is contained in several works. These include the *Nihāyah, Durrat al-tāj*, and the *Sharḥ ḥikmat al-ishrāq* (*Commentary Upon the Theosophy of the Orient of Light of Suhrawardī*). In order to appreciate the significant contribution of Quṭb al-Dīn to this field, it is necessary to recall that after Ibn al-Haytham (d.430/1039) there was a relative lack of interest in optics among Muslims. The optical writings of even a scientist of the stature of al-Ṭūsī were found to be much inferior to those of Ibn al-Haytham. It was Quṭb al-Dīn who, inspired by the metaphysics of light of Suhrawardī, played an important role in bringing about new vigour in optical studies in the seventh/thirteenth century.

Quṭb al-Dīn, who became acquainted with Ibn al-Haytham's *Kitāb al-manāẓir* (*Optical Thesaurus*) in the course of his travels, applied the optical knowledge contained in that work to the phenomena of the rainbow. He was the first to give a qualitatively correct explanation of the cause of the rainbow.[52] Kamāl al-Dīn al-Fārisī (d. 1320), who studied at Marāghah under Quṭb al-Dīn, seized upon his teacher's leading ideas and developed them in detail in his extensive commentary upon Ibn al-Haytham entitled *Tanqīh al-manāẓir* (*The Revision of the Optics*).[53] He carried out an experiment based upon those ideas. The experiment confirmed Quṭb al-Dīn's theory of the primary rainbow and led to the discovery of the cause of the second bow. At about the same time in the West, an independent research carried out by Theodoric of Freiberg (based also on Ibn al-Haytham's *Optics*) produced smiliar conclusions. Both Kamāl al-Dīn and Theodoric failed, however, to give a satisfactory explanation of the colors of the rainbow.[54]

The significance of Quṭb al-Dīn in optics may be summarized as follows: First, he helped to revive the optical teachings of Ibn al-Haytham. Second, he and his students, particularly Kamāl al-Dīn, laid the foundation of a separate science of the rainbow (*qaws qazaḥ*). This science appeared in the classification of the sciences composed in the century of Quṭb al-Dīn's death.[55] Through his *Sharḥ ḥikmat al-ishrāq*, Quṭb al-Dīn also helped to popularize the *ishrāqī* theory of vision propounded by Suhrawardī.[56]

Next to mathematics, the philosophical science on which Quṭb al-Dīn wrote most was metaphysics. There are about half a dozen metaphysical treatises, including the *Durrat al-tāj*,[57] attributed to him. Most of them are in the form of commentaries. The most important of these is the *Sharḥ ḥikmat al-ishrāq*.[58] In the Islamic world,it is the most popular commentary upon Suhrawardī's *Ḥikmat al-ishrāq* (*The Theosophy of the Orient of Light*), eclipsing even the commentary of Shahrazūrī, Suhrawardī's disciple and first commentator. It has been used as a standard text of *Ishrāqī* (illuminationist) philosophy in the traditional schools of Persia and India until today.[59] Since Quṭb al-Dīn's commentary has been accepted through the centuries as the most authentic interpretation of the doctrines of Suhrawardī, it is possible to speak of the former as the most eminent representative of the *Ishrāqī* school after the founder himself.

Historically, Quṭb al-Dīn's link with the *Ishrāqī* school is still a mystery. It is not known under whom he received instruction in *Ishrāqī* doctrines. According to certain oral traditions,[60] Quṭb al-Dīn is believed to have met one or more disciples of Suhrawardī who were known to have gone underground to spread their master's teaching after he was put to death.[61] These sources further maintain that before Quṭb al-Dīn left Marāghah he had already been exposed to Suhrawardī's teaching. If this claim is true, then, very likely, it was Quṭb al-Dīn himself who introduced al-Ṭūsī to Suhrawardī.[62]

Quṭb al-Dīn's identification with *Ishrāqī* philosophy is of significance for our study. I will attempt to show in the next chapter that certain features of his classification of the sciences are directly influenced by *Ishrāqī* teaching.

The other commentaries of Quṭb al-Dīn, which we have categorized as metaphysical works, include the following: (1) *al-Sharḥ wa'l-ḥāshiyat 'ala'l-ishārāt wa'l-tanbīhāt (Commentary and Glosses Upon the Ishārāt)*, (2) *Sharḥ al-najāt (Commentary upon the Najāt)*, (3) *Sharḥ kitāb rawḍat al-nāẓir (Commentary upon the "Rawḍat al-nāẓir")* and (4) *Ḥāshiyat 'alā ḥikmat al-'ayn (Glosses Upon the "Ḥikmat al-'ayn")*. The first two treatises are important commentaries upon two of Ibn Sīnā's philosophical works, namely the already mentioned *Ishārāt* (which Quṭb al-Dīn read with al-Ṭūsī) and the *Najāt* (*Book of Deliverance*). These philosophical works of Ibn Sīnā deal in great detail not only with metaphysical sciences but also with natural philosophy. Quṭb al-Dīn's commentaries upon them may also then be regarded as works on natural philosophy. One of the main features of these commentaries is that they seek to portray Ibn Sīnā's philosophy mainly through the Suhrawardian interpretation – that is, on the basis of the *Ishrāqī* teaching.

The *Sharḥ kitāb rawḍat al-nāẓir* is Quṭb al-Dīn's commentary upon al-Ṭūsī's *Rawḍat al-nāẓir*. It deals mainly with questions of ontology. The *Glosses Upon the "Ḥikmat al-'ayn"* is the first of many commentaries upon the well-known *Ḥikmat al-'ayn* of Najm al-Dīn Dabīrān al-Kātibī al-Qazwīnī.

Properly speaking, the *Durrat al-tāj* should have been placed under a different category, namely the category of encylopedic works since it deals with various branches of knowledge including strictly religious sciences. It is essentially a philosophical and scientific work. As a philosophical work it is predominantly metaphysical since its major theme deals with the excellence of wisdom (*ḥikmah*). Written on the model of Ibn Sīnā's *Kitāb al-Shifā'*, it is generally considered the outstanding Persian encyclopedia of Peripatetic philosophy.[63] It is the most important source for our study of Quṭb al-Dīn's classification of the sciences. The classification is given in the introduction of that work together with a detailed discussion of the meaning and significance of knowledge and wisdom.

Apart from the introduction, the encyclopedia consists of five books dealing with logic, metaphysics, natural philosophy, mathematics and theodicy, as well as a four-part conclusion on religion and mysticism.[64] The materials for each book are drawn from the works of various authors. Viewed as a metaphysical treatise, it was mainly influenced by the writings of Ibn Sīnā and Suhrawardī. Quṭb al-Dīn's link with the school of *'irfān* (gnosis) of Ibn 'Arabī is reflected in his treatment of Sufism. His main source in this case is the *Manāhij al-ibād ila'l-ma'ād* of Sa'd al-Dīn al-Farghānī, a disciple of Rūmī and al-Qunawī.

Another major category of Quṭb al-Dīn's philosopical writings deals with medicine which he considers a branch of natural philosophy. His most voluminous work belongs to this category. This is the five-volume *Sharḥ kullīyāt al-qānūn* already mentioned. The above work came to be regarded by many students of Islamic medicine as the most thorough and profound commentary upon the *Canon*.[65] The other medical works of Quṭb al-Dīn comprise a treatise on leprosy (*Risālat fi'l-baraṣ*), a commentary upon Ibn Sīnā's '*Canticum*' (*Sharḥ al-urjūzah*), and a treatise on medical ethics.[66] The whole corpus of his medical works played an important role in the propagation of Ibn Sīnā's medical teachings not only in Persia but also in the Indian subcontinent especially from the ninth/fifteenth century onward.

Almost all of Quṭb al-Dīn's philosophical works deal with what he calls theoretical philosophy. So far as is known he wrote only one treatise on practical philosophy. This is the *Risālat dar 'ilm-i akhlāq* (*Treatise on Ethics*) which he wrote in Persian and which is apparently lost. But judging from his treatment of ethics in the *Durrat al-tāj* it seems that in this domain too he was greatly influenced by the writings of Ibn Sīnā.

10.2.2 Religious Works

In the religious sciences,[67] Quṭb al-Dīn is credited with over a dozen treatises. His most important works in this domain deal with the science of Qur'anic interpretation. According to Ḥājjī Khalīfah, an eleventh/seventeenth century bibliographer, Quṭb al-Dīn wrote a voluminous commentary upon the Qur'ān,[68] entitled *Fatḥ al-mannān fī tafsīr al-Qur'ān* (*The Triumph of the All Bounteous in Commentary on the Qur'ān*). Quṭb al-Dīn also wrote the *Fī mushkilāt al-tafāsīr* (*On the Difficulties in Interpretations*) in which he sought to show that, correctly interpreted, no Quranic passage contradicts another.

Another treatise by Quṭb al-Dīn on Qur'anic interpretation is his commentary upon al-Zamakhsharī's *al-Kashshāf 'an ḥaqā'iq al-tanzīl* (*The Unveiler of the Mysteries of Revelation*). The next important category of his religious works is concerned with the science of traditions. His works in this domain include the *Jam' al-uṣūl* (*Encyclopedia of Principles*) and the *Sharḥ al-Sunnah* (*Commentary upon the Traditions*).

The following religious treatises by Quṭb al-Dīn also deserve to be mentioned:

(1) the *Sharḥ al-mukhtaṣar (Commentary on the Abridgement)*
This is a commentary on the abridged version of Ibn Ḥājib's (d.646/1249) *Muntahā al-su'āl wa'l-amal fī 'ilmay al-uṣūl wa'l-jadal (All That Can Be Asked or Hoped Concerning the Twin Sciences of the Principles of Jurisprudence and Dialectics)*. The *Muntahā* was a famous epitome of Mālikī law (though Quṭb al-Dīn himself was a Shāfi'ī).

(2) *Intikhāb-i Sulaymānīyah (Solomon's Choice)*
This is an abridged Persian translation of al-Ghazzālī's *Iḥyā'* made for Maḥmūd Bey, the son of the Amīr Muẓaffar al-Dīn to whom Quṭb al-Dīn dedicated his *Ikhtiyārāt-i muẓaffarī (Muẓaffarī Selections)*.

(3) *Shajarat al-īmān (The Tree of Faith)*
This work, also known as *'Ahd-nāmah (Book of Will and Testament)*, deals with supererogatory religious observances.

(4) *Risālah fī taḥqīq al-jabr wa'l-qadar (Treatise on the Reality of Fate and Predestination)*
Quṭb al-Dīn is known to have written a few treatises on the sciences of language. These too belong to the category of religious works since, like al-Ghazzālī, he classifies the sciences of language under religious knowledge. As an example one may mention his *Miftāḥ al-miftāḥ (The Key to the Key)*, the earliest commentary on the popular textbook of Arabic grammar and style, *Miftāḥ al-'ulūm (The Key to the Sciences)* of Sirāj al-Dīn al-Sakkākī (d.626/1228-29)[69]

10.3 General Significance of His Works

The foregoing survey of Quṭb al-Dīn's writings shows that his primary intellectual engagement was in the field of philosophical sciences. Thus the philosophical sciences enjoy a dominant position not only in his classification of knowledge but also in his intellectual life – as clearly demonstrated both by his lifelong education and the body of works he produced. But he was also a respected scholar of the religious sciences. On account of the unusual breadth of his knowledge in his time, he was conferred various honorific titles. The eighth/fourteenth century historian, Abu'l Fidā', gave him the title of *al-Mutafannin* (Master in many sciences)

The spread of Quṭb al-Dīn's intellectual influence was made possible both through his students and his writings. He is known to have numerous outstanding students.[70] One of them, Kamāl al-Dīn al-Fārisī, has been mentioned in connection with the rainbow theory. One may also mention Quṭb al-Dīn al-Rāzī, the author of the *Muḥākamat* (*The Critical Decisions*). In this work, al-Rāzī conducts a "trial" of the relative merits of the commentaries of Naṣīr al-Dīn al-Ṭūsī and Fakhr al-Dīn Rāzī upon the *Ishārāt* of Ibn Sīnā. As for Quṭb al-Dīn's writings, the most influential were the *Sharḥ kullīyāt al-qānūn* (in medicine), the *Nihāyah* (in astronomy) and *Sharḥ ḥikmat al-ishrāq* (in metaphysics and general philosophy). The general influence of his writings has been summarized by Nasr:

> His writings were also one of the influential intellectual elements that made possible the Safavid renaissance in philosophy and the sciences in Persia, and his name continued to be respected and his works studied in the Ottoman and the Mogul empires.[71]

ENDNOTES

Chapter 10

[1]Except for the various articles written by E. Wiedemann and S.H. Nasr, there is no significant treatment yet of the life and works of Quṭb al-Dīn in any European language. Nasr's article is the most recent and also the most comprehensive. See his "Quṭb al-Dīn al-Shīrāzī," *DSB*, XI, 247-53. For Wiedemann's articles, see "Zu den optischen Kenntnissen von Qutb al-Dīn al-Schirazi" in *Archv fur die Geschichte der Narturwissenschaften und der Technik*, 3(1912), 187-93; "Ueber dei Gestalt, Lage und Bewegung der Erde Sowie philsophisch-astronomische Betrachtungen von Quṭb al-Dīn al-Schirazi," *ibid*, 395-422; "Ueber eine Schrift ueber die Bewegung des Röllens und die Beziehung Zwishen dem Geraden und den Gekruemmten, von Quṭb al-Dīn Mahmūd b. Mas'ūd al-Schirazi, in *Sitzungsberichte der Physikalisch-medizimischen Sozietat in Erlangen*, 58-59 (1926-1927), 219-224; and *Encyclopedia of Islam*, 2nd ed., V, 547-8. For a more specific reference to Quṭb al-Dīn as a scientist, see G.Sarton, *An Introduction to the History of Science*, Baltimore, 1941, II, 1017-20; also S.H.Nasr, *Science and Civilization in Islam*, p.56.

The most reliable source for the biography of Quṭb al-Dīn is his own work *Sharh kullīyāt al-qānūn (Commentary upon the Principles of the "Canon of Ibn Sīnā")*. The original text of this important work has not yet been published. However, the substance of the autobiography is contained in L. Leclerc, *Histoire de la médicine arabe*, Paris, 1876, II, 129-30. Among the traditional biographies of Qutb al-Dīn, we may mention those of Ibn Shuhbah (*Ṭabaqāt al-Shāfi'iyah*), al-Subkī (*Ṭabaqāt al-Shāfi'iyah al-kubrā*, VI, 248); al-Suyūṭī (*Bughyat al-wu'āt*); and Abu'l-Fidā', *Annales Moslemici*, ed. J.J. Reiske, Copenhagen, 1794, V,63,243.

The best modern account of Quṭb al-Dīn's life and works is the Persian article by M. Minovi, "Mullā Quṭb Shīrāzī," in *Yād-nāma-ye īrānī-ye Minorsky*, Teheran, 1969, pp. 165-205. Another important account, also in Persian, is given by S.M. Mishkāt in the introduction to his edition of Quṭb al-Dīn's *Durrat al-tāj li ghurrat al-dībāj fi'l-hikma (Pearls of the Crown, the Best Introduction to Wisdom)*, Teheran, 1938-41, pt. I, Vol. I. (This work will be cited as *Durrat al-tāj*).

[2]This Suhrawardī is the author of the famous Sufi treatrise, *'Awārif al-ma'ārif* and the nephew of the founder of the Suhrawardīyah Order. He came from the same place as Suhrawardī, the founder of the illuminationist (*Ishrāqī*) school of Islamic philosophy, and whose greatest commentator was Quṭb al-Dīn himself. On the former Suhrawardī, see A. Schimmel, *Mystical Dimension of Islam*, p.245.

[3]See E. Wiedemann, *Encyclopedia of Islam*, p. 547

[4]There are two kinds of *khirqah*: the *khirqah irādah* which the aspirant gets from the master to whom he has sworn allegiance and who is responsible for his progress; and the *khirqah tabarruk* which he may obtain from different masters with whom he has lived or whom he has visited during his journeys – if the master considered him worthy of receiving some of his blessings. See A. Schimmel, *op. cit*; p. 102. Quṭb al-Dīn was said to have received the second kind of *khirqah* from a number of Sufi masters including his father and a certain Najīb al-Dīn 'Alī al-Shīrāzī. See *Durrat al-tāj*, p. ج

[5]*ibid*, p. ج

[6]According to traditional sources, he was a Shāfi'ite scholar who taught at the Niẓāmīyah school in Baghdad. See *al-Ḥawādith al-jāmi'ah*, Baghdad (1251 A.H), quoted by S.M. Mishkāt (ed.) *Durrat al-tāj*, p. ج

[7]We infer this from the fact that al-Kabshī was a Shāfi'ite professor at the Niẓāmīyah.

[8]Al-Suyuṭī, *op. cit.*, has al-Rukshawī instead of al-Bushkānī; E. Wiedemann (*op. cit.*, p. 547) has al-Rushkānī.

[9]Apart from the *Canon*, the works of Ibn Sīnā upon which Rāzī wrote commentaries are *Ishārāt wa'l-tanbīhāt* and *'Uyūn al-ḥikmah*. See S.H. Nasr, "Fakhr al-Dīn Rāzī," *A History of Muslim Philosophy*, I, p. 644. Rāzī's *Mabāhith al-mashriqīyah*, being a commentary on Peripatetic philosophy in general, may also be regarded as a commentary on Ibn Sīnā inasmuch as the latter is the leading representative of that philosophy.

[10]For a detailed description of the Marāghah observatory, see A. Sayili, *The Observatory in Islam*, Ankara, 1960.

[11]It was al-Ṭūsī himself who was said to have influenced Hülagü to establish the Marāghah observatory.

[12]On the significance of al-Ṭūsī as the most important successor to Ibn Sīnā in the whole domain of the arts and sciences and philosophy, see S.H. Nasr, *Science and Civilization in Islam*, pp. 54-56, 321-22; see also S.M. Afnān, *Avicenna*, London, 1958, p. 244.

[13]See E.G. Browne, *Literary History of Persia*. III, 18.

[14]S.H. Nasr, *op. cit.*, p.81.

[15]B. Husain Siddiqi, "Naṣīr al-Dīn Ṭūsī," *A History of Muslim Philosophy*, I, p. 565.

[16]E.G. Brown, *op. cit.*, II, 485.

[17]See E.S. Kennedy, "The Exact Sciences in Iran under the Seljūqs and Mongols," in *The Cambridge History of Iran*, V, 670; also O. Neugebauer, "Studies in Byzantine Astronomical Terminology," *Trans. American Philosophical Society*, L(1960), pt. 2, p. 28

[18]S.H. Nasr, "Quṭb al-Dīn al-Shīrāzī," *op. cit.*, p.249.
[19]*Durrat al-tāj*, p. ژ
[20]*Ibid*, p. س-ژ
[21]The *Ilkhānid Zīj*, originally composed in Persian and then translated into Arabic, was completed in 670/1274. Although Quṭb al-Dīn played a major role in the observations made at Marāghah which led to its composition, his name was not mentioned in its introduction by al-Ṭūsī (see below).
[22]S.H. Nasr, *op. cit.*, p. 248.
[23]*Durrat al-tāj*, p. س . It is very unlikely, however that the work was written during his stay in Isfahan since it was completed around 680/1281.
[24]*Ibid*, p. ش
[25]'A. Jāmī, *Nafahāt al-uns*, ed. M.Tauhīdīpūr, Tehran, 1336/1957, quoted by A. Schimmel, *op. cit.*, p. 313.
[26]On the details of this correspondence, see W.C. Chittick, "Mysticism vs. Philosophy in Earlier Islamic History: The al-Ṭūsī, al-Qunawī Correspondence," *Religious Studies*, 17(1981), 82-104.
[27]A Schimmel, *op. cit.*, p. 312. Konya's literary language was Persian although its population spoke partly Greek – there being a strong Christian substratum in the former city of Iconium – and partly Turkish.
[28]E. Wiedemann, *op. cit.*, p. 547. (Cf. *The Cambridge History of Iran*, ed. J. A. Boyle, 1968, V, 365).
[29]In 659/1260, not long after Quṭb al-Dīn left his native city Shīrāz, the Mamlūks of Egypt defeated the Mongols at 'Ain Jālūt (Goliath's Spring) near Nazareth. This Mamlūk victory stemmed the tide of Mongol conquest of the Islamic world. See, for example, A.S. Ahsan, "The Fall of Baghdad," in *A History of Muslim Philosophy*, II, 790-91. Aḥmad Takūdār's attempt to establish friendly relations with the Sulṭān of Egypt was opposed by his fellow princes, who favored resumption of hostilities with the Mamlūks (*The Cambridge History of Iran*, V, 365).
[30]For a general study of the medical life and works of Ibn al-Nafīs and his significance in the history of medicine, see A.Z. Iskandar, "Ibn al-Nafīs," *DSB*, pp. 602-6; also M. Meyerhof, "Ibn an-Nafīs (XIIIth Cen.) and his theory of the lesser circulation," in *Isis*, 1935, 23, 100-20. On Ibn al-Quff, see G. Sobhy, "Ibn'l Kuff, an Arabian Surgeon of the VII Century Alhigra," in *Journal of the Egyptian Medical Association*, 1937, 20, 349-57.
[31]On these commentaries and other medical works of Ibn al-Nafīs, see M. Meyerhof, *op cit.*, pp. 112-15; also S.H. Nasr, *Islamic Science*, p.181
[32]S.H. Nasr, "Quṭb al-Dīn al-Shīrāzī," *op. cit.*, p. 249.
[33]*Durrat al-tāj*, p. ص
[34]On Rashīd al-Dīn (the author of the first world history, *Jāmi' al-tawārīkh*) and his cultural significance, see, for example, *The Cambridge History of Iran*, vol. V, where references to him are made in a number of chapters. See also E.G. Browne, *Arabian Medicine*, Westport, Connecticut, 1983, Hyperion rep. ed., pp. 103-9.
[35]He was said to have designed and built Shenb, a suburb west of Tabrīz in 696/1297. On his architectural achievements, see D.N. Wilber, *The Architecture of Islamic Iran: Īl-Khānid Period*, Princeton, 1955, p. 17.
[36]See S.S. Hassan and M.Á. Chaghatai, "Muslim Architecture in Later Centuries," *A History of Muslim Philosophy*, II, 1090.
[37]See E.G. Browne, *op. cit.*, pp. 105-6.
[38]E. Wiedemann, *op. cit.*, p. 547.
[39]The most complete list and also the best account of the works of Quṭb al-Dīn to have been given so far are those of J.T. Walbridge III. See his *Philosophy of Quṭb al-Dīn*

Shīrāzī: A Study in the Integration of Islamic Philosophy, PhD thesis, Harvard University (May, 1983), appendix C, pp. 235-276. Walbridge's account contains useful information concerning the manuscripts evidence of the works attributed to Quṭb al-Dīn. See also M. Minovi, *op. cit.*; *Durrat al-tāj*, pp. اج-م ; and S.H. Nasr, *op. cit.*, pp. 248-49.

[40]*Durrat al-tāj,* I,71

[41]Nasr (*op. cit.*) has listed a total of twenty six treatises by Quṭb al-Dīn on the philosophical sciences.

[42]E. Wiedemann, *op. cit.*

[43]This new treatment of the problem of planetary motion was initiated by al-Ṭūsī who gave a severe criticism of the Ptolemaic planetary theory in his *Tadhkirah (Memorial of Astronomy)*. For a discussion of this new development in planetary theory, see E.S. Kennedy, "Late Medieval Planetary Theory," in *Isis* 1966, 57, 365-78; also his "The Exact Sciences in Iran under the Seljūqs and Mongols" in *The Cambridge History of Iran*, V, 668-70. See also S.H Nasr, *Islamic Science*, p. 109.

[44]This work was composed in 683/1284 shortly after the *Nihāyah* (around 680/1281). Both works were written on the model of al-Ṭūsī's *Tadhkirah.*

[45]E.S. Kennedy, *op. cit.*, p. 669; also S.H. Nasr, "Quṭb al-Dīn al-Shīrāzī," *op. cit.*, p. 251.

[46]E.S. Kennedy, *op. cit*

[47]*Ibid*, p. 670. For description of the Marāghah and Ibn al-Shāṭir models, see the various articles of Kennedy in *Isis*: 1957, 48, 428-32; 1959, 50, 227-35; 1962, 55, 492-9; and 1966, 57, 208-19.

[48]The *Sharḥ al-tadhkirah* also contains a commentary upon the *Bayān maqāṣid al-tadhkirah* of Muḥammad ibn 'Alī al-Himādhī.

[49]*Al-Zīj al-sulṭānī* has also been attributed to Muhammad ibn Mubārak Shams al-Dīn Mīrak al-Bukhārī. See S.H. Nasr, *op. cit.*, p. 249. On the two astronomical tables of Quṭb al-Dīn, see E.S. Kennedy, *A Survey of Islamic Astronomical Tables*, Philadelphia, 1956.

[50]It was analyzed by E. Wiedemann in his "Ueber eine Schrift ueber die Bewegung des Rollens und die Beziehung zwischen dem Geraden und dem Gekruemmten von Quṭb al-Dīn Maḥmūd b. Mas'ūd al-Schirazi," in *Sitzungsberichte der Physikalisch-medizinischen Sozietat in Erlangen*, 1926-1927, 58-59, 219-24.

[51]E. Wiedemann, "Kuṭb al-Dīn Shīrāzī," *op. cit.*, p. 547.

[52]Quṭb al-Dīn put forward the theory that the rainbow is produced by the behavior of rays of sunlight falling upon spherical droplets of water and that this behavior is a combination of refractions and reflections after the ray has entered the drop. The *primary bow* was explained by him to be due to two refractions and one internal reflection, and the *secondary bow* to two refractions and two internal reflections of solar rays in minute spherical drops of water suspended in the air. See M. Abdur Rahman Khan, "Physics and Mineralogy," in *A History of Muslim Philosophy*, II, 1295; S.H. Nasr, *Islamic Science*, p. 142; also E.S. Kennedy, "The Exact Sciences in Iran under the Seljūqs and Mongols," *op. cit.*, p.675.

[53]E.S. Kennedy, *op. cit.*, p. 675. The *Tanqīh al-manāzir* has been published in two volumes in Hyderabad-Dn. (1347-8 A.H). On Kamāl al-Dīn's contribution to optics, see E. Wiedemaann, "Zur Optik von Kamal al-Din, "*Archiv fur die Gesh der Naturwissenschaften...*, III (1910-12), 161-77.

[54]See C.B. Boyer, *The Rainbow from Myth to Mathematics*, New York – London, 1959, chap. V.

[55]S.H. Nasr, "Quṭb al-Dīn al-Shīrāzī'," *op. cit.*, p. 250

[56]According to the *Ishrāqī* theory, one can only have vision of a lighted object. What

happens in an act of vision is that the soul of the observer surrounds the object and is illuminated by its light. This act of illumination of the soul (*nafs*) in presence of the object is what the *Ishrāqis* called vision. Hence for them, physical vision partakes of the illuminative nature of all knowledge. See S.H. Nasr, *Three Muslim Sages*, p. 68; also his "Shihāb al-Dīn Suhrawardī Maqtūl," *A History of Muslim Philosophy*, I, 386-7.

[57]The *Durrat al-tāj* is one of the only two works of Quṭb al-Dīn which have been printed. The other is *Sharḥ ḥikmat al-ishrāq* (see below). One of the five books of the *Durrat al-tāj* deals with metaphysics. But this fact alone does not enable us to regard it as a metaphysical treatise. See our explanation below for its inclusion among the metaphysical treatises.

[58]A major study of this work has been carried out recently. See J.T. Walbridge, *op. cit.*

[59]*Ishrāqī* philosophy is also taught in the Faculty of religious science or "Theology" of Tehran University. There is, in fact, a special chair for *Ishrāqī* philosophy at that University. See S.H. Nasr, *Three Muslim Sages*, p. 81.

[60]I am indebted to Prof. Nasr for the following views concerning the association of Quṭb al-Dīn with the *Ishrāqī* school.

[61]Suhrawardī died in 587/1191 during his imprisonment by Ṣalāḥ al-Dīn al-Ayyūbī (the famous Saladin). The latter was said to have yielded to the demand of the doctors of the law for the execution of Suhrawardī on the grounds of propagating doctrines against the tenets of the faith. See S.H. Nasr, *Three Muslim Sages* p. 57.

[62]At Marāghah, Quṭb al-Dīn was not just a student of al-Ṭūsī. He was an outstanding scholar in his own right. There was mutual influence during their long association.

[63]See S.H. Nasr, "Quṭb al-Dīn al-Shīrāzī," *op. cit.*, p. 249.

[64]The introduction and the books on logic, metaphysics, and theodicy were published by S.M. Mishkāt, Tehran (1938-41); S.H. Tabasi, Tehran (1938-44), has published the book on mathematics excluding certain portions on geometry. See *Ibid.*

[65]See S.H. Nasr, *Islamic Science*, p. 182.

[66]This work on medical ethics is entitled *Risālat fī bayān al-ḥājat ila'l-ṭibb wa ādāb al-aṭibbā' wa waṣāyāhum (Treatise on the Explanation of the Necessity of Medicine and of the Manners and Duties of Physicians)*.

[67]As will be shown in the next chapter, the religious sciences understood by Quṭb al-Dīn are in the main identical with those of al-Ghazzālī. However, the two thinkers differ regarding the place of religious sciences within the total classification of knowledge.

[68]According to Ḥājjī Khalīfah, the commentary is a forty-volume work. Twenty-nine volumes are known to survive in the Esad Efendi Collection at Suleymaniye Library, Istanbul. See J.T Walbridge, *op. cit.*, p. 268. See also *Durrat al-tāj*, p. ات

[69]Quṭb al-Dīn dedicated the commentary to the poet Humām al-Dīn Tabrīzī (d. 714/1314-15), a student of al-Ṭūsī. Humām al-Dīn was responsible for collecting the book of panegyrics presented to Quṭb al-Dīn a few years before his death. See J.T. Walbridge, *op. cit.*, p. 272.

[70]For a list of well-known students of Quṭb al-Dīn, see *ibid*, pp. 231-34.

[71]S.H. Nasr, "Quṭb al-Dīn al-Shīrāzī," *op. cit.*, p. 253.

CHAPTER 11

QUṬB AL-DĪN'S CLASSIFICATION OF THE SCIENCES

In the *Durrat al-tāj*, Quṭb al-Dīn presents the following classification of the sciences:[1]

A. The philosophical sciences (*'ulūm ḥikmīy*)

These are divided into the theoretical (*naẓarīy*) and the practical (*'amalīy*):

(1) Theoretical philosophical sciences consist of
metaphysics
mathematics
natural philosophy
logic

(2) Practical philosophical sciences constist of
ethics
economis
politics

B. The non-philosophical sciences (*'ulūm ghayr ḥikmīy*)

These are termed the religious (*dīnīy*) sciences if they are according to, or fall under the teachings of the *Sharī'ah* (revealed law). Otherwise, they are called the non-religious (*ghayr dīnīy*) sciences.

Religious sciences may be classified in two different ways:

(1) Classification into transmitted (*naqlīy*) sciences and intellectual (*'aqlīy*) sciences

(2) Classification into the sciences of fundamentals (*uṣūl*) and the science of branches (*furū'*)

11.1 "Ḥikmat" as the Basis of Classification

A key concept in Quṭb al-Dīn's classification is *ḥikmat* (philosophy). The distinction between *ḥikmat* and non-*ḥikmat* forms of knowledge constitutes the most fundamental basis of his classification. Some exposition of Quṭb al-Dīn's views concerning *ḥikmat* is therefore necessary if we are to understand the philosophical foundation of his classification.

According to Quṭb al-Dīn, it is a view held by all Muslims that *ḥikmat* (wisdom) constitutes the highest and most noble form of knowledge. In the *Durrat al-tāj*, he cites numerous Qur'anic verses to show that Muslim belief in the pre-eminence of *ḥikmat* finds explicit and strong support in Islamic revelation.[2] We do know, however, that what precisely constitutes *ḥikmat* and what distinguishes it from non-*ḥikmat* is a matter of contention among Muslims. There are differences of opinions among Muslim scholars concerning this question because neither the Qur'ān nor the prophetic ḥadīths has given an explicit answer to it. For this reason, Quṭb al-Dīn made quite clear that in his understanding of *ḥikmat* he was following the tradition of the *ahl ma'rīfah* (lit: people of true knowledge).[3] By this group, I think he means the philosophers. A number of studies have shown that the philosophers who exercised the greatest influence upon Quṭb al-Dīn's philosophical ideas were the Peripatetic Ibn Sīnā and Suhrawardī of the *Ishrāqī* (illuminationist) philosophical school.[4]

As understood in the Islamic philosophical tradition, *ḥikmat* is not the name of one particular science or discipline but rather a generic noun representing several sciences. In Quṭb al-Dīn's classification, *ḥikmat* is identified with the theoretical philosophical sciences comprising of metaphysics, mathematics, natural science, and logic and with the practical philosophical sciences comprising of ethics, economics, and politics. Quṭb al-Dīn mentions two main characteristics of *ḥikmat* which distinguish it from non-*ḥikmat*. The first concerns the timeless and universal nature of *ḥikmat*. *Ḥikmat* refers to that form of knowledge which remains one and the same for all times and cultures.[5] The second pertains to *ḥikmat*'s essentiality. As Quṭb al-Dīn defines it, *ḥikmat* is knowing things "as they really are" as well as acting truthfully and correctly to the best of one's ability so that in realizing such knowledge and action the human soul attains its perfection.[6]

Knowledge of the essences or true nature of things is what constitutes the philosophical sciences. This knowledge is timeless because essences, natures, realities or quiddities[7] of things refer to the eternal aspects of the universe. The knowledge is arrived at through a continous process of conceptions (*taṣawwur*) and assents (*taṣdīq*) in which definitions of increasing perfection interacted with the experience of concrete things. In philosophical knowledge complete definitions[8] are produced which signify the essences of things in question and the movement from known to unknown objects of assent is by means of scientific proof which is certain in nature.[9] What is central to the philosophicl sciences is the knowledge of universals. What Quṭb al-Dīn meant was that the knowledge of universals constitutes the basis of the conceptual structure of our knowledge of the external world. Universals are known through the particulars. Although in external reality, says Quṭb al-Dīn, things are particular and concrete, the intellect knows their realities in terms of universals. Quṭb al-Dīn maintains the view that the universals like genus, species, difference and accidents possess an objective reality of their own. Universals are not arbitrary constructions in the human mind but correlates of the intelligibility of the external world.[10]

11.2 Divisions of "Ḥikmat"

Since *ḥikmat* is the knowledge of things as they really are, its division into the diffferent branches reflects the division of all things or existents. Existents are of two kinds: the first is that whose existence is not contingent upon the volitions of man; in the second kind the existence of a being is contigent upon man's volitions. Sciences of the former are called theoretical philosophy (*ḥikmat naẓarīy*); those of the latter practical philosophy (*ḥikmat 'amalīy*).

11.2.1 Theoretical Philosophy and Its Divisions

Quṭb al-Dīn classifies existents which are objects of theoretical philosophy according to the degrees of their immateriality or separation from matter. There are three classes of existents:[11]

(1) Those which are completely separate from matter such as God, intellects (*'uqūl*) and souls (*nufūs*)

(2) Those which cannot exist except in association with matter but which can be known without reference to matter. Examples are numbers and geometrical entities like squares, triangles, spheres and circles.

(3) Those which are not separate from matter and which cannot be known except in association with matter. These are the natural substances: minerals, plants, and animals The above three classes of existents give rise to the three fundamental divisions of theoretical philosophy, namely, metaphysics (*'ilm mā ba'd al-ṭabī'ah*), mathematics (*'ilm riyāḍī*), and natural science (*'ilm ṭabī'ī*) respectively. This was, in fact, al-Fārābī's threefold classification of theoretical philosophy. Like al-Fārābī, Quṭb al-Dīn maintains the view that of the three divisions, metaphysics occupies the highest position, mathematics the intermediate, and natural science the lowest.[12] In their subdivision of ech of these sciences, the two thinkers differ. While al-Fārābī simply spoke of their respective parts, Qutb al-Dīn went to distinguish between two kinds of parts, namely the major and the minor, a division which he also applied to the religious sciences. The major parts are comprised of knowledge of the principles (*uṣūl*) or roots of the science in question and the minor parts knowledge of its branches.[13]

11.2.1.1 **Metaphysics**

Quṭb al-Dīn's metaphysics consists of two major parts, that is divine science (*'ilm-i ilāhī*)[14] and first philosophy (*falsafah-i ūlā*)[15], and at least three minor parts. The three minor parts mentioned by him are the science of prophethood (*nubuwwah*), the science of religious authority (*imāmah*), and eschatalogy. His divine science and first philosophy are identical to the whole (i.e., the three parts) of al-Fārābī's metaphysics in respect of subject matter although there is a defference in terminological usage.[16] The inclusion of the science of prophethood, the science of religious authority, and eschatology as branches of metaphysics was a feature of al-Ghazzālī's but not al-Fārābī's classification. Unlike al-Ghazzālī, however, Quṭb al-Dīn eliminates from the domain of metaphysics theurgy (*'ilm nairanjāt*), oneiromancy (*'ilm ta'bīr*), and magic (*'ilm ṭalismāt*). Quṭb al-Dīn views these sciences as branches of natural science.

11.2.1.2 **Mathematics**

Quṭb al-Dīn divides mathematics into four major branches and nine minor ones:

A. **Major branches**

- (1) Geometry
- (2) Arithmetic
- (3) Astronomy
- (4) Music

B. **Minor branches**

- (1) Optics
- (2) Algebra
- (3) The science of weights
- (4) Surveying[7]
- (5) The science of calculation[18]
- (6) Mechanical Engineering
- (7) The science of the balance[19]
- (8) The science of astronomical tables and calendars
- (9) The science of irrigation[20]

Quṭb al-Dīn's division of mathematics differs from al-Fārābī's in several aspects. Of the nine sciences listed by Quṭb al-Dīn as minor branches, only optics, the science of weights and mechanical engineering appear in al-Fārābī's classification but these sciences are treated there on the same level with geometry, arithmetic, astronomy, and music. Another difference is that Quṭb al-Dīn completely ignores the division of each mathematical science into its theoretical and practical parts.

Apparently, Quṭb al-Dīn's concept of *minor branches* of mathematics is based upon the consideration that those branches are subdivisions of the major branches. Optics, the science of weights, surveying, mechanical engineering, the science of the balance, and the science of irrigation were considered to have branched out of geometry. Algebra and the science of calculation were subdivisions of arithmetic; the science of astronomical tables and calendars a subdivision of astronomy. The big increase in the number of branches of mathematics in Quṭb al-Dīn's classification, compared to al-Fārābī's mathematics, does not point to any significant difference in their mathematical philosophy. It merely reflects the historical development of Muslim mathematics. Historical studies of Islamic science confirm that all the nine minor branches mentioned by Quṭb al-Dīn were studied in his time as separate, independent sciences.

11.2.1.3 **Natural Science**

Quṭb al-Dīn's division of natural science may be summarized as follows:

There are eight major branches of natural science, namely, the sciences of:

(1) Natural things that are heard
(2) The nature of simple and composite bodies
(3) Generation and corruption of bodies
(4) Meteorology
(5) Mineralogy
(6) Botany
(7) Zoology
(8) Psychology

The minor branches of natural science include:

(1) Medicine
(2) Judicial astrology
(3) Agriculture
(4) Physiognomy
(5) Oneiromancy
(6) Alchemy
(7) Natural magic or science of talismans
(8) Theurgy

There are major differences between al-Fārābī's and Quṭb al-Dīn's division of natural science. Quṭb al-Dīn's natural science is more comprehensive. Except for two minor features, the eight major branches of his natural science are almost identical to the whole of al-Fārābī's natural science. Quṭb al-Dīn lists psychology as a separate science. In al-Fārābī the subject-matter of psychology was still not separated from that of zoology. The second feature pertains to Quṭb al-Dīn's reorganization of the discipline of meteorology. Al-Fārābī divides the subject-matter of Aristotle's meteorology into two parts, which he makes the fourth and the fifth branches of natural science.[21] With Quṭb al-Dīn, these two branches were combined to form a single science of meteorology. In my view, he had done so for the following reason. He defines meteorology as the science of the causes of atmospheric phenomena and related terrestrial phenomena, like thunder, lightning, rain, snow, and earthquake. The causes of these phenomena are to be sought in the knowledge of the properties of the four elements, that is, fire, air, water, and earth, in their relation to each other as simple bodies and as mixtures in compound bodies; in other words, in the knowledge of those things which al-Fārābī identifies with the subject-matter of the fourth and fifth branches of his natural science. Quṭb al-Dīn considers the two parts inseparable.

As in mathematics, the minor branches of natural sciense were so named by Quṭb al-Dīn because they were considered to have branched out of its major branches. Not a single one of these "minor branches" appears in al-Fārābī's classification. Al-Fārābī, we saw, views these sciences as practical arts or occult sciences, which fall outside the scope of his classification. In contrast, Quṭb al-Dīn's conception of *ḥikmat* enables him to accomodate these siences into his classification.

11.2.1.4 **Logic**

In accordance with his conception of *ḥikmat*, Quṭb al-Dīn treats logic as a branch of theoretical philosophy. For him, logic is a form of *ḥikmat*, because like the other parts of philosophy it deals with the natures of things. Unlike the other philosophical sciences, however, logic does not treat of the natures of things as they exist either externally, mentally, or as they are free from both modes of existence.[22] Rather, logic is concerned with these natures as they exist in the mind, but insofar as they are subjects or predicates, universals or particulars, essential or accidental and other "states" of such nature. Quṭb al-Dīn maintains that logic is not only a part of philosophy but also a tool of its other parts.

In his division of the science of logic, Quṭb al-Dīn follows the traditional Muslim Peripatetic division into the nine books of the Organon.

11.2.2 Practical Philosophy and Its Divisions

Quṭb al-Dīn accepts and defends Aristotle's threefold division of practical philosophy into ethics, economics, and politics. He maintains that these three sciences constitute independent disciplines. The nature of these disciplines is not religious but philosophical. Quṭb al-Dīn, therefore, differs from both al-Fārābī and al-Ghazzālī in defining the positions of these three sciences in the classification of knowledge.

According to Quṭb al-Dīn, practical philosophy is primarily concerned with the principles of voluntary human actions and works which are good and virtuous and by means of which human beings attain their livelihood in this world and perfection in the hereafter. Principles of good works and virtuous actions, whether individual or collective, are of two types. The first type Quṭb al-Dīn calls natural principles.[23] By this he means those principles which do not change with time and which remain unaffected by changes in ways of life or life styles. The second kind of principles is "situational" in nature. Human affairs are said to be situational when they involve changes dictated by the needs of new situations and circumstances.

Knowledge of "natural principles" of human actions is derived from observations and reflections upon human society, made by men of learning over the ages.[24] This knowledge falls under *ḥikmat* because the natural principles, besides being immutable or timeless, are universally applicable to human societies of all places and times. Quṭb al-Dīn did not consider knowledge of "situational principles" as *ḥikmat*. The reason he gives is that these principles refer either to manners and customs (*ādāb wa rusūm*) collectively agreed upon by a society or group or to the divine laws (*nawāmis ilāhī*) that originated from prophets or religious leaders. Manners and customs and divine laws are here understood as matters which have to do with particular peoples and particular times. Accordingly, Quṭb al-Dīn places the study of situational principles of human actions under the religious sciences or more precisely under *fiqh*.

The threefold division of practical philosophy was based upon a similar division of human acts. According to Quṭb al-Dīn, there are three types of human acts:

(1) Individual acts
(2) Collective acts at the level of the home or family
(3) Collective acts at the level of town, state, and country

Individual acts are acts which pertain to a single individual alone. Ethics (*tahdhīb-i akhlāq*, lit: moral refinement) is concerned with the natural principles of individual acts, namely the virtues or the states of the soul by which an individual does good works. Economics (*tadbīr-i manāzil*, lit: governance of the household) pertains to the second division of human acts. The basis of traditional economics is the principles of household association. Through this science one knows how man ought to conduct the governance of his household – which is common to him, his wife, his children, his servants, and his slaves – so as to lead to a well-ordered life that enables him to gain happiness.

Politics (*siyāsat mudun*) deals with the third division of human acts. It is concerned with the principles of political association.[25] In contrast to both al-Fārābī and Ibn Sīnā, Quṭb al-Dīn did not consider the study of prophecy and the divine Law as integral parts of political science.[26]

11.3 Non-Philosophical or Religious Sciences

In the *Durrat al-tāj*, the non-philosophical sciences are regarded as synonymous with religious sciences. According to Quṭb al-Dīn, religious sciences are either (1) *naqlī* (transmitted), (2) *'aqlī* (intellectual) or (3) both *naqlī* and *'aqlī*.[27] By *naqlī* sciences he means those sciences which could only be established through evidences that are heard or transmitted from relevant authorities. As an example, he mentions the science pertaining to acts of worship like prayer and fasting. Reason cannot establish the religious principle whereby a Muslim is required to fast on the last day of the month of Ramaḍan but is forbidden to do so the following day. Scientifically speaking, says Quṭb al-Dīn, the two consecutive days are hardly distinguishable from one another. The sayings of the Prophet alone decide on this matter.

By *'aqlī* sciences Quṭb al-Dīn means the sciences which can be established by the human intellect, regardless of whether there is *naqlī* evidence or not. For example, knowledge of the existence of God and knowledge of the reality of prophethood can be rationally demonstrated. As for the third category of religious sciences, it may be regarded as a sub-category of the second. The third category refers to those sciences which are established by both the intellect and transmitted sources.

Quṭb al-Dīn uses the above categorization of religious sciences as the basis of his division of those sciences into two parts, namely (a) the sciences of fundamental principles of religion (*'ilm-i uṣūl-i dīn*) and (b) the science of branches of religion (*furū'-i dīn*). He calls those sciences *furū'* which could not be established without *naqlī* evidence. Sciences of fundamental principles of religion are those which can be established by human reason irrespective of whether there is *naqlī* evidence or not.

11.3.1 Sciences of Fundamental Principles of Religion

Quṭb al-Dīn divides this science into four parts:

(1) Knowledge of the unique Essence of God[28]
(2) Knowledge of divine Attributes[29]
(3) Knowledge of God's works[30]
(4) Knowledge of prophethood and the divine message and wisdom associated with it[31]

11.3.2 Sciences of Branches of Religion

Quṭb al-Dīn gives a lengthy division of this science, which may be summarized as follows:

A. The science which is considered as the goal (*maqṣūd*). It is comprised of four parts:

1. The science of the Book (i.e., the Qur'ān)

This science is further divided into twelve parts, namely the sciences of:

(1) Recitation[32]
(2) *Wuqūf*[33]
(3) Qur'anic lexicography
(4) Desinential inflection (*'ilm-i i'rāb*)[34]
(5) Circumstances of the revelation of Qur'anic verses[35]
(6) The abrogating (*nāsikh*) and the abrogated (*mansūkh*)
(7) Hermeneutic interpretation[36]
(8) History[37]
(9) Deduction of meanings of the Qur'ān[38]
(10) Guidance and moral counselling
(11) Semantics (*'ilm-i ma'ānī*)[39]
(12) Syntax and style (*'ilm-i bayān*)[40]

2. The science of *ḥadīths*[41]
3. The science of the principles of jurisprudence
4. Jurisprudence

B. The science of literature

This science consists of twelve parts, namely the sciences of:

1. Idiomatic expressions
2. Word compositions
3. Etymology
4. *'ilm-i i'rāb*
5. Semantics
6. Literary criticism (*'ilm-i bayān*)
7. Prosody
8. *'ilm-i qawāfī*[43]
9. Letter writing
10. Poetry writing
11. Calligraphy
12. Discourse (*'ilm-i muḥāḍarat*)

Quṭb al-Dīn's enumeration of the sciences of branches of religion shows that there is some overlapping between the Qur'ānic sciences and the sciences of literature. Three sciences appear as branches of both the sciences of the Qur'ān and the science of literature. The three sciences are *'ilm-i i'rāb*, semantics (*'ilm-i ma'ānī*), and *'ilm-i bayān*. Quṭb al-Dīn's description of these three sciences is inadequate to explain why there is an overlapping. An explanation that may be offered concerning this question is that the Arabic language is the central element in both the science of the Qur'ān and the science of literature. The three sciences mentioned happen to be the pillars of the Arabic language. They pertain to the structure of that language, which is the language of the Qur'ān as well as the vehicle of literary expression among the Muslim peoples of Quṭb al-Dīn's time.

The science of literature, like the sciences of the Qur'ān and *ḥadīths*, is viewed as a transmitted science. Included in this science are the linguistic sciences enumerated by al-Fārābī.

ENDNOTES

Chapter 11

[1]See *Durrat al-tāj*, Pt. I, Vol. I. 71-98
[2]The verse he repeatedly quoted is the following: He (Allāh) giveth wisdom unto whom He will, and he unto whom wisdom is given, he truly hath received abundant good (*The Qur'ān*, 2:269). See *ibid*, for example, pp, 71, 73.
[3]*Ibid*, p. 72
[4]The most recent and also the most thorough of these studies is that of John Tuthill Walbridge III, *op. cit*.
[6]*Durrat al-tāj*, I, I, 72.
[6]*Ibid*.
[7]The terms *dhāt* (essence), *ṭabī'ah* (nature), *ḥaqīqah* (reality), and *māhīyah* (quiddity) appear to be used interchangeably by Quṭb al-Dīn. I have used here the terms as spelt in Arabic and not in their Persian form as used by Quṭb al-Dīn.
[8]Quṭb al-Dīn defines a complete definition as

> "an expression indicating the quiddity of the thing by correspondence (*muṭābaqah*). It is composed of the genus and difference.....The genus contains all the common essentials and the difference all distinguishing essentials if the genus and species are composite. Just as the creation of the thing outside is incomplete without the creation of all its parts, so too is its creation in the mind – i.e., its complete conception – without the creation of all the essentials in the mind."

See Walbridge, *op. cit.*, p. 139.
[9]This "certain" proof refers to the method of *burhān* (demonstration) in the same sense used by al-Fārābī and Ibn Sīnā.
[10]Walbridge, *op. cit.*, p. 139

[11]*Durrat al-tāj,* I, I, 73.
[12]*Ibid.* Quṭb al-Dīn's idea of mathematics as intermediate between metaphysics and natural science is slightly different from that of al-Fārābī. Quṭb al-Dīn's idea is based solely on ontological consideration as indicated by his threefold division of existents.
[13]Quṭb al-Dīn did not explain how the two parts are related to each other. See my explanation below under the discussion of mathematics and natural science.
[14]Quṭb al-Dīn defines divine science as the science dealing with God and existents that are "close to Him" such as intellects and souls, through which He created other existents. *Durrat al-tāj,* I, I, 74.
[15]Quṭb al-Dīn's First Philosophy deals with universals. It is divided into two parts: (1) matters common to all concepts (*mafhūmāt*), such as existence, quiddity, unity, necessity, eternity, cause, and substance, along with their contraries and related terms; (2) the classes of accidents. Walbridge, *op. cit.*, p. 110.
[16]In al-Fārābī, the term *al-'ilm al-ilāhī* is used to refer to the whole of metaphysics, embracing Quṭb al Dīn's *'ilm-i ilāhī* and First Philosophy.
[17]Surveying (*'ilm-i misāḥah*) is the science of areal measurements pertaining primarily to lands.
[18]The Arabic term *'ilm al-jam' wa'l-tafrīq* means literally the science of the operations of "combining" and "separating." The operations of "combining" refer to addition and multiplication and the operations of "separating" to substraction and division.

By Quṭb al-Dīn's time, these four fundamental arithmetical operations were applied not only to whole numbers but also to rational numbers and surds.
[19]The term used by Quṭb al-Dīn is *'ilm-i awzān wa mawāzīn.* See *Durrat al-tāj,* I, I, 75. This science deals primarily with centers of gravity and ways of applying the balance in order to measure the specific weights of different substances. See S.H. Nasr, *Science and Civilization in Islam,* pp. 139-44.
[20]As a mathematical science, the science of irrigation (*'ilm-i naql-i miyāh*) deals with the measurement of water canals.
[21]As indicated in chapters five and six, al-Fārābī's fourth branch of natural sciences deals with the principles of the reactions which the four elements undergo in order to form compounds. His fifth branch deals with the common properties of composite bodies.
[22]According to Quṭb al-Dīn, quiddity could be considered on three levels: (1) quiddity conditioned on something (*māhīyah bi-sharṭ-i shay'*); (2) quiddity abstracted from all external connections but existent in the mind; and (3) quiddity conditioned on nothing (*māhīyah bi-sharṭ-i lā shay'*). The first level is quiddity as it exists externally as a concrete thing; the second is mentally existent quiddity; and third is quiddity which has no existence in objects or minds. Walbridge, *op. cit,.* pp. 121-2.
[23]*Durrat al-tāj,* I,I, 81.
[24]*Ibid.*
[25]By political association Quṭb al-Dīn means ways of sharing things between different individuals and groups in order to bring about cooperation for the benefits of man. *Ibid,* p. 80.
[26]With Quṭb al-Dīn prophecy is treated under metaphysics and the divine law under the religious sciences.
[27]*Durrat al-tāj,* I, I, 83.
[28]This is the knowledge that God alone is Necessary Being (*wājib al-wujūd*). All other beings are contingent beings which have need of Creator in order to exist. *Ibid,* p. 85.
[29]Divine Attributes are of two kinds. One is Attributes of transcendence (*tanzih*) or majesty (*jalāl*). With respect to these Attributes, God is absolutely free of contingency, substance, accident and change and absolutely transcends space and time. The other is

Attributes of perfection (*kamāl*). Examples are Life, Knowledge, Power, Will, Hearing, Seeing, Speech, Mercy and Forgiving.
[30]Quṭb al-Dīn maintains that knowledge of creation leads to the knowledge of divine Attributes, especially of God's Power and Wisdom.
[31]According to Quṭb al-Dīn, prophethood is necessary because the human intellect cannot comprehend everything especially pertaining to religious questions, a good example of which is worship.
[32]This science (*'ilm-i qirā'at*) is concerned with the different styles of recitation of the Qur'ān, based upon the tradition of the Prophet.
[33]This is the science of correct "punctuation." An incorrect beginning or ending of a verse may affect the interpretation of its meaning.
[34]According to Quṭb al-Dīn, this knowledge is one of the necessary conditions for the correct interpretation (*tafsīr*) of the Qur'ān.
[35]The Qur'ān was revealed in diverse circumstances over a period of about twenty-three years. Knowledge of these specific circumstances is useful because it enables one, for example, to infer general ethico-legal principles.
[36]This science (*'ilm-i ta'wīl*) deals with the intended or hidden meaning of a Qur'anic verse, as opposed to its literal meaning.
[37]As a branch of the science of the Qur'ān, history (*'ilm-i qiṣaṣ*) deals with the religious lives and missions of prophets mentioned in the Qur'ān, especially with their trials and tribulations in the face of great and often violent opposition against the divine message of which they claimed to be bearers. The science of history also deals with the fall of mighty nations and peoples of the past. As mentioned in the Qur'ān, their fall was as a result of their disobedience and corruption of the religious and moral life divinely ordained for them. The main aim of this science, says Quṭb al-Dīn, is to arrive at the appropriate moral lessons for the people of the present times. *Durrat al-tāj*, I, I 90.
[38]This science is concerned with the derivation of ideas in general from word combinations in general. Among these ideas are those which constitute law. Quṭb al-Dīn points out the importance of this discipline to jurists and scholars in the field of principles of jurisprudence. *Ibid.*
[39]This science, made necessary by the peculiar nature of Arabic speech, is concerned with the rules which enable us to know the very situations that apply to word combinations, as far as they affect meanings.
[40]This discipline is concerned with the knowledge of how a single idea or meaning may be expressed by different forms of speech or word combinations.
[41]Quṭb al-Dīn includes in this science knowledge of the companions of the Prophet and of their successors as well as knowledge of *ḥadīth* transmissions.
[42]The term used by Quṭb al-Dīn is *'ilm-i abniyah* or *'ilm-i taṣrīf*.
[43]This is the science of rhymes relating to poetry.

CONCLUSION

The Philosophical Bases of the Three Classifications: Similarities and Differences

My study of three Muslim classifications of the sciences – composed by al-Fārābī, al-Ghazzālī, and Quṭb al-Dīn al-Shīrāzī – shows that these classifications are at once based on philosophical ideas that are common to all Islamic intellectual schools, and ideas which are specific to the intellectual and religious world-view of its author and of the school he represents. There are two dominant ideas that shape the underlying philosophical basis of each classification. One is the idea of the hierarchy and the unity of the sciences. Another is the idea of the distinction between religion and philosophy. The latter idea is also related to the distinction between revelation and reason.

The general idea of hierarchy of reality is shown (chapter two) to be rooted in Islamic revelation. The Qur'ān and *ḥadīths* contain numerous references to such ideas as the hierarchy of creation, hierarchy of believers and knowers, hierarchy of witnesses of divine unity, and the hierarchic structure of the Qur'ān itself. Although the general idea of hierarchy is accepted by all three thinkers, it is in al-Fārābī that the idea receives its most comprehensive and detailed treatment. The idea of hierarchy permeates al-Fārābī's philosophical thought. There is the hierarchy of the cognitive faculties of the human soul in his psychology; the hierarchy of syllogistic proofs and the corresponding degrees of certainty in his epistemology and logic; the hierarchy of existents (*mawjūdāt*) in his metaphysics; the hierarchy of virtues and goods in his political philosophy; and many other secondary kinds of hierarchy. On the basis of these different kinds of hierarchy al-Fārābī formulates the idea of the hierarchy of the sciences.

Since al-Fārābī's classification is the first to be studied and is also the concentration of this study, the hierarchy of the sciences is established through him. There are three fundamental bases of hierarchically ordering the sciences: the methodological, the ontological, and the ethical. The methodological basis is derived from al-Fārābī's hierarchical ordering of proofs, arguments, and modes of knowing things; the ontological basis, from his hierarchically ordered view of the universe; and the ethical basis, from his hierarchical ordering of human needs, goods, and goals. These three bases are related to three main aspects of the sciences. The ontological basis is related to the subject-matters of the sciences; the methodological to the methods and modes of knowing the objects of study; and the ethical to the aims and goals of the sciences.

As general principles, the three criteria of establishing the hierarchy of the sciences are accepted by all three thinkers. However, in their specific formulation of each criteria these thinkers differ from each other. Consequently, their hierarchies of the sciences are not identical. Moreover, as related to their classifications of the sciences, we do not find the same kind of emphasis given to the three criteria. The difference in their specific formulation of each criteria and in the emphasis given to it is closely related to their specific intellectual and religious world-view, especially pertaining to the distinction between religion and philosophy or between revelation and reason.

In all three classifications the distinction between philosophical and religious sciences is clearly pronounced, although the terminologies used for the two groups of sciences are different. Al-Ghazzālī distinguishes between *shar'īyah* (religious) sciences and *'aqlīyah* (intellectual or rational) sciences. The latter he also calls *ghayr shar'īyah* (non-religious) sciences. Quṭb al-Dīn distinguishes between *'ulūm ḥikmīy* (philosophical sciences) and *'ulūm ghayr ḥikmī* (non-philosophical sciences). The latter he treats as synonymous with religious sciences, since he is concerned with the sciences cultivated in a civilization possessing a *sharī'ah* (revealed law). In al-Fārābī's classification, however, no specific terminology is used, because the two groups of sciences are not mentioned by name. But his enumeration of the religious sciences of *kalām* and *fiqh* immediately follows that of the philosophical sciences, namely mathematics, natural sciences, metaphysics, and political science.

Even the terminologies used by the authors are indicative of their respective philosophical attitude toward the two groups of sciences. Al-Ghazzālī's use of the term *ghayr shar'īyah* (non-religious) for the intellectual sciences means that, for him, the *shar'īyah* sciences are paramount and serve as the basis for naming all other sciences. Similarly, Quṭb al-Dīn's use of the term *ghayr 'ulūm ḥikmīy* for the religious sciences means that the philosophical sciences serve as his basis for comparison with the other sciences.

Al-Fārābī's classification gives greater prominence to the philosophical sciences. In fact, its main aim was to make logic and the philosophical sciences better known and more generally accepted among Muslims. The classification constitutes also al-Fārābī's attempt at projecting a superior image of the philosophical sciences in relation to the religious sciences. The only religious sciences included in his classification are *kalām* and *fiqh* but these are only briefly treated. These sciences are in fact subordinate to his political science. The important religious science of *uṣūl al-fiqh* (principles of jurisprudence) is incorporated into political science. Al-Fārābī appeals to methodological ground to support his view that the philosophical sciences are superior to the religious sciences. Philosophy employs the most excellent method of reasoning and proof, namely the demonstrative method, whereas the religious sciences at best employ the dialectical method.

In maintaining that philosophy is superior to religion, al-Fārābī represents the general view of the Muslim Peripatetic school of philosopher-scientists. In the perspective of this school, philosophy and religion are two approaches to the same truth. Philosophy is the better approach, since it affords certain knowledge of revealed truth, whereas religion at best affords "approximate" certainty. What is, in fact, contrasted is not philosophy, understood as a rational system formulated independently of intellection and revelation, and religion, understood as a total revealed tradition. Rather, the distinction is envisaged in the context of one and the same revealed tradition. This fact is cleary reflected in al-Fārābī classification. His political science deals with the same revealed doctrines and practices treated by *kalām* and *fiqh*. But the former treats the doctrines and practices at the level of philosophy, while *kalām* and *fiqh* treat them at the level of religion.

In al-Fārābī, the distinction between religion and philosopy is not formulated in terms of the distinction between revealed faith and reason. Rather, the distinction is primarily conceived in terms of a contrast between two types of reasoning or between two ways of accepting revealed truths. Al-Fārābī and his school of philosophers, unlike al-Ghazzālī, do not project revelation and reason as two mutually exclusive domains. They place great reliance on the use of reason, with logic as its tool, to reach transcendent truths. Through the use of reason man may attain pure intellection (*ta'aqqul*) or intellectual intuition, which is described as the summit of philosophical experience. For al-Fārābī, revelation understood as the prophet's intellectual vision of spiritual realities is of the same nature as the pure intellection of the philosophers. For this reason, the question of a fundamental distinction between sciences based on revelation and sciences based on reason does not arise in al-Fārābī.

The philosophical sciences themselves admit of degrees of excellence. According to all the three criteria of hierarchically ordering the sciences, metaphysics is the most excellent philosophical science. Natural science occupies the lowest position among the philosophical sciences. According to the ontological and ethical bases, mathematics and political science emerge as a kind of intermediate sciences between natural science and metaphysics. As intermediate sciences, they are closely related on the one hand with natural science and on the other with metaphysics. According to the methodological basis, the position of mathematics and political science is less clear. Al-Fārābī seems to consider the geometrical method as of being the same degree of excellence as the demonstrative method employed in metaphysics. And he implies that natural science and political science are of the same status with respect to the methodological criteria.

Although in the hierarchy of the sciences mathematics and political science are intermediate between natural science and metaphysics, their order of enumeration in al-Fārābī's classification is different. Al-Fārābī has ordered the sciences in their order of instruction (*tartīb al-ta'līm*) or learning. The first philosophical science recommended to be studied is mathematics, or more precisely arithmetic and geometry, since they deal with entities that are the easiest to be grasped by the human mind.

In his overall treatment of the positions of the various sciences in the hierarchy of the sciences al-Fārābī has given more emphasis to the methodological basis than to the other two bases. However, in discussing the methodological basis his primary concern is not with the positions of the philosophical sciences in relation to each other. His focus of comparison is between philosophy and the religious sciences of *kalām* and *fiqh*.

In contrast, al-Gazzālī's classification gives greater prominence to the religious sciences. The religious sciences are superior to the philosophical sciences, because the former are said to be based on revealed teaching, while the latter are based on reason. Al-Ghazzālī, in both capacities as a *mutakallim* and a Sufi, emphasize on the negative aspect of reason as veil and limitation and its inability to reach tanscendent truths. He argues that the fact that the metaphysical sciences of the philosophers are plagued with errors and inconsistencies clearly shows that their "philosophic" or demonstrative method is inapplicable to the domain of transcendent truths.

Al-Ghazzālī identifies philosophy with purely rational truths or human wisdom rather than with revealed *ḥikmah* (wisdom). Consequently, he seeks to redraw the boundaries of philosophy to make it "legitimate" before the eyes of orthodoxy defined by *kalām*. His "legitimate" philosophy becomes reduced to the logical, mathematical and the natural sciences. He considers these sciences as either religiously and philosophically neutral or are not in conflict with revealed doctrines. Since he views the philosophers' political and ethical sciences to be based on borrowed doctrines from revelation, he incorporates these sciences into the religious group. In his classification of the sciences given in the *Iḥyā'*, he identifies non-religious or intellectual sciences with his "legitimate" philosophy.

Al-Ghazzālī's division of knowledge into the religious and the intellectual sciences is related to the *mutakallimūn's* conception of the relationship between revelation and reason. Revelation and reason are conceived as two mutually exclusive sources of knowledge, namely of the religious and intellectual sciences respectively. But this view of revelation and reason is problematic. Insofar as al-Ghazzālī was committed to this view, he appears not to be consisent in his position as to where the line should be drawn between the religious and intellectual sciences. He could not remain faithful to his own definitions of the religious and the intellectual sciences when actually enumerating these two groups of sciences. He attributes certain knowledge comprised in some of the intellectual sciences like medicine and astronomy to the prophets. In *al-Risālat al-ladunīyah*, in contrast to his classification in the *Iḥyā'*, he extends his intellectual sciences or "legitimate" philosophy to include metaphysics.

As a *mutakallim* al-Ghazzālī defends the distinction between the religious and the intellectual sciences. But as a Sufi, he realizes that the distinction could only have a limited validity. He maintains that at the level of gnosis or *ma'rifah* of the Sufis the distinction between the religious and the intellectual ceases to exist. "Presential knowledge" (*'ilm al-ḥuḍūrī*) of the Sufis is at once religious and intellectual. In *al-Risālat al-ladunīyah*, written from a Sufi perspective, he maintains that most of the branches of religious knowledge are intellectual in the opinion of him who knows them, and most of the branches of intellectual knowledge belong to the religious code in the opinion of him who understands them. In the light of this view of al-Ghazzālī, we see wisdom in al-Fārābī's approach of distinguishing the philosophical and religious sciences. Al-Ghazzālī presents the division of knowledge into the presential and the attained as more fundamental and universal than the division into the religious and the intellectual.

Al-Ghazzālī with his jurist background and moralist standpoints gives great emphasis to the ethical basis of ordering the sciences. He applies this basis to both the religious and the intellectual sciences. This application results in his idea of the division of knowledge into the *farḍ 'ayn* and the *farḍ kifāyah* sciences, and the idea of the distinction between praiseworthy and blameworthy sciences. Al-Ghazzālī maintains the supremacy and the priority of the *farḍ 'ayn* over the *farḍ kifāyah* sciences. The *farḍ 'ayn* sciences are praiseworthy in an absolute sense, while the *farḍ kifāyah* sciences are praiseworthy only within a certain limit, which he calls "the limit of sufficiency." The *farḍ kifāyah* sciences themselves are of various degrees of excellence. The concept of praiseworthy sciences is more general than the concept of *farḍ 'ayn,* since the former embraces the religious and the intellectual sciences, whereas *farḍ 'ayn* refers to certain religious sciences only.

In his discussion of the ethico-legal basis of the hierarchy of the sciences al-Ghazzālī pays more attention to the religious sciences, whereas al-Fārābī was more interested in the ethical status of the philosophical sciences. Al-Fārābī and al-Ghazzālī share the view that the science of God is the most useful of the sciences; in al-Ghazzālī's terminology, this science is "praiseworthy in an absolute sense."

Al-Ghazzālī's classification of the sciences has not entirely resolved the "problem of the boundary" between the religious and the intellectual. However, his idea that certain kinds of knowledge are at once religious and intellectual was taken up by Quṭb al-Dīn and applied to his classification of the sciences. In Quṭb al-Dīn we also find another perspective of looking at the problem of distinction between the religious and the philosophical sciences.

Coming after al-Fārābī and al-Ghazzālī, Quṭb al-Dīn was able to undertake a synthesis of the classifications of his predecessors. Quṭb al-Dīn divides the sciences into *ḥikmat* (philosophical) and *ghayr-ḥikmat* (non-philosophical) sciences. The use of the term *ḥikmat* for "philosophy" is significant. After al-Ghazzālī's powerful attack on the philosophers, philosophy in the Islamic world took on a new form that was less rationalistic and more in harmony with the Qur'anic world-view. Al-Ghazzālī in fact paved the way for the spread of the Illuminationist (*ishrāqī*) school of philosophy, with which Quṭb al-Dīn is identified. Quṭb al-Dīn referred more than once to the Qur'anic basis of *ḥikmat*. His philosophy is illuminationist philosophy (*ḥikmat dhoqī*) based upon supra-rational experience or illumination of the intellect but at the same time makes the best use of discursive reasoning. Presential knowledge is a fundamental basis of Quṭb al-Dīn's *ḥikmat*.

In his classification, Quṭb al-Dīn's emphasis is on the philosophical sciences. He presents them as sciences which are the same for all times and cultures or civilizations, and as sciences of the natures of things that comprise the immutable aspects of the universe. The non-philosophical sciences are the sciences associated with the *Sharī'ah*. These sciences are not the same for all times and cultures, since God has revealed different *sharī'ahs* for different branches of humanity and in different epochs of history. Quṭb al-Dīn, thus, does not envisage the distinction between religion and philosophy in terms of a distinction between revelation and reason. In this, he inherits the Farabian distinction that philosophy is what belongs to all humanity and religion to a particular branch of humanity.

An important feature of Quṭb al-Dīn's classification is his division of the religious sciences into (1) transmitted, (2) intellectual, and (3) both transmitted and intellectual sciences. Quṭb al-Dīn categorically states that there are sciences which are established by both the intellect and transmitted sources. This is equivalent to saying that there are revealed teachings which may be established independently by reason. The sciences of such kinds of teachings Quṭb al-Dīn categorizes as the sciences of fundamental principles of religion. The sciences of the branches of religion, including linguistic science and literature, belong to the category of transmitted sciences. Quṭb al-Dīn thus offers a novel solution to the problem of the boundary between the religious and the intellectual sciences. In Quṭb al-Dīn's classification scheme the sciences of the fundamental principles of religion correspond to metaphysics and certain parts of practical *ḥikmat*.

According to all three classifications, the highest knowledge is the knowledge of God. It is for the sake of the knowledge of God that all other forms of knowledge are sought. Moreover, knowledge of all things other than God must be conceptually or organically related to the knowledge of God. This idea together with the view that all knowledge comes ultimately from the same source constitute the idea of the unity of knowledge shared by the three authors.

SELECTED BIBLIOGRAPHY

General

'Abd al-Rāziq, M., *Tamhīd li-tārīkh al-falsafat al-islāmīyah*, Cairo, 1959.

Anawati, M.M., and Gardet, L., *Introduction à la théologie musulmane, essai de théologie comparée* Paris: J. Vrin, 1948.

Arberry, A. J., *Revelation and Reason in Islam*, London, 1957.

Arnold, T. W., *The Caliphate*, Oxford, 1924.

Atiyeh, G. N., *Al-Kindī, The Philosopher of the Arabs*, Rawalpindi: Islamic Research Institute, 1966.

Averroes, *Tahāfut al-tahāfut*, trans. S. van den Bergh, London: Luzac and Co., 1954.

Badawī, 'A., ed., *Arisṭū 'ind al-'arab*, Cairo, 1947.

____________, ed., *Rasā'il falsafīyat li'l-Kindī wa'l-Fārābī wa, ibn Bājjah wa ibn 'Adī*, Benghazī: University of Libya Press, 1973.

Barthold, W., *Turkestan Down to the Mongol Invasion*, trans. and rev. by author and Gibb, H.A.R., London: Luzac and Co., 1928.

Brown, E.G., *A Literary History of Persia*, 4 vols., London: T. Fisher Unwin, 1902-1924, and Cambridge: Cambridge University Press, 1951.

Burckhardt, T., *Alchemy, Science of the Cosmos, Science of the Soul*, Baltimore, 1974.

The Cambridge History of Iran, ed., J. Boyle, Cambridge, 1968, vol. 5.

Carra de Vaux, B., *Les penseurs de l'Islam*, 5 vols., Paris: P. Geuthner, 1921-1926.

Corbin, H., with Nasr, S.H., and Yahya, O., *Histoire de la philosohpie islamique*, Paris, 1964.

de Boer, T.J., *The History of Philosophy in Islam*, trans., Jones, E.R., London: Luzac and Co., 1965.

Dictionary of Scientific Biography, ed. C. Gillispie, New York: Charles Scribner's Sons., 1970 on.

Dunlop, D.M., *Arabic Science in the West*, Karachi: Pakistan Historical Society, 1958.

____________, *Arabic Civilization up to 1500 A.D.*, Beirut and London, 1971.

The Encyclopaedia of Islam, 2nd edn., Leiden and London, E.J. Brill, 1960 on.

Fakhry, M., *A History of Islamic Philosophy*, New York and London: Columbia University Press, 1970.

Gauthier, L., *Introduction à l'étude de la philosophie musulmane*, Paris: Editions Ernest Leroux, 1900.

Ghoraba, H., "The dilemma of religion and philosophy in Islam", *Islamic Quarterly* (London), 2: 241-51 (1955); 3: 4-15 (1955); 3: 73-87 (1956).

Gilson, E., *History Of Christian Philosophy in the Middle Ages*, New York: Random House, 1955.

Gyekye, K., *Arabic Logic: Ibn al-Ṭayyib's Commentary on Porphyry's Eisagoge*, Albany: SUNY Press, 1979.

Hourani, G., ed., *Essays on Islamic Philosophy and Science*, Albany: SUNY Press, 1975.

Husik, I., *History of Medieval Jewish Philosophy*, New York, 1976.

Ibn Abī Usaibi'ah, *'Uyūn al-anbā fī ṭabaqāt al-aṭibbā'*, Beirut, 1963.

Ibn Khaldūn, *The Muqaddimah*, trans. Rosenthal, F., 3 vols., New York: Pantheon Books, 1958; London and Henley: Routledge and Kegan Paul, 1986.

Ibn Khallikān, *Ibn khallikān's Biographical Dictionary*, trans. de Slane, W.M., Paris: Oriental Translation Fund, 1842-1871; New York and London: Johnson Reprint Corporation, 1961.

Ibn al-Nadīm, *The Fihrist of al-Nadīm*, ed. and trans., Dodge, B., 2 vols., New York and London: Columbia University Press, 1970.

Ibn al-Qifṭī, *Akhbār al-ḥukamā'*, Leipzig: Dieterich, 1903; Cairo: Sa'adah Press, 1326 (1908).

Ibn Rushd, *Averroes' Commentary on Plato's Republic* trans. Rosenthal, E.I.J., Cambridge, 1956.

Iqbal, M., *The Development of Metaphysics in Persia*, London: Luzac, 1908.

Izutsu, T., *The Structure of Sabzawarian Metaphysics*, Tehran.

Ivry, A.L., *Al-Kindī's Metaphysics*, Albany, 1974.

Le Strange, G., *The Lands of the Eastern Caliphate*, New York, 1966.

Lerner, R. and Mahdi, M., eds., *Medieval Political Philosophy: A Source-book*, Toronto: The Free Press of Glencoe, 1963.

Lloyd, A.C., "Neoplatonic Logic and Aristotelian Logic," *Phronesis*, 1: 1 (November 1955); 1: 2 (1956).

Madkour, I., *Fi'l-falsafat al-islāmīyah*, Cairo: Dār Iḥyā' al-kutub al-'arabīyah, 1947.

____________, *L'Organon d'Aristote dans le monde arabe*, Paris, 1934.

Makdisi, G., "Muslim Institutions of Learning in Eleventh Century Baghdad," *Bulletin of the School of Oriental and African Studies*, 22: 1-56 (1961).

____________, *The Rise of Colleges: Institutions of Learning in Islam and the West*, Edinburgh: Edinburgh University Press, 1981.

Marmura, M., ed., *Islamic Theology and Philosophy*, Albany: SUNY Press, 1984.

Morewedge, P., ed., *Islamic Philosophical Theology*, Albany: SUNY Press, 1979.

Munk, S., *Mélanges de philosophie juive et arabe*, Paris: J. Vrin, 1927; and 1955 edn.

Nasr, S.H., *Three Muslim Sages*, Cambridge: Harvard University Press, 1964.

____________, *An Introduction to Islamic Cosmological Doctrines*, Boulder: Shambhala Publications, Inc., 1978; previously, Cambridge: Harvard University Press, 1964.

____________, *Islamic Studies, Essays on Law and Society, The Sciences, Philosophy and Sufism*, Beirut: Librairie de Liban, 1967.

____________, *Science and Civilization in Islam*, Cambridge: Harvard University Press, 1968.

____________, *Islamic Science: An Illustrated Study*, London: World of Islam Festival Publishing Co., Thorsons Publisher Ltd., 1976.

____________, *Islamic Life and Thought*, Albany: SUNY Press, 1981.

____________, "Reflections on Methodology in the Islamic Sciences", *Hamdard Islamicus*, 3: 3 (1980), 3-13.

Nicholson, R., *A Literary History of the Arabs*, London, 1907.

O'Leary, DeLacy, *Arabic Thought and Its Place in History*, London 1922.

____________, *How Greek Science Passed to the Arabs*, London; Routledge and Kegan Paul, 1949.

Pearson, J.D., *Index Islamicus*, Cambridge, 1958 on.

Peters, F.E., *Aristotle and the Arabs: The Aristotelian Tradition in Islam*, New York: New York University Press, and London: University of London, 1968.

Pines, S., "Some Problems of Islamic Philosophy," *Islamic Culture* 11: 66-80 (1927).
Rahman, F., trans., *Avicenna's Psychology*, London: Oxford University Press, 1952; Westport, Connecticut: Hyperion Press, 1981.
————, *Prophecy in Islam: Philosophy and Orthodoxy*, Chicago: University of Chicago Press, 1958 and repr. 1979.
Rescher, N., *Studies in the History of Arabic Logic*, Pittsburgh University Press, 1963.
————, *The Development of Arabic Logic*, Pittsburgh: Pittsburgh University Press, 1964.
————, *Studies in Arabic Philosophy*, Pittsburgh: Pittsburgh University Press, 1966.
Rosenthal, E.I.J., *Political Thought in Medieval Islam*, Cambridge, 1958.
Rosenthal, F., *A History of Muslim Historiography*, Leiden, 1951.
————, *Knowledge Triumphant*, Leiden, 1970.
————, *The Classical Heritage in Islam*, trans. from German, E. Marmorstein and J. Marmorstein, London: Routledge and Kegan paul Ltd., 1975.
————, "On the Knowledge of Plato's Philosophy in the Islamic World," *Islamic Culture*, 14: 387-422 (1940).
Ross, W.D., ed., *The Works of Aristotle translated into English*, 12 vols., Oxford, 1910-52.
Sarton, G., *Introduction to the History of Science*, 3 vols., Baltimore: The Williams and Wilkins Co;, 1927-1948.
Schimmel, A., *Mystical Dimensions of Islam*, Chapel Hill, 1975.
Schuon, F., *Logic and Transcendence*, trans. P.N. Townsend, London: Perennial Books Ltd., 1975.
————, *Dimensions of Islam*, trans. P.N. Townsend, New York, 1969; London: George Allen & Unwin, 1970.
————, *Islam and the Perennial Philosophy*, trans. J.P. Hobson, London: World of Islam Festival Publishing Co., and Thorsons Publishers Ltd., 1976.
————, *Sufism: Veil and Quintessence*, Bloomington: World Wisdom Books, 1982.
al-Shahrastani, M., *al-Milal wa'l-niḥal*, London, 1892.
Sharif, M.M., ed. *A History of Muslim Philosophy*, 2 vols.,

Wiesbaden: Otto Harrassowitz, 1963.

Sheikh, M. Saeed, *Islamic Philosophy*, London: The Octagon Press, 1982.

Siraj al-Din, A., *The Book of Certainty*, London, 1952.

Stern, S., "Notes on al-Kindī's Treatise on Definitions," *Journal of the Royal Asiatic Siciety*, London (1959), 32-43.

Stern, S., Hourani, A. and Brown, V. eds., *Islamic Philosophy and the Classical Tradition: Essays Presented to Richard Walzer*, Oxfod, 1972.

Tibawi, A., *Arabic and Islamic Themes*, London, 1970.

____________, *Islamic Education*, London, 1972.

Tritton, A.S., *Materials on Muslim Education in the Middle Ages*, London, 1957.

von Grunebaum, G.E., *Medieval Islam*, Chicago, 1953.

____________, *Islam: Essays in the Nature and Growth of a Cultural Tradition*, London: Routledge and Kegan Paul, 1955.

____________, ed., *Logic in Classical Islamic Culture*, Wiesbaden: Otto Harrassowitz, 1970.

Walzer, R., *Greek into Arabic: Essays on Islamic Philosophy*, Cambridge: Havard University Press, 1962.

Watt, W.M., *Islamic Philosophy and Theology*, Edinburgh: Edinburgh University Press, 1962.

____________, *The Formative Period of Islamic Thought*, Edinburgh: Edinburgh University Press, 1973.

____________, *The Majesty that was Islam*, London, 1976.

Wolfson, H.A., "The Internal Senses in Latin, Arabic and Hebrew Philosophic Texts," *Harvard Theological Review*, 28: 69-133 (1935).

____________, *Crescas' Critique of Aristotle: Problems of Aristotle's Physics in Jewish and Arabic Philosophy*, Cambridge: Harvard University Press, 1929.

____________, *The Philosophy of the Kalam*, Cambridge (Mass.) and London: Harvard University Press, 1976.

Al-FĀRĀBĪ
(Primary Sources)

al-Fārābī, *Catalogo de las ciencias*, ed. and trans. A.G. Palencia,

Madrid: La Facultad de Filosofia y Letras, University of Madrid, 1932 and 1953 edn.

——————, *Epistola Sull' Intellecto*, trans. F. Lucchetta Padova, 1974.

——————, *Al-Fārābī's Commentary and Short Treatise on Aristotle's de Interpretatione*, trans., F.W. Zimmermann, London: Oxford University Press, 1981.

——————, *Al-Fārābī, Deux Ouvrages inédits sur la rhétorique*, pt. 1, ed. and trans. J. Langhade and M. Grignaschi, Beyrouth, 1971.

——————, *Al-Fārābī: Fuṣūl al-madanī (Aphorisms of the Statesman)*, ed. and trans. D.M. Dunlop, Cambridge: Cambridge University Press, 1961.

——————, *Al-Fārābī: Idées des habitants de la cité vertueuse*, trans. R.P. Jaussen, Y. Karam and J. Chlala, Cairo: L'institut Francais D'archéologie Orientale, 1949.

——————, *Al-Fārābī's Philosophische Abhandlungen*, ed. F. Dieterici, Leiden, 1890 (also 1896 edn.), comprising the following:

Kitāb al jam' bain ra'yai al-ḥakīmain Aflāṭūn al-ilāhī wa Arisṭūṭālīs
Fī aghrāḍ al-ḥakīm fī kull maqālat min al-kitāb al-mausūm bi'l-ḥurūf
Maqālat fī ma'ānī al-'aql
'Uyūn al-masā'il
Risālat fuṣūṣ al-ḥikam
Risālat fī jawāb masā'il su'ila 'anhā
Fī mā yasiḥḥ wa lā yasiḥḥ min aḥkām al-nujūm
Risālat fī mā yanbaghī an yuqaddam qabla ta'allum al-falsafah
Qiṭ'ah min tarjamat al-Fārābī wahya mākhūdhah min tārīkh al-ḥukamā'

——————, *Al-Fārābī's Philosophy of Plato and Aristotle*, trans., M. Mahdi, Ithacaa: Cornell University Press, 1969.

——————, *Al-Fārābī's Short Commentary on Aristotle's Prior Analytics*, trans. N. Rescher, Pittsburgh: Pittsburgh University Press, 1963.

——————, *Iḥṣā' al-'ulūm*, ed. 'U. Amīn, Cairo: Dār al-Fikr al-'Arabī, 1949.

——————, *Kitāb al-alfāẓ al-musta'malah fi'l-manṭiq (Utterances*

Employed in Logic), ed. M. Mahdi, Beirut: Dar el-Mashreq Publishers, 1968.

__________, *Kitāb al-ḥurūf (The Book of Letters): Commentary on Aristotle's Metaphysics*, ed. with introduction and notes, M. Mahdi, Beirut: Dar el-Mashreq Publishers, 1969.

__________, *Kitāb al-khaṭābah*, ed. J. Langhade, Beirut, 1971.

__________, *Kitāb al-madīnat al-fāḍilah*, ed. F. Dieterici, Leiden, 1895.

__________, *Kitāb al-millah wa nuṣūṣ ukhrā*, ed. M. Mahdi, Beirut: Dar el-Mashreq Publishers, 1968.

__________, *Kitāb al-siyāsat al-madanīyah (al-Fārābī's Political Regime)*, ed. F. Najjar, Beyrouth: Imprimerie Catholique, 1964.

__________, *Rasā'il (Ten Tracts)*,. Hyderabad: Dāirat'ul Ma'ārifi'l Osmania, 1345 A.H., comprising the following:

Risālat fī ithbāt al-mufāraqāt
Maqālat fī aghrāḍ mā ba'd al-ṭabī'ah
Taḥṣīl al-sa'ādah
Al-Da'āwī al-qalbīyah
Sharḥ risālah zainūn al-kabīr al-yūnānī
Kitāb al-siyāsat al-madanīyah
Kitāb al-fuṣūṣ
Risālat fī faḍīlat al-'ulūm wa'l-ṣinā'āt
Risālat fī masā'il mutafarriqah

__________, *Risālah fi'l-'aql*, ed. N. Bouyges, Beirut: Imprimerie Catholique, 1938.

Arberry, A.J., ed. and trans., "Fārābī's Canons of Poetry," *Rivista degli Studi Orientali*, (Rome), 17: 266-78 (1938).

Dunlop, D.M., ed and trans., "Al-Fārābī's Introductory Sections on Logic," *The Islamic Quarterly* (London), 2: 264-82 (1955).

__________, ed. and trans., "Al-Fārābī's *Eisagoge*," *The Islamic Quarterly* (London), 3: 117-38 (1956-57).

__________, ed. and trans., "Al-Fārābī's Introductory *Risālah* on Logic, "*The Islamic Quarterly* (London), 3: 224-35 (1956-57).

__________, ed., and trans, "Al-Fārābī's Paraphrase of the *Categories* of Aristotle," *The Islamic Quarterly* (London), 4: 168-97 (1957-58); 5: 21-54 (1959).

d'Erlanger, R., trans., *La Musique Arabe*, vols. I, II, Paris: Paul Geuthner, 1930-1935.

Farmer, H.G., *Al-Fārābī's Arabic – Latin Writings on Music*, Glasgow: The Civic Press, 1934.

Gibson, E., trans. "De Intellectu d'alFarabi," *Archives d'Histoire Doctrinale et Littéraire du Moyen Age*, Paris, 4: 126-41 (1926).

Hyman, A., trans. "The Letter Concerning The Intellect," *Philosophy in the Middle-Ages*, eds. A. Hyman and J.J. Walsh, New York: Harper and Row Publishers. 1967.

Mahdi, M. trans., "Al-Fārābī against Philoponus," *Journal of Near Eastern Studies*, 26: 223-60 (1967).

————, trans. "Plato's Laws," *Medieval Political Philosophy: A Source-book*, eds. R. Lerner and M. Mahdi, Toronto, Canada: The Free Press of Glencoe, 1963, 83-94.

Najjar, F.M. trans., "On Political Science, Jurisprudence and Dialetical Theology," *ibid.*, 24-30.

————, trans., "The Political Regime", *ibid*, 31-57.

Rosenthal, F., *The Classical Heritage in Islam*, trans. from German, E. Marmorstein and J. Marmorstein, London: Routledge and Kegan Paul Ltd., 1975, 54-55.

Sayili, A. and Lugal, N., eds., and trans., "Abū Naṣr al-Fārābī's Article on Vacuum," *Türk Tarih Kurumu Basimevi* (Ankara). 15 (1): 21-36 (1951).

Wiedemann, E., trans., "Zur Alchemie bei den Arabern," *Journal fur praktische chemie*, N.F., 76: 115-22 (August 1907).

Secondary Sources

'Abd al-Rāziq, M., *Failasūf al-'arab wa'l-mu'allim al-thānī*, Cairo, 1945.

Afshar, I., ed., *Essays on Fārābī*, Tehran: Central Library and Documentation Center, University of Tehran, 1975.

al-Andalusī, Ṣā'id ibn Aḥmad, *Ṭabaqāt al-umam*, Najaf, 1967.

Atēs, A., "Fārābī'nin Eserlerinin Bibliografyasi," *Türk Tarih Kurumu Bellsten* (Ankara), 15(57): 175-92 (1951).

al-Bayhaqī, Ẓahir al-Dīn, *Tārīkh ḥukamā' al-Islām*, Damascus, 1946.

————, *Tatimmah ṣiwān al-ḥikmah*, Lahore, 1935.

Blumberg, H., "Al-Fārābī's Five Chapters on Logie," *Proceedings of the American Academy for Jewish Research*, 6 115-21 (1934-1935).

de Boer, T.J., *The History of Philosophy in Islam*, trans. E.R. Jones, London: Luzac and Co., 1961.

Carra de Vaux, B., "al-Fārābī," *The Encyclopaedia of Islam*, eds., M. Th. Houtsma, A.J. Wensinck, vol. II, Leiden: E.J. Brill, 1911-1938, 57-59.

__________, 'Fārābī" *Hastings Encyclopedia of Religion and Ethics* (New York), 5: 757-59 (1914).

Cunbur, M., *Fārābī Bibliografyasi*, Ankara, 1973.

Dunlop, D.M., *Arab Civilization up to 1500 A.D.* Beirut and London, 1971.

Efros, I., "Palquera's *Reshit Hokmah* and al-Fārābī's *Iḥṣā' al-'ulūm*," *Jewish Quarterly Review*, New Series, 25: 227-35 (1935).

Fackenheim, E.L., "Al-Fārābī: His Life, Times and Thought," *Middle Eastern Affairs*, 2: 54-59 (1951).

Fakhry, M., *A History of Islamic Philosophy*, New York and London: Columbia University Press, 1970.

Farmer, H.G., "Clues for the Arabian Influence on European Musical Theory," *Journal of The Royal Asiatic Society* (London), 1925, 61-80.

__________, "The Influence of al-Fārābī's *Iḥṣā' al-'ulūm* (De Scientiis) on the Writers on Music in Western Europe," *ibid.* 1932, 561-92.

Georr, K., "Fārābī est-il l'auteur de fucūc-al-hikam?" *Revue des Etudes Islamiques*, 15:31-39 (1941-1946).

__________, *Bibliographie Critique d'al-Fārābī*, Paris, 1964.

Gilson, E., "Les Sources gréco-arabes de l'augustinisme avicennisant," *Archives d'Histoire Doctrinale et Littéraire du Moyen Age*, Paris, 4: 5-149. (1929).

Gundisalvo, D., "Classification of the Sciences," trans. N. Clagett and E. Grant, in *A Source Book in Mediaeval Science*, ed. E. Grant, Harvard University Press, 1974, 59-76.

al-Hashim, J., *Al-Fārābī*, Beirut, 1960.

Lowenthal, A., "Al-Fārābī," *The Jewish Encyclopedia* (New York), vol. I, 1901 (repr. 1912), 374-75.

Madkour, I., *La Place d'al-Farabi dans l'ecole philosophique musulmane*, Paris, 1934.

__________, "Naẓarīyah al-nubūwah 'ind al-Fārābī," *al-Risālah* (Egypt), 1936, 1731-34, 1783-86, 1830-32, 1869-71, 1913-15, 1944-96; 1937, 8-10, 59-60, 90-91.

__________, "Al-Fārābī," *A History of Muslim Philosophy*, ed.,

Sharif, M.M., Wiesbaden: O. Harrassowitz, vol. I, 1963-1966, 450-68.
Mahdi, M., "Al-Fārābī," *History of Political Philosophy*, ed. L. Strauss and J. Cropsey, Chicago, 1963, 160-80.
————, "Science, Philosophy and Religion in al-Fārābī's *Enumeration of the Sciences*," *The Cultural Context of Medieval Learning*, ed. J.E. Murdoch and E.D. Sylla, Dordrecht, 1975, pp. 113-147.
————, "Al-Fārābī", *Dictionary of Scientific Biography*, ed. C.G. Gillispie, New York: Charles Scribner's Sons, 1970, vol. IV.
————, "Remarks on al-Fārābī's *Attainment of Happiness*," *Essays on Islamic Philosophy and Science*, ed. G.F. Hourani, Albany: SUNY Press, 1975, pp. 47-65.
Maḥmūd, A., *Al-Fārābī*, Cairo, 1944.
Maḥfūz, H.A., *al-Fārābī fi'l-marāji' al-'arabīyah*, Baghdad, 1975.
————, and Yasīn, *J.A., Mu'allafāt al-Fārābī*, Baghdad, 1975.
Massignon, L., "Sur le texte original arabe du *De Intellectu* d'al-Farabi" *Archives d'Histoire Doctrinale et Littéraire du Moyen Age*, 4: 151-52 (1929).
Najjar, F., "Al-Fārābī on Political Science," *The Muslim World*, 48: 94-102 (1958).
————, "Fārābī's Political Philosophy and Shi'ism," *Studia Islamica*, 14: 57-72 (1961).
Plessner, M., "Al-Fārābī uber Medizin, eine ubersehene und seine neuentdeckte Quelle," *XXI Congress Internazionale di Storia, Medicina*, 1970, 1533-1539.
Rahman, F., "L'Intellectus Acquisitus in al-Fārābī," *Giornale Critico della Filosofia Italiana*, 7: 351-57 (1953).
Rescher, N., *Al-Fārābī: An Annotated Bibliography*, Pittsburgh: Pitsburgh University Press, 1962.
————, "Al-Fārābī on Logical Tradition," *The Journal of the History of Ideas*, 24: 127-32 (1963).
Rosenthal, E.I.J., "The Place of Politics in the Philosophy of al-Fārābī," *Islamic Culture* (Hyderabad-Deccan), 29(3): 157-78 (1955).
Salman, D.H., "The Medieval Latin Translations of al-Fārābī's Works," *The New Scholasticism*, 13: 245-61 (1939).
Sherwani, H.K., "Al-Fārābī's Political Philosophy," *Proceedings of*

the 9th All-India Oriental Conference, 1937, pp. 337-60.

___________, "Al-Fārābī's Political Theories," *Islamic Culture*, 12: 288-305 (1938).

Steinschneider, M., *Al-Fārābī, des arabischen Philosophen Leben und Schriften*, St. Petersburg, 1869.

Stern, S.M., "Al-Mas'ūdī and the Philosopher al-Fārābī" *Al-Mas'ūdī Millenary Commemoration Volume*, eds. S. Maqbul and A. Rahman, Aligarh, 1960, pp. 28-41.

Strauss, L., "Quelques remarques sur la science politique de Maimonide et de Farabi," *Revue des Etudes Juives*, 100: 1-37 (1936).

___________"Fārābī's Plato," *Louis Ginzberg Jubilee Volume of the American Academy for Jewish Research* (New York), 1945, pp. 357-93.

___________, "How Fārābī read Plato's *Laws*," *Mélanges Louis* Massignon, vol. III, Damascux, 1957, pp. 319-44.

Wāfī, Al 'Abd al-Wḥid, "AlFārābī," *Turāth al-incanīyah* (Egypt), 2: 569-82.

Wali-ur-Rahman, M., "Al-Fārābī and His Theory of Dreams," *Islamic Culture*, 10: 137-51 (1936).

___________, "The Psychology of al-Fārābī,: *Islamic Culture*, II: 228-47 (1937).

___________Walzer, R., "Al-Fārābī's Theory of Propechey and Divination," *Journal of Hellenic Studies*, 57 (1957).

___________, "Al-Fārābī," *Encyclopaedia of Islam*, 2nd edn., vol, II, 77-91.

Wright, O., "Al-Fārābī: Music," *Dictionary of Scientific Biography*, vo. IV.

Zimmermann, F.W., "Some observations on al-Fārābī and Logical Tradition," *Islamic Philosophy and the Classical Tradition*, eds., S.M. Stern, A. Hourani, and V. Brown, Oxford, 1972, pp. 517-46.

AL-GHAZZĀLĪ

(Primary Sources)

Al-Ghazzālī, *The Alchemy of Happiness*, trans. C. Field, Lahore, 1979.

——— *The Book of Knowledge*, trans. N.A. Faris, Lahore, 1962.

——— *The Confessions of al-Ghazzālī*, trans. C. Field, London, 1909.

——— *The Faith and Practice of al-Ghazzālī*, trans. W.M. Watt, London, 1953.

——— *The Foundations of the Articles of Faith*, trans. N.A. Faris, Lahore, 1974.

——— *Freedom and Fulfillment: An Annotated Translation of al-Munqidh min al-dalāl and Other Relevant Works of al-Ghazzālī*, trans. R.J. McCarthy, Boston, 1980.

——— *al-Iqtiṣād fi'l-i'tiqād*, Cairo, 1327/1901.

——— *Maqāsid al-falāsifah*, ed. Sulayman Dunya, Cairo, 1964.

——— *Mishkāt al-anwar (the Niche for Lights)*, trans. W.H. Gairdner, Lahore, 1952.

——— *Mi'yār al-'ilm*, ed. Sulyman Dunya, Cairo, 1964.

——— *al-Munqidh min al-ḍalāl*, ed. F.Jabre, Beyrouth, 1959.

——— *Tahāfut al-falāsifah (The Incoherence of the Philosophers)*, trans. S.A. Kamali, Lahore, 1958.

Smith, M., trans., "Al-Risālat al-ladunīyah," *Journal of the Royal Asiatic Society*, London, 1938, Part II, pp. 177-200; Part III, pp. 353-74.

Ṭībāwī, A.L., "Al-Ghazālī's Tract on Dogmatic Theology," *Islamic Quarterly*, 9: 65-122 (1965).

Secondary Sources

'Abd el-Jalil, J.M., *Autor de la sincérité d'al-Ghazzali*, Damascus, 1956.

'Ali, Syed Nawab, *Some Moral and Religious Teachings of al-Ghazzali*, Lahore, 1968.

Badawī, 'A., *I'tirāfāt al-ghazālī aw kayfa 'arrakha al-ghazālī nafsahu*, Cairo, 1945.

Bousquet, G.H., *Iḥyā' 'Ouloūm ed-dīn ou vivification des sciences de la foi*, Paris, 1955.

Bouyges, M., *Essai de chronologie des oeuvres de al-Ghazali*, Beyrouth: Imprimerie Catholique, 1959.

Carra de Vaux, B., *Al-Ghazzālī*, Paris, 1902.

Corbin, H., "The Ismā'īlī Response to the Polemic of Ghazālī," in S.H. Nasr (ed.), *Ismā'īlī Contributions to Islamic Culture*, pp. 69-98.

Faris, N.A., *"Iḥyā' 'ulūm al-dīn* of al-Ghazzālī," *Proceedings of the American Philosophical Society*, 71: 15-19 (1939).

Gairdner, W.H.T., "Al-Ghazali's *Mishkāt al-anwār* and the Ghazali-Problem," *Der Islam*, 5: 121-53 (1914).

Gardner, W.R.W., *An Account of al-Ghazālī's Life and Works*, Madras, 1919.

Hourani, G.F., "The Chronology of Ghazālī's Writings," *Journal of the American Oriental Society*, 79: 225-33 (1959).

Jabre, F., *La biographie et l'oeuvre de Ghazālī reconsidérées à la lumière des Ṭabaqāt de Sobki*, 1954.

____________, *La notion de la certitude selon Ghazālī dan ses origines psychologiques et historiques*, Paris 1058.

____________, *La notion de la ma'rifa chez Ghazālī*, Beyrouth, 1958.

Laoust, H., *La politique de Gazālī*, Paris, 1970.

MacDonald, D.B., "The Life of al-Ghazzālī with Special Reference to His Religious Experiences and Opinions", *Journal of The American Oriental Society*, 20: 71-132 (1899).

____________, The Name al-Ghazzālī", *Journal of the Royal Asiatic Society*, 1902, 18-22.

Marmura, M., "Ghazālī and Demonstrative Science," *Journal of the History of Philosophy*, 3: 2 (Oct. 1965), 183-204.

____________, "Ghazālī's Attitude to the Secular Sciences and Logic," *Essays on Islamic Philosophy and Science*, ed. G. Hourani, SUNY Press, 1975, pp. 100-11.

Mubarak, Z., *Al-akhlāq 'ind al-ghazzālī*, Cairo, 1924.

Palacios, A., *Algazel, Dogmatica, moral y ascetica*, Zaragosa, 1901.

____________, *La espiritualidad de Algazel y su sentido cristiano*, Madrid-Granada, 1931-41, 4 vols.

Sherif, M.A., *Al-Ghazālī's Theory of Virtue*, Albany: SUNY Press, 1975.

Smith, M., *al-Ghazālī the Mystic*, London, 1944.

Umaruddin, M., *The Ethical Philosophy of al-Ghazzālī*, Lahore, 1970.

al-'Uthmān, 'A., *Sīrat al-ghazālī wa-aqwāl al-mutaqaddimīn fīhi*, Damascus, 1960.

Watt, W.M., "The Authenticity of the Works Attributed to al-Ghazzālī," *Journal of the Royal Asiatic Society*, 1952, 24-45.

————, "A Forgery in al-Ghazzālī's *Mishkāt*?" *Journal of the Royal Asiatic Society*, 1949, 5-22.

————, *Muslim Intellectual: A Study of al-Ghazālī*, Edinburgh, 1963.

Wensinck, A.J., *La Pensée de Ghazzālī*, Paris, 1940.

Zwemer, S.M., *A Moslem Seeker After God: Showing Islam at Its Best in the Life and Teachings of al-Ghazali, Mystic and Theologian of the Eleventh Century*, New York, Chicago, London and Edinburgh, 1920.

QUṬB AL-DĪN AL-SHĪRĀZĪ

(Primary Sources)

Quṭb al-Dīn al-Shīrāzī, *Durrat al-tāj li-ghurrat al-dībāj fi'l ḥikma (Pearls of the Crown, the Best Introduction to Wisdom)*, 8 vols. in 2; vol. I edited by Sayyid Muḥammad Mishkāt; vol. II edited by Sayyid Ḥasan Mishkan Tabasi, Tehran: Majlis, 1317-24.

————, *Sharḥ Ḥikmat al-Ishrāq*, ed. Asad Allāh Harati, Tehran, 1313-15.

Secondary Sources

Chittick, W.C., "Mysticism vs. Philosophy in Earlier Islamic History: the al-Ṭusī, al-Qunawī Correspondence," *Religious Studies*, 17: 87-104 (1981).

Iqbal, 'Abbās, " 'Allamah-yi Quṭb al-Dīn Shīrāzī," *Majallah-yi Armaghan*, 16: 659-668 (1311 A.H).

Kennedy, E.S., "Late Medieval Planetary Theory," *Isis*, 57: 365-78 (1966).

Leclerc, L., *Histoire de la médicine arabe*, Paris: Libraire des Societes Asiatiques, 1876.

Minovi, M., "Mullā Quṭb al-Shīrāzī," *Yād-Nāmah-ye īrānī-ye Minorsky*, Theran, 1969, 195-205.

Mir, Muhammad-Taqi, " 'Allamah-yi Quṭb al-Dīn Shīrāzī," *Khirad wa Kurshish*, 2: 451-65 (1349 A.H.)

Mishkāt, Sayyid Muhammad, Introduction to *Durrat al-tāj* by Quṭb al-Dīn al-Shīrāzī.

Nasr, S.H., "Quṭb al-Dīn al-Shīrāzī," *Dictionary of Scientific Biography*, II: 247-53.

Qurbani, Abu'l-Qasim, "Quṭb al-Dīn Shīrāzī," *Rah-Numa-yi Kitāb*, II: 429-33.

Saliba, G., "The Original Source of Quṭb al-Dīn al-Shīrāzī's Planetary Model," *Journal for the History of Arabic Science* 3: 3-18 (1979).

Walbridge III, J.T., *"The Philosophy of Quṭb al-Dīn Shīrāzī: A Study in the Integration of Iṣlamic Philosophy*, PhD dissertation, Harvard University, 1983.

Wiedmann, E., "Quṭb al-Dīn al-Shīrāzī," *Encyclopedia of Islam*, 2nd edn., 2: 1166-67.

____________, "Ueber eine Schrift ueber die Bewegung des Rollens und die Beziehung Zwishen dem Geraden und den Gekruemmten, von Quṭb al-Dīn Maḥmūd b. Mas'ūd al-Schirazi," *Sitzungs-berichte der physikalisch-medizinischen Sozietat in Erlangen*, 58-59: 219-24 (1926-27).

____________, "Zu den optischen Kenntnissen von Quṭb al-Dīn al-*Schirazi, "Archv fur die Geschicte der Naturwissenschaften und der Technik*, 3: 1876-93 (1912).

INDEXS

Names

A
al-'Abbādī, 'Abd al-Ḥamīd, 33n36
'Abbāsid Caliphate, 11, 156, 162, 173
'Abbāssid Empire, 155
'Abd al-Karīm al-'Uthmān, 172n6
'Abd al-Mu'min, 239
'Abd al-Rāziq, M., 6n4
Abdur Rahman Khan, M., 247n52
Abraham (prophet), 163, 169
Abrahamic tradition, 29
Abū 'Abdallāh al-Nātilī, 14
Abū Bakr ibn al-'Arabī, 177n75
Abū Bishr Mattā ibn Yūnus. *See* Mattā ibn Yūnān
Abū Firās, 20
Abū Ḥātim al-Rāzī, 199n35, 200n61
Abū Isḥāq Ibrāhīm ibn 'Abdullāh al-Baghdādī, 26
Abū Ja'far Muḥammad ibn al-Qāsim al-Karkhī, 19
Abu'l Fatḥ Naṣr ibn Ibrāhīm al-Maqdisī al-Nābulusī. *See* al-Nābulusī,
Abū'l-Faraj, 20
Abu'l-Fidā', 244-5
Abū Naṣr al-Ismā'īlī, 157
Abū Sa'id al-Sīrāfī, 34n41
Abū Ya'qūb al-Sijistānī, 199n35, 200n61
Adam (prophet), 170
Afnān, S.M., 245n12
Afshar, I., 35n55
Aḥmad al-Ghazzālī, 157, 162, 177n67
Aḥmad Takūdar, 234-6
Ahsan, A.S., 246n29
'Ain Jālūt (Goliath's Spring), 246n29
Aleppo, 20, 31n1, 33n25, 36n70, 156
Alexander of Aphrodisias, 118n41
Alexandria, 15-6, 18, 116n6
Alexandrian school, 22, 130
'Alī, the fourth Caliph, 132n6
'Alī al-Wāḥidī al-Naisābūrī, 224n37
'Alī ibn 'Īsā, 19
Almagest (of Ptolemy), 14, 25
Alp-Arslān, 156, 173n10
al-Āmidī, Abu'l-Ḥasan, 11, 32n16, 36n70
Amīn, 'Uthmān, 6n4, 36n57, 38n84, 40n111
Amīr Muẓaffar al-Dīn, 243
Analytica Posteriora, 16-7
Anatolia, 26, 233-5
Antioch, 15
Arabic, 13, 16-7, 21, 23, 31, 35, 36n59, 82, 92, 96, 131, 132n6, 133n13, 134, 160, 172n6, 177, 179n121, 192, 220, 236-7, 246, 260, 262
Arabs, 81

Arghun, 235-6
Aristotle, 10, 18-9, 24-5, 27-8, 30, 31n3, 34n40, 38, 39n97, 40n109, 48, 57, 60, 65n22, 66, 82, 92, 106, 108, 116n6, 117, 119n53, 129, 135, 139-40, 142-5, 149, 186-7, 200, 255-6
Aristotelian, 22, 27-8, 35n55, 48, 50, 69, 96, 100, 102, 126, 129, 133n10, 204
Aristotelianism, 187
  Muslim, 161
Arnold, T., 41n119
Ascalon, 37n77
al-Ash'ari, Abu'l-Ḥasan, 184, 198-9
Ash'arism, 156-7, 173n13, 198
Ash'arites, 156-7, 173, 182, 198, 201n77
Asia Minor, 156
Asīl al-Dīn, 232
Asin Palacios, M., 178
al-'Attas, S.M.N., 92n43, 201n76
Atēs, Ahmet, 23, 37n80, 38n87
Athens, 10
Atiyeh, G.N., 133n10
Averroes, 150, 54. *See also* Ibn Rushd
Avicenna, 40n112. *See also* Ibn Sīnā

B
Bābakīyah, 191
Badawī, 'Abd al-Raḥmān, 39n95, 177n75, 190
Baghdad, 2, 10-1, 15-9, 21, 32n16, 33n27, 34n41, 35n56, 37n73, 155-6, 159-61, 163-4, 173-4, 176-7, 229, 233, 245
al-Baghdādī, 190
al-Baghdādī, Muḥammad ibn al-Sukrān, 233
al-Baghdādī (Abū Isḥāq Ibrāhīm ibn 'Abdullāh). *See* Abū Isḥāq Ibrāhīm
Bahā' al-Dīn Muḥammad al-Juwaynī, 233
Bahā' al-Dīn Pahlawān, 234
al-Baqarī, 163
al-Bāqillānī, 174n22, 184, 190, 198, 199n23
Barhebraeus, 230
Barthold, W., 32n8
al-Basṭāmī, 162
Bātinism, 190, 192
Bāṭinites (Bāṭinis), 163, 190, 192-3,
al-Bayhaqī, 31n4, 37n77, 196n6
Beer, G., 176n44
Being, Necessary, 261n28
al-Bīrūnī, 32n7, 237
Bodge, B., 32n8
Bouyges, M., 67n44, 169, 174n21, 175-6, 177n82, 178
Boyer, C.B., 247n54
Boyle, J., 175n31, 246n28
Brockelmann, C., 34n36
Brown, E.G., 33, 37n74, 173n7, 245-6
Bukhārā, 13-4
al-Bukhārī, 224n37
Būyid (Buwaihid), 155-6
Byzantines, 156, 231

C
Canon of Medicine (Ibn Sīnā's), 229-30, 232, 235, 242, 245
Categories (of Aristotle), 22, 35n55, 121, 135, 140
Caucassus, 156
Chaghatai, M.A. 246n36
Chaldeans, 81
Chejne, A.G., 148n12, 225n56
Chittick, W., 246
Christians, 82
Clagett, N., 135n37
Constantinople, 18
  Imperial Professor at, 36n61
Copernicus, 231, 238
Corbin, H., 162, 176n61, 179n121, 201
Cunbur, Mujgan, 37n80, 38, 39n94

D
Damascus, 11, 19-21, 32, 36n70, 37, 163-4, 177
Daniel of Morlay, 25
De Boer, 15, 31n6, 33n26, 35n47, 117n21
de Slane, W. MacGukin, 30n1, 31n6
Dieterici, F., 40, 65n18, 67n37
Dionysius Thrax, 36n59
Divine Essence, 64n2
Divine Intellect, 79, 91n26, 147n2
Divine Justice, 64n3
Divine Knowledge, 200
Divine Names and Qualities, 64n2,

149n42, 198
Divine Presence, 194, 196
Divine Truth, 195
Divine Unity, 91n35, 210
Dome of the Rock, 163
Dominicus Gundissalinus, 176n44
Dunlop, D.M., 10, 19, 32n11, 36-7, 41, 66n24, 68, 118n30, 134n26
Durr-al-tāj (of Quṭb al-Dīn al-Shīrāzī), 236, 238-42, 245-50, 257, 260

E
Eche, Y., 173
Egypt, 20, 37n73, 81, 156, 174, 235, 246
d'Elanger, Baron R., 39n101
Elements of Euclid, 14, 25, 147n7
Enoch (prophet), 215
Esmail, A., 175n31
Euclid, 25, 39n100, 147n7
European, 5, 41n124, 238, 244

F
Fakhr al-Dīn Rāzī, 201n65, 230-2, 244-5
Fakhr al-Mulk, 164
Fakhry, M., 15, 35n45, 40n108, 199n20
al-Fakurī, H., 174n22
Fārāb, 10-1, 13, 32n8, 33n27
Farmer, H.G., 25, 39
al-Fārābī, 1-5, 9-35, 38-151, 158, 160-1, 185-8, 200, 204-5, 207, 219-22, 224n43, 225n69, 236, 239, 252-7, 260-1, 263-6, 268-9
Farabian, 96, 269
Faris, N.A., 65n14, 178n101, 225n72
al-Fārisī, 'Abd al-Ghāfir, 165, 172n6, 174n19, 178n89
Fārīyāb, 32n8
al-Fārmadhī, 155, 158, 172n5, 175
al-Faruqi, I.R., 225n44
al-Faruqi, L.L., 225n44
Fāṭimid Caliph, 201
Fāṭimid caliphate, 157, 174, 192
Fāṭimid dynasty, 156
Fāṭimid Imāms, 201
Fāṭimids, 156, 162, 173n13, 174
Field, C., 177n84, 225n44
Firdawsi, 155, 178n90
First Being, 97-8, 102-3
First Cause, 56, 71-2, 80, 82, 89n5, 96-8, 107, 114-5
First Heaven, 98
First Principle, 97, 107
First Truth, 82
Frank, R.M., 34n39, 174n25
Furlani, G., 175n32
Fuṣūl al-madanī, 19, 28-9
Fuṣūṣ al-ḥikam, 27, 40

G
Gabriel (archangel), 56
Gabrieli, F., 41n121
Gairdner, W.H.T., 169, 178, 199n31
Galston, M.S., 68, 93, 117n11, 132n4, 135n39
Gardet, L., 133n10
Gardner, W.R.W., 159, 172n6, 175n36
Gatje. H., 36n59
Gauthier, L., 178n100
Genghis Khan, 230
Georr, K., 27, 37n80, 40n112
Gerard of Cremona, 25, 35n55
Ghāzān Khān, 235-6
Ghaznavids, 155
al-Ghazzālī, 1-2, 4-5, 9, 12, 22, 29, 44, 65, 107, 117n20, 145, 147, 155-225, 230-1, 236, 243, 248n67, 252, 256, 263-9
Gibb, Sir Hamilton, 64n3
Gillispie, C.G., 31n3
Gilson, E., 38n93
Goldstein, B.R., 117n20
Goldziher, I., 169, 173n13
Good, the Supreme, 109, 115
Grant, E., 135n37
Greek, 13, 17-8, 24-5, 28, 33n26, 35n56, 36, 57, 81-2, 92n48, 133n13, 134n26, 186, 231, 246
Gundisalvo, Domingo 134n37, 135n37
Gyeke, K., 36n61

H
Hachem, H., 178n101
Ḥājjī, Khalīfah., 175n39, 242, 248n68
al-Ḥākim, (Fāṭimid Caliph), 201n71

Hamadān, 164, 176n59, 177n83
Hamdānids, 20, 37n77
Harrān, 15-6, 18
Hārūn al-Rashīd (Caliph), 17
al-Ḥasan al-Baṣrī, 201n70
al-Ḥasan ibn Ṣuwār, 35n55
Hasan, A., 223n16
Hassan, M.R., 172n4, 173n10
Hassan, S.S., 246n36
Heath, Sir Thomas L., 39n100
Hebrew, 21, 24, 29, 35n55, 38-9, 41, 176
Hebron, 163
Hegel, 19, 36n67
Hermetic-Pythagorean, 186
Heraclius, 36n61
Ḥikmat al-ishrāq, (of Suhrawardī), 240
Hodgson, M.G.S., 175n31
Homā'ī, J., 172n1
Homes, H.A., 177n84
Hourani, G.F., 175n37, 176, 177n77, 193n33
Huart, C., 33n36
al-Hujwīrī, 155, 172n2
Hülagü, 230, 234-5
Humām al-Dīn Tabrīzī, 248n69
Ḥusain, Ṭāhā, 33n36
Hye, M.A., 198n13

I

Ibn 'Abbās, 45, 193
Ibn Abī Usaibi'ah, 11, 13, 15-6, 19-20, 24, 31-3, 34n40, 35-8, 40
Ibn 'Aqīl, 33, 34n36, 173
Ibn 'Aqnin, 25
Ibn 'Arabī, 90n11, 167, 233, 241
Ibn 'Asākir al-Dimashqī, 172n6
Ibn al-Athīr, 177
Ibn Buṭlān, 67n31
Ibn al-Furāt, 19
Ibn Ḥājib, 243
Ibn Ḥawqal, 32n8
Ibn al-Haytham, 237, 239
Ibn Ḥazm, 148n12, 215, 225n56
Ibn Ja'far, Qudāmah, 33n36
Ibn al-Jawzī, Abu'l-Faraj, 172n6
Ibn Khālawayh, 20
Ibn Khaldūn, 9, 11-2, 32n19, 33n23, 174n23, 183-4, 205, 223n10
Ibn Khallikān, 11, 13, 16, 18, 20, 30n1, 31-3, 35n52, 36-7, 38n91, 172n6, 175n39, 178n87
Ibn Killis, 174n15
Ibn Muqlah, 19
Ibn al-Nadīm, 32n9
Ibn al-Nafīs, 235, 246
Ibn al-Quff, 235, 246n30
Ibn al-Rāwandī, 34n38
Ibn Rushd, 29, 35n55, 40n110, 117n20, 126, 144, 150n55, 161, 176n47, 178
Ibn al-Sarrāj, 16, 35n50
Ibn Shākir, 231
Ibn al-Shāṭir, 238, 247
Ibn Shuhbah, 229, 245
Ibn Sīnā, 9, 14, 25, 27, 29, 40n113, 41n118, 52, 67, 75, 95, 116n5, 117n20, 126, 134n30, 145, 147, 160-1, 182, 185-8, 199n33, 204, 221, 225n69, 229-32, 235, 239, 241-2, 244-5, 250, 257, 260n9
Ibn Ṭibbon, 23, 41
Ibn Ṭufayl, 166, 178n100
Ibn Wahb al-Katib, 33n36
Ibrāhīm ibn 'Adī, 31n1
Ibrāhīm al-Marwazī, 15-6, 18
Iconium (city), 246n27
Idrīs (prophet), 215
Iḥṣā' al-'ulūm, 6n4, 21-2, 27, 30, 36n57, 43, 55, 61-2, 81, 121, 125, 132, 134, 140, 142, 145, 148-9
Iḥyā' 'ulūm al-dīn, 164-6, 171, 176n47, 177-9, 190, 193, 203, 210-1, 213, 224, 243, 267
Ikhshīdid (dynasty), 19-20
Ikhwān al-Ṣafā', 120n88, 186, 189
Ilkhānid Zīj, 232-3, 246
Ilkhānids, 234-5
Inati, Shams C., 134n30
Isaac Albalag, 174n44
Isagoge, 22, 38n85
Isbijāb, 10
Isfahan, 233, 246
Iskandar, A.Z., 246n30
Islam, 2-3, 9-13, 15, 22, 28-30, 32-3, 40, 47, 57, 64, 69, 76, 80, 82, 124, 126, 132-3, 155, 167, 171, 190, 201, 219, 223, 230, 233-4, 236

Ismāʻīlīs, 4, 158, 162, 190, 192-4, 199, 201
Ismāʻīlism, 158, 173n13, 189-92, 199n37, 200n61, 201
Ismāʻīlites, 156
Israeli, Isaac, 67n32
Izutsu, T., 40n114

J
Jabre, Father, 163, 169, 177n75, 179
Jackson, A.V.W., 172n3
Jalāl al-Dīn Rūmī. *See* Rūmī
Jāmī, 172n5, 234, 246n25
al-Jarra, 174n22
Jerusalem, 163, 177n79
Judah Nathan, 176n44
al-Junayd, 162
Jurjān, 157
al-Juwaynī (theologian), 156-8, 173n12, 174
al-Juzjānī, 9

K
Kamāl al-Dīn Abī'l-Khayr al-Kāzirūnī, 229
Kamāl al-Dīn al-Fārisī, 239, 244, 247n53
Kamāl al-Dīn Rāfiʻī, 234
Kamali, S.A., 175n43, 176
Kāzrūn, 229
Kennedy, E.S., 238, 245n17, 247
al-Khaṭṭābī, 17, 31n3
Khurāsān, 15, 21, 32n8, 155, 164, 173n10, 233
Khurramiyah or *Kurramidīnīyah* (Bāṭinī movement), 191
al-Kindī, 1, 10, 21, 23, 30, 32n7, 126, 133n10
Konya, 233-4, 246
Kraus, P., 201n65
Kubesov, A., 39n100
al-Kundurī, ʻAmid al-Mulk, 156, 173
al-Kurdī, 175n40
Kuyel, M. Turker, 31n6, 34n36

L
Lane, E.W., 117n9
Langhade, 31n6, 93n64
Laoust, Henry, 162, 177
Latin West, 9
Leclerc, L., 244
Lees, W.N., 172n5
Leibniz, 81
Lerner, R., 41n118, 118n43, 150, 151n72, 225n69
Lovejoy, Arthur O., 95
Lucchetta, Francesca, 39n94

M
MacDonald, D.B., 169, 172, 173n11, 175n27, 224n44
Madkour, I, 17, 31n5, 33n22, 34n38, 35n57, 37n77, 66n23, 117n6, 135n41
Mahdi, M., 6n4, 13, 15, 17, 26, 31, 32n10, 33, 34n41, 35-6, 37n73, 39n97, 40-1, 68n49, 89, 90n12, 118n43, 133n12, 150, 151n72, 225n69
Maḥfūẓ, H.A., 31, 32n12, 33n26, 34n40, 35-8, 39n107, 40n109
Maḥmūd Bey, 243
Mahoney, E.P., 116n2
Maimonides, 23, 29, 40n110, 41, 150n61, 174n23, 184
Makdisi, G., 14, 32n18, 34, 173
Malik-Shāh, 156, 162
Mamlūk Sulṭān of Egypt, 234, 246
al-Ma'mūn (Caliph), 17
al-Manṣūr Sayf al-Dīn Qalā'ūn, 234
Maqbul, S., 31n3
al-Maqdisī, 33n36
Marāghah, 230-3, 238-40, 245-8
Marāyā, 35n55
Margoliouth, D.S., 34n41, 135n46
Marmorstein, E., 6n5
Marmorstein, J., 6n5
Marmura, M.E., 174n25, 199n33
Masergasan, 17
Mashhad, 155
Mashkar, Muhammad Javad, 37n78
Massignon, L., 39n93
al-Masʻūdī, Abu'l-Ḥasan, 31n3
Mattā ibn Yūnān, 16-9, 34n41, 35n55, 38n88, 131
Mawlawī of Anatolia, 26
Mazardāran, 157
McCarthy, R.J., 163, 172, 173n11, 174n19, 175, 176n60, 177-9, 197n1, 198-202, 223n13, 225
Mecca, 156-7, 163, 181

Medina, 156-7
Mediterranean, 156
Mercury (planet), 231, 238
Merv (Marw), 15-8
Meyerhof, M., 197n6, 246
Minovi, M., 245, 247
Mir Dāmād, 30
Mishkāt, S.M., 245, 248
Mishkāt al-anwār, 165-71, 179, 185, 225-6
Mizān al-'amal, 165-6, 199n33, 203-4, 217
Mogul (empire), 244
Molaṭya, 234
Mongols, 2, 234, 246
Morewedge, P., 34n39, 116n2
Morris, J.W., 132n5
Moses ibn Ṭibbon, 39n100, 41n117
Mosul, 20
Muḥammad Ḥassan Hitu, 174n21
Muḥammad ibn 'Alī al-Himādhi, 247n48
Muḥammad ibn Mubārak Shams, al-Dīn Mīrak al-Bukhārī, 247n49
Muḥammad Sa'd al-Dīn, 235
al-Muḥāsibī, 162
Muḥsin-i Fayḍ, 179n121
Mu'īn al-Dīn Parwānah, 234
Mujīr al-Dīn, 177n79
Muller, 178n100
Munk, S., 41n117
al-Munqidh min al-ḍalāl, 9, 158-9, 165-9, 171-2, 175, 177n77, 179, 181, 188, 190, 192-3, 200, 205
The *Muqaddimah* (of Ibn Khaldūn), 32n19
al-Muqtadir (Caliph), 16, 18
Murdoch, J.E., 6n4
Murtaḍā, S., 173n6, 175n39
al-Mustaẓhir (Caliph), 158, 162, 175n28
al-Mustaẓhirī 161, 175, 177, 190-3, 201,
al-Mu'taḍid (Caliph) 16,
al-Mutanabbi, 20
al-Mutaqqī (Caliph), 19
Mu'tazilite, 198, 201n77
Muwaffaq al-Dīn Ya'qūb al-Sāmarrī, 235

N
al-Nābulusī, Abu'l-Fatḥ Naṣr ibn Ibrāhīm al-Maqdisī, 163-4
Nadir, A.N., 41n123, 66n26
Naficy, Sa'id, 33n31
Naishapur, 157-9, 163-5, 197n6
Najīb al-Dīn 'Alī al-Shīrāzī,, 245n4
Najjar, F., 38n92, 41, 65n21, 89n6, 117, 119n65, 120
Najm al-Dīn Dabīran al-Kātibī al Qazwīnī, 233, 241
Nanji, A., 175n31
al-Nashshār, 34n36
Naṣīr al-Dīn al-Ṭūsī, 2, 230-4, 237-41, 244-8
Nāṣiri-i Khusraw, 199n35, 200n61
Nasr, Seyyed Hossein, 2, 5-6, 20, 27, 30, 31n6, 32-3, 36n70, 37, 38n89, 39n106, 40-1, 64n3, 89n1, 90n18, 91, 92n44, 116n5, 120n88, 126, 132n6, 147n2, 151n70, 160, 173n15, 174, 175n31, 176-7, 179, 197, 199-202, 223, 232, 244-8, 261
Naṣr I ibn Aḥmad, 13
Nazareth, 246n29
Neoplatonic, 48, 51, 96, 118n27, 166
Neoplatonists, 170
neo-Pythagoreanism, 190
Nestorian, 17-8
Neugebauer, O., 245n17
New Persian, 13, 35n56
Nicholson, R.A., 172n2
Nicomachus, 186
Niẓām al-Mulk, 155-7, 162, 164, 172n4, 173-4
Niẓāmiyah, 157, 164, 173-4, 233, 245
Nuḥ ibn Asad, 10

O
O'Leary, De Lacy, 35n53
Olympiodorus, (school of), 36n61
Ottoman (empire), 244
Organon (Aristotle's), 17, 22, 35n56, 57, 129-30, 135n40, 256,
Otrar, 32n8

P
Palencia, Gonzales, 17, 35n57, 148n10
Peripatetic, 2, 10, 23, 118n27, 182, 186,

231-2, 250, 256, 265
Persia, 17, 27, 155, 158, 233-4, 240, 242, 244
Persian, 5, 10, 13, 164, 229, 236, 238, 241-3, 245-6, 260
Peters, F.E., 15, 35n46
Philoponus, John, 39n97, 66n29
Philoppowski, 41n117
Physics (Aristotle's), 19, 148n22
Pines, S., 27, 40
Plato, 27-30, 41, 82, 92n48, 106, 113, 117, 119n53, 187, 200
Plessner M., 39n99
Plotinian, 95
Plotinus, 28, 116n6, 118n41
Poetics (Aristotle's), 22, 122
Poggi, V.M., 175n32
Porphyry, 22
Prior Analytics (Aristotle's) 22, 34n38, 122, 135
Ptolemaic (model), 238, 247
Ptolemy, 25, 177n20
Pythagoras, 186-7, 189, 200
Pythgorean, 25, 182
Pythagoreanism, 182, 190, 200n61

Q
Qānūn fi'l-ṭibb (of Ibn Sīnā), 299
Qarāmiṭah (Carmathians), 192
Qarāmites, 192
Qazvīn, 233
al-Qazwīnī. *See* Najm al-Din Dabiran Kātibī
al-Qā'im (Caliph), 173n8
al-Qifṭī, 16, 19, 31n4, 35n51, 36n64, 38n93
al-Qirqisānī, 33n36
Quadrivium, 25, 137
Quasem, M.A., 65n6, 176n49, 201n78, 202
Qumsha'ī, Ilāhī, 27
al-Qunawi, Ṣadr al-Dīn, 233-4. 241
Qur'ān, 11-2, 28, 44-5, 64-5, 75, 107, 119n57, 133n10, 157, 169, 178, 183, 191-4, 198n15, 199, 201-2, 208, 212, 214, 223, 236, 242, 250, 259-60, 262-3
Quṭb al-Dīn al-Rāzī, 244
Quṭb al-Dīn al-Shīrāzī, 1-5, 225n69, 229-64, 268-70
Quwairī, 15-6, 18

R
al-Rādhkānī, Aḥmad, 157
al-Rāḍī (Caliph), 19
Rāfiḍites, 173n11
Rahman, A., 31n3
Rahman, F., 27, 40n112, 67n34, 68n63, 80, 90, 91n28, 92n42, 219, 225
Rashīd al-Dīn Faḍlallāh, 235-6, 246
al-Rāzī (Muhammad Zakariyā'), 26,
Reiske, J.J., 245
Republic (Plato's), 29
Rescher, N., 19, 23, 27, 31n6, 34n38, 35, 36n68, 37n80, 38n83, 39n107, 41n124, 43, 64n1, 93n58
Rhetoric (Aristotle's), 22, 36n67, 122
al-Risālat al-laduniyah, 165-8, 186-7, 203-5, 209, 217, 223, 267-8
Rosenfeld, B.A., 39n100
Rosenthal, E.I. J., 41n120, 120n83, 150n54
Rosenthal, F., 5, 6n4, 32n19, 36n67, 132
Rūmī, 21, 233-4, 241

S
al-Ṣabbaḥ, 158, 201
Ṣadrā, Mullā, 126, 179n121
Sa'd al-Dīn al-Farghānī, 241
Safavid (s), 30, 244
Ṣalāḥ al-Dīn al-Ayyūbī, 248n61
Salman, D.H., 41
Sāmānid, 10-1, 13
Sambursky, S., 118n41
Sarton, G., 244n1
Sassanids, 13
Sayf al-Dawlah, 20-1, 33n25, 37n74
Sayili, A., 38n82, 41n122, 245n10
Ṣā'id ibn Aḥmad al-Andalusī, 23, 31n4, 38n 91, 41n117, 133n8
Schimmel, A., 245-6
Scholastics, 41n124
Schoun, F., 28, 40n116, 64n2, 65n11, 66n24, 77, 81, 91, 92n51, 119n64, 198n20, 200, 226
al-Shāfi'i (Shāfi'ites), 147, 156 173n11, 206, 243, 245

Seljūq, the first Great, 156
Seljūq Empire, 156-7, 173n13
Seljūqs, 155-6. 164, 234
al-Shādhilī, Abu'l-Ḥasan, 20
Shādhilīyah Order, 20
Shāh Waliallāh, 1
al-Shahrastānī, 174n22, 184
Shahrazūrī, 240
Shams al-Dīn Juwaynī, 234
Shams al-Dīn Muḥammad ibn
Aḥmad al-Kabshī, 229, 233, 245n7
Sharaf al-Dīn al-Mas'ūdī, 201n65
Sharaf al-Dīn Zakī al-Bushkānī, 229
Sharif, M.M., 31n5, 33n31, 172, 174, 179n121
Sharī'ah, 22, 29, 32n20, 33, 44-5, 47, 65n20, 115, 133n10, 146, 151n70, 162, 171, 191-2, 201. 207, 215, 249, 264, 269
Sheikh, M.S., 172n6
Shemtob ben Falaquera, 29
Shenb, 246n35
Sherif, M.A., 6n3, 166, 176n55, 178, 199n33, 222n3, 223n14
al-Shiblī, 162
Shihāb al-Dīn Abū Ḥafṣ 'Umar al-Suhrawardī, 229, 245n2
Shī'ism, 2, 156, 179n121, 191-2, 194, 201
Shī'ite (s), 1, 155-7, 173n11, 193-4, 201
Shīrāz, 229-32, 246
Siddiqi, B. Husain, 245n15
al-Sijistānī, Abū Sulaimān, 32n7
Sirāj al-Dīn-al-Sakkākī, 243
Sīwās, 234
Smith, Huston, 93n51
Smith, M., 172n6, 174n18, 178, 202n90
Sobhy, G., 246n30
Socrates, 119n53, 200
Sogdian, 10, 35n56
Stagirite, 25, 66n28
Steinschneider, M., 31n5, 38n80, 39
Stephanus of Alexandria, 36n61
Stern, S.M., 31n3
Le Strange, G., 32n8
Strauss, L., 27, 38n92, 41, 150n61
al-Subkī, 157, 172n6, 174, 175n39, 229, 245
Sufis, 1-2, 4, 25-6, 80, 155, 157-60, 162-5, 170-1, 175-6, 179n116, 181-2, 187 , 193-7, 200-1, 202n87, 210. 219, 223n22, 229, 233, 245, 267-8
Sufism, 155, 158, 162-3, 167, 169, 171, 177, 190-2, 195-6, 201n77, 229, 241
al-Suhrawardī, 161, 232, 239-41, 245, 248, 250
Suhrawardīyah Order, 245n2
Sulayman Dunya, 222n1
Sulṭān of Naishapur, 155
Sulṭān-Qal'a, 17
Sunni (s), 1, 155-6, 173n13, 189, 192-4
Sunnism, 2, 191
Suter, H., 39n100
al-Suyutī, 245
Sylla, E.D., 6n4
Syriac, 13, 17, 22, 33n26, 35n56, 36n59, 38, 82

T

Ṭābarān, 178n90
Tabasi, S.H., 248n64
Ṭabāṭabā'ī, 201n73
Tabrīz, 234-6, 246n35
Tahāfut al-falāsifah, 160-2, 165-6, 175-6, 185, 188, 198n9, 200
Ṭanṭāwī, 25
Taqī Khān, 173n10
Taqizadeh, S.H., 132n6
Tāqūt al-Hamawī, 172n6
Ṭarkhān, 10
Tauhīdīpūr, M., 246n25
Ta'līmism, 161
Ta'līmites, 4, 158-9, 161-2, 175, 176n59, 177n83, 181, 189-91, 193
Theodoric of Freiberg, 239
Tibāwī, A.L., 173n15, 174n15, 177
Topics (Aristotle's), 22, 34n38, 60, 122
Transoxiana, 10-1, 32n8, 201
Ṭughrul-Beg, 155-6, 173
Turkestan, 10
Turkish, 10-1, 13, 23, 31n6, 35n56, 155, 177n84, 246
Ṭūs, 155, 157, 164-5, 172n3, 176n59

U

'Umar Khayyām, 182, 197
Ummayyad, 163
Umaruddin, M., 6n3, 173n7

V
Vajda, G., 33n36, 34n38
Van den Burgh, S., 176n47
van Ess, Josef, 34
von Grunebaum, G.E., 34

W
Wāfī, A.A., 31n5, 33
Walbridge, J.T, 246n39, 247-8, 260-
al-Walīd, 'Alī ibn Muḥammad ibn, 162, 190-1, 201
Wali-ur-Rahman, M., 41n123, 66n25, 132n6
Walzer, R., 11, 15, 27, 31, 32n13, 34n43, 35n55, 38n83, 66n29, 67n39, 116, 118n27
Wasīj, 10
Watt, M., 36n65, 159, 165-71, 172n6, 173, 174n22, 175n38, 177-9
Wiedemann, E., 30, 41n122, 237, 244-7
Wilber, D.N., 246n35
Wolfson, H.A., 67, 174, 198n17
Wright, O., 39n102

Y
Yaḥyā ibn 'Adī, 31n1
Yasin, J.A., 38n80, 39n107
Yemen, 162
Yūhannā ibn Ḥaylān, 15-8, 35

Z
al-Zabīdī, 172n6, 179n121
al-Zamakhsharī, 242
al-Zarkashī, 34n36
Zimmermann, F.W., 15, 17, 35-6, 133n13, 134-5
Zwemer, S.M., 172

Subjects

A
absolute good, 106-7
abstraction, 51-3, 57
degrees of, 51
accident (s), 71, 185, 251, 261
ādāb al-jadal, 14
ādāb al-kalām, 14
'ādāh, 211, 223n26
agriculture, 55, 110, 119n49, 126, 133n7, 141, 254
ahl al-bāṭin, 201
ahl al-jāhilīyah, 116
ahl al-manṭiq wa'l-burhān, 158
ahl al-ma'rifah, 250
ahl al-taqlīd, 183, 198n13
ahl al-ra'y wa'l-naẓar, 182
'aks, 93n58
'ālam al-shahādah, 221
alchemical art. See art
'ālam al-khayāl, 51

'ālam al-malakūt, 221
'ālam al-mulk, 221
'ālam al-shāhādah, 221
alchemical art. *See* art
alchemy 30, 43, 126, 132n6, 141, 209, 220, 254
al-alfāẓ al-dāllah, 127
alfāẓ mufradah, 121
alfāẓ murakkabah, 121
algebra, 253-4
analogy, 183, 198n11
angel of revelation, 73, 75
angelic world. *See* world
angelology, 142, 149n44
angel (s), 45, 65, 73-4, 96, 98, 115, 119n47, 196
animals,
  non-rational, 99, 100, 105-6, 115, 149n29
  rational, 99, 149n29
animate principles. *See* principles
animate substance. *See* substance
anteriority, 96-7
anthropology, 145
al-'aqīdah, 198
'aql, 48, 50, 77, 90n18, 179, 188
al-'aql al-'amalī. *See* intellect, practical
al-'aql bi'l-fi'l. *See* intellect, actual
'aql bi'l-quwwah. *See* intellect, potential
al-'aql al-fa''āl. *See* Intellect, Active *or* intellect, active
al-'aql al-hayūlānī. *See* intellect, material
al-'aql al-juz'ī, 78
al-'aql al-kullī, 78, 168-9, 179
al-'aql al-munfa'il. *See* intellect, actual
al-'aql al-mustafād. *See* intellect, acquired
al-'aql al-naẓarī. *See* intellect, theoretical
'aqlīyah, 203, 222n3, 264
arbāb al-aḥwāl, 162, 195
archetypes, 147n2
  Platonic, 78
architecture, 126
arithmetic, 12, 14, 89, 103, 112, 122, 137, 139, 147n4, 148n16, 205, 209, 213-4, 253-4, 266
art (s), 46-7, 49, 51, 54, 66, 80, 82, 108, 113-4, 126, 133n7, 139, 144, 235-6, 245
  alchemical, 132n6
  demonstrative, 129-30
  dialectical, 133n7
  ethical, 150n63
  excellent, 144
  the master, 144
  medical, 150n58
  non-syllogistic, 126, 133n7
  philosophic, 122, 129
  poetical, 133n7, 135n40
  practical, 110, 112-3, 116, 123, 126, 138-39, 141, 150n55, 255
  rhetorical, 133n7, 135n40
  sophistical, 133n7, 135n42
  syllogistic, 122, 125, 130, 133n7
art of arts, 82
art of dialectic, 122
art of poetry, 122, 130. *See also* poetry
art of rhetoric, 122, 130
al-asbāb al-thawānī. *See* Second Causes (second causes)
al-aṣḥāb al-aqwāl, 162
assent, 58-61, 68n56, 79, 84-86, 88, 196, 251
astrology, 26, 138, 209, 214-16, 225
  judicial (*aḥkām al-nujūm*), 26, 122, 137-38, 148, 214, 254
astronomical system, 117n20
astronomical tables, 238
astronomy, 40, 46-7, 62, 101-3, 122, 137, 209, 214, 223n12, 230-1, 236-9, 244, 253-4, 267
  mathematical, 137-9, 214, 224n43
  theoretical and practical, 138
atomism, 184
awā'il al-yaqīn, 58
axioms of certainty, 58-61
āyat al-kursī. *See* Verse of the Throne
'ayn al-qalb. *See* eye of the heart
a'ẓam. *See* magnitude (s)

B

badanīyah, 104
bāṭin, 191-3, 201, 211
al-bāṭinīyah, 181, 190-1, 201
being (s)
  absolute incorporeal, 97-98, 123, 142

celestial, 150n53
divine, 150n53
eternal, 100
immaterial, 96-7
incorporeal, 98
intermediate, 101
metaphysical, 103, 105, 111, 114
natural, 104
second, 98-9
spiritual, 104, 115, 125
terrestrial, 99, 104, 115
theoretical, 110
voluntary, 103-4
being *qua* being, 142
bid'ah, 183
bi'l-ṭab' (by nature), 96
bodies
artificial, 139
celestial, 96-7, 99, 101-2, 105, 118, 137-8
composite, 140-1, 254, 261n21
compound, 123, 255
generated, 140
heavenly, 117-8, 122-3, 138
heterogeneous, 140-1, 149
homogenous, 140-1, 149
natural, 99-100, 103, 123, 139-41, 148n23, 149
sensible, 97
simple, 123, 140, 148n23, 254-5
terrestrial, 96-7, 99, 102, 105
botany, 141, 254
al-burhān, 22, 85, 185, 260n9. *See also* demostration; demonstrative method
al-burhān al-yaqīnī (certain method), 81

C
calligraphy, 259
causality, 198n20
cause (s),
accidental, 96
actual, 96
Aristotelian, 61, 72, 80, 103, 117n13, 139
celestial, 118n27
efficient, 68n60, 96-7, 103, 148n19
essential, 96
final, 68n60, 96-7, 103, 148n19
first, 62
formal, 68n60, 96-97, 103, 148n19
immediate, 98, 215
material, 68n60, 96-7, 103, 148n19
potential, 96
proximate, 96
remote, 96
second, 96, 98
secondary, 72
ultimate, 62
cause of all causes, 215
celestial matter. *See* matter
certainty, 58-9, 160, 189, 196
approximate, 58-60, 86, 88, 125, 265
degrees of, 58, 84, 87
necessary, 59-60, 85
certainty at times, 59, 85, 89, 93n60
chemistry, 141
civilization, Islamic, 30, 126
classes of knowers, 159-60, 181-82, 196
classification of assents, 59, 86
classification of beings, 96-8
classification of internal senses, 67
classification of knowers, 4, 182
classification of the sciences, 1, 4-5, 9, 32n7, 43, 48, 55, 66n28, 95, 107, 112, 121, 133n10, 147, 159, 161, 165, 189, 203, 207, 219, 236, 239, 240-1, 249, 263, 267-8
classification of virtues, 107
climatic zones, 123, 138
common sense (faculty of), 50, 52-3, 66n29, 67n32, 90n22
community, Islamic, 192
composition, 101
metric, 128
conception, 58-60, 84, 251
perfect, 68n56
contemplation, 195
contemplative knowledge. *See* knowledge
cosmogony, 142, 149n44,
cosmography, 237
cosmological scheme, Plotinian, 95
cosmology, 142, 149n44, 186
cosmos, 50, 116, 141, 222
created order, 109

D

al-dahriyūn, 200
dalīl, 84
ḍarurīyāt, 159
definition (s)
  essential elements of, 97
  perfect, 58, 60, 84
demonstration, 57, 79, 81-2, 85-6, 130, 134n37, 135, 139, 142, 185-7, 189, 260n9
demonstrative method. *See* method
demonstrative premises. *See* premises
demonstrative proof. *See* proof
demonstrative reasoning. *See* reasoning
demonstrative syllogisms. *See* syllogism (s)
devices
  arithemtical, 123
  ingenious, 137, 139
  mechnical, 123
  optical, 123, 138
al-dhākirah (memory), 54
dhawq, 170, 185, 196-7, 200
dhikr (s), 175, 195
dialectical arguments, 85
dialectical method. *See* method
dialectical proof. *See* proof
dialectical theology. *See* theology
dīn, 81, 92
al-dīniy (religious), 237, 149
disciplines, Islamic, 126
discourse
  rational, 77
  syllogistic 77-8, 91n32
discursive thought, 78
divine acts, 80, 111
divine attributes, 44, 111, 117n17, 169-71, 197, 209, 258, 261n29, 262
divine essence, 44, 111, 197, 209
divine intellect. *See* intellect
divine justice, 116
divine law, 29, 45, 71, 109, 256-7, 261n26
  Greek, 29
  divine legislation, 65n20
  divine name, 44
  divine nature, 117n17
  divine principle, 43-4
  divine revelation. *See* revelation
divine science. *See* science (s)
divine unity, 170, 207, 213, 217, 223n22, 224n39
divine works, 44, 197, 221
dreams
  false, 132n6
  interpretation of, 30, 43, 126, 132n6
  patho-genetic, 132n6
  true, 132n6
al-dunyawīyah, 217
dynasty,
  Fāṭimid, 156
  Seljūq, 156

E

economics, 114, 145, 237, 249-50, 256-7
ecthesis, 93n58
elements
  the four, 99-100, 117n27, 118, 140, 148n23, 255, 261n21
  mixtures of the, 149n29
emanation or effusion, 95, 98
empiricists, 196
engineering, 103, 112, 123, 147n1
  mechanical, 147n1, 253
engineering sciences. *See* sciences
entelechy, 100
epistemology, 4, 24, 43-4, 126, 182, 210, 263
eschatology, 207, 252
esoteric, 30, 166-9, 171, 190-2, 194, 202, 206, 210, 219. *See also al-bāṭin*
esoteric function, the Prophet's, 191
esotericism, 190-2
ether, 117n27, 118
ethical basis, 95, 106, 115, 207, 263, 266, 268
ethical philosophy. *See* philosophy
ethical sciences. *See* sciences
ethico-legal basis, 47, 216, 268
ethics, 4, 27-8, 38, 47, 114, 124, 145, 161, 171, 189, 222, 237, 242, 249-250, 256-7
  medical, 242, 248n66
  Sufi, 171
etymology, 259
excellence, 96, 119n66, 211, 266, 268
excellence of deliberation, 111
excellence of the practical intellect, 111

excellence of the theoretical intellect, 110
exegesis, 12
existence, 58-9, 63, 82, 89n5, 96-100, 118, 142, 251, 255, 261
  actual, 99
  potential, 99, 102
  predicate, 58, 60
  simple, 58
exoteric, 166-9, 171, 191-2, 210
exoterism, 167
experimentation, 205
expressions
  composite, 121-2, 127-8
  significant, 127
  simple, 121, 127-8
eye of the heart, 195

F
faculties
  animal, 53
  cognitive, 49-50, 67n35, 74, 77
  internal, 54
faculties of knowing, 158
faculties of the soul, 48, 66n28, 103, 105-6, 142
faculty
  appetitive, 49, 74, 90n24, 103, 109, 112, 118n28
  deliberative, 71, 108, 110, 112
  mental, 78
  imaginative, 49-51, 54, 57, 60, 62, 66n29, 67n35, 68n68, 73-4, 77, 90, 102-3, 118n28
  rational, 49-50, 53-57, 60, 66, 75, 77, 92n45, 100, 103-5, 108-9, 188
  sensitive, 49-51, 54, 62, 68n68, 74, 102-4, 118n28
  vegetative, 49, 103
faculty of choice, 108-9
faculty of common sense. *See* common sense
faculty of compositive animal imagination, 52, 54
faculty of compositive human imagination, 51
faculty of estimation, 51–2. *See also* *wahm*
faculty of growth, 118n28, 141
faculty of memory, 51, 53-4, 61
faculty of motion, 118n28
faculty of nutrition, 141
faculty of representation, 51-52, 67n34
faculty of reproduction, 141. *See also* reproductive faculty
faculty of skill, 108, 111
faḍā'il, 104
al-faḍā'il al-'amaliyah, 108
al-faḍā'il al-fikriyah, 108
al-faḍā'il al-khuluqiyah, 109
al-faḍā'il al-naẓariyah, 108
al-faḍā'il al-nuṭuqiyah, 108
al-faḍā'il al-ṣinā'iyah, 108
faḍl, 96
falāsifah, 4, 71, 80, 174n22, 181, 186
falsafah, 2, 157-8, 174n25, 184-6, 188-9, 206. *See also* philosophy
falsafah-i ūlā, 252
al-falsafat al-'amaliyah, 143. *See also* philosophy, practical
al-falsafat al-madaniyah, 145, 150n63
al-falsafat al-naẓariyah, 143. *See also* philosophy, theoretical
al-falsafat al-siyāsiyah, 145, 150n63. *See also* philosophy, political
falsification, 88
fanā', 195
farḍ 'ayn, 206-7, 210-3, 216
farḍ kifāyah, 184, 199n21, 206-7, 210-11, 213-4, 216, 220
fardāniyah, 170
fayḍ, 95, 98. *See also* emanation or effusion
faylasuf, 12
fikriyah, 104, 108
fiqh, 12-3, 83, 125-6, 133n10, 146-7, 155-7, 183, 198n11, 222, 257, 264-6. *See also* jurisprudence
first philosophy. *See* philosophy
form (s) (*ṣūrah*), 50-2, 55-7, 61-3, 66n28, 67n34, 96-7, 99-102, 117n25, 118, 127-8, 137, 196
  imaginative, 57
  intelligible, 56, 62-3
  non-sensible, 52
  sensed, 52-3
  sensible, 67n34
  universal, 55
forms of reasoning, 130

G
geodesy, 237
generation of intellects. *See* intellect (s)
generation of the heavens. *See* heavens
geography, 236-7
geology, 141
geomancy, 132n6
geometrical justice, 116
geometrical method. *See* method
geometrical properties on light. *See* light
geometry, 46-7, 62, 89, 103, 112, 122, 137-8, 147-8, 209, 214, 236, 238, 248, 254, 266
  practical, 122, 148n10
  theoretical, 122, 138-9, 147n7, 148n16
gharībah, 126
ghayr al-dīnīy, 237, 249
ghayr al-ḥikmī, 236, 264
gnosis, 28, 71, 75, 115, 160-1, 219, 232, 234, 241, 267. *See also ma'rifah*
(the) good, 109, 111, 115, 143
  absolute, 111
  intellectual, 109, 116
  material, 116
  moral, 109, 115-6
  natural, 115
  spiritual, 116
government (s), 145
  corrupt, 145
  virtuous, 145, 150n59
grammar, 12-4, 16, 34n41, 125, 128, 131, 133
  universal, 131
grammarians, 14, 124-5, 128. *See also al-naḥwiyūn*

H
al-ḥadd al-awsaṭ, 84. *See also* middle term
al-ḥadd al-tamm, 58. *See also* definition (s), perfect
ḥadīth (s), 22, 44-5, 157, 193, 201n79, 202, 207, 224n37, 236, 250, 259-60, 262n41, 263
ḥakīm (sage), 27
happiness, 55, 65, 70-2, 89n1, 93n52, 107, 111, 113, 115-6, 119n65, 123, 143, 150n55, 258
  presumed, 114
  realization of, 144
  supreme, 71-2, 79, 106-7, 143
  true, 71, 103, 111, 114-5, 143-4
ḥaqīqat al-nubuwwah, 196
ḥaqq al-yaqīn, 195
ḥaram, 65n20
al-ḥawāss al-bāṭinah (internal senses), 51. *See also* senses, internal
hay'ah, 101, 118n33
heaven of archetypes, 120n88
heaven of the moon, 98
heavenly bodies. *See* bodies
heavens, 45, 80, 101, 116, 237
  mathematical forms of the, 101
hierarchy of beings, 46, 72, 80, 95, 97, 99, 101-2, 116, 149n29, 150n50
hierarchy of believers and knowers, 45, 263
hierarchy of creation, 45, 263
hierarchy of faculties of the human soul, 50, 64, 69, 263
hierarchy of intelligences and souls, 95
hierarchy of knowledge, 46
heirarchy of modes of knowing, 182
hierarchy of the philosophical sciences, 89
hierarchy of the premises of a syllogism, 85, 89
hierarchy of the principles of bodies, 99
hierarchy of reality, 44, 263
hierarchy of the sciences, 1, 4, 25, 43-7, 83, 95, 97, 115, 124, 126, 139, 263-4, 266, 268
hierarchy of second beings, 98
hierarchy of separate intelligences, 56, 67n42
hierarchy of syllogistic proofs, 85, 130, 263
hierarchy of virtues, 263
hierarchy of witnesses of divine unity, 45, 263
ḥikmah (*ḥikmat*), 28, 71, 81, 110-1, 186, 241, 250-1, 225-6, 267, 269-70
ḥikmat dhoqī, 269
ḥikmat-i 'amalīy, 237, 251
ḥikmat-i naẓarīy, 237, 251
al-ḥikmī, 236
hīmiyā', 132n6

ḥissīyāt, 159
holy spirit, 56
ḥuḍūri, 203
ḥukamā', 75
ḥulūl, 202
human anatomy, 150n58
human body, 105, 141, 149n38, 150n58
human community, 145
human intellect. *See* intellect
human language, 128, 220
human mind, 73, 76, 84, 100, 139, 147n2, 251, 266
human society. *See* society
human soul. *See* soul
human virtues. *See* virtues
human will, 99, 104, 144
human wisdom. *See* wisdom
humanities, 145
ḥuṣūlī, 203

I
'ibādah, 211, 223n26
al-'ibārah, 22
al-iftirāḍ, 93n58
ijmā', 183, 208, 223n10
ijtihād, 194
ikhtilāṭ (mixtures of the elements) 149n29
ikhtiyār, 103, 109
al-ilahīyūn, 200
ilhām, 167, 218
illuminationist theosophy, 232, 269
'ilm, 5, 43, 83-4, 170, 185, 236
'ilm al-'adad, 122
'ilm aḥkām al-nujūm, 122
'ilm aḥwāl al-qalb, 217
'ilm al-akhlāq, 208
'ilm al-athqāl, 123
'ilm al-fiqh, 43, 123, 145. *See also* jurisprudence
'ilm al-firāsah, 132n6
'ilm al-ḥadīth, 12
'ilm al-handasah, 122
'ilm al-hay'ah, 101
al-'ilm al-ḥikmī, 204
'ilm al-ḥiyal, 123, 147n1
'ilm ḥuḍūrī, 195, 268. *See also* presential knowledge
'ilm ḥuṣūlī, 195. *See also* knowledge, attained or acquired
al-'ilm al-ilāhī, 83, 97, 123, 261. *See also* science, divine
al-'ilm al-ladunī, 202n89, 204, 218
'ilm al-lisān, 121
'ilm mā ba'd al-ṭabī'ah, 252. *See also* metaphysics
al-'ilm al-madanī, 28, 83, 100, 103, 123
'ilm al-manāẓir, 101, 122. *See also* optics
'ilm al-manṭiq 121. *See also* logic
'ilm al-mu'amalah, 210
'ilm al-mukāshafah, 178n101, 195, 204, 210
'ilm al-mūsīqā, 123. *See also* music
'ilm al-mūsīqā al-'amalīyah, 123. *See also* music, practical
'ilm al-mūsīqā al-nazarīyah, 123. *See also* music, theoretical
'ilm niranjāt, 253
'ilm al-nujūm, 40, 122, 137. *See also* astronomy
'ilm nujūm al-ta'līmī, 122. *See also* astronomy, mathematical
al-'ilm al-ṭabī'ī, 99, 123, 252
'ilm ta'bīr, 253
'ilm ṭalismāt, 253
'ilm ṭarīq al-ākhirah, 210
'ilm al-tawḥīd 207
'ilm-i abniyah, 262n42
'ilm-i awzān wa mawāzīn, 261
'ilm-i bayān, 259-60
'ilm-i i'rāb, 259-60
'ilm-i ma'ānī, 259-60
'ilm-i misāḥah, 261
'ilm-i muḥāḍarat, 259
'ilm-i naqli-i miyāh, 261
'ilm-i qawāfī, 87, 259
'ilm-i qirā'at, 262
'ilm-i qiṣaṣ, 262
'ilm-i taṣrīf, 262n42
'ilm-i ta'wīl, 262n36
imagination
  animal, 53, 67n36
  compositive, 53
  deliberative, 61
  rational, 53, 61
  sensitive, 53
Imām, 150n59, 191-2, 194, 201
Imām, Infallible, 158, 190, 193

al-Imām al-ma'ṣūm, 192
Imāmah, 162, 192-3, 252
induction, 60, 84, 88
  ordinary, 60, 88
  scientific, 60-1, 88
Intellect, Universal, 169
intellect (s), 48-51, 57-9, 61-3, 65n22, 66, 68-9, 71, 75, 77-9, 91n35, 98, 102, 105, 107, 118n36, 129, 149n38, 179, 196, 206, 251, 262, 269
  acquired, 62-3, 68n68, 70, 75-6
  active, 51, 56-7, 63-4, 66n30, 67n42, 69-74, 76, 90n22, 96, 98-101, 117n21, 147n2, 149n38
  actual, 56, 62-3, 89n4, 102, 118n36
  cosmic, 72
  heavenly, 99
  material, 67n40
  passive, 70
  potential, 51, 56-7, 62-3
  practical, 108, 110-2, 204
  prophetic, 48, 64, 69
  separate, 98, 101-3, 105
  theoretical, 71, 105, 108, 110-1, 204
  transcendent, 149n38
intellection, 62, 75-8, 81, 86, 91n26, 265-5
  cosmic, 72, 91n26
  degrees of, 62
intellectual good. *See* the good
intellectual history, Islamic, 2, 161, 219, 232
intellectual intuition. *See* intuition
intellectual perfection. *See* perfection
intellectual sciences. *See* sciences
intellectual tradition. *See* tradition
intellectual vision, 113, 266
intelligence, 49, 66, 202n87, 209, 229
intelligences, separate, 98, 105, 114-5
intelligibles, 49, 55-8, 66, 73-4, 79-80, 98, 100-1, 128-9, 131
  actual, 51, 55
  composite, 122
  first, 77
  mathematical, 101
  metaphysical, 101
  natural, 71, 101, 144, 150n53
  potential, 51, 55, 62-3
  primary, 57-62, 67n46, 85, 110
  pure, 62-3, 68n68
  secondary, 78
  simple, 121
  theoretical, 110
  voluntary, 71-2, 119n51, 144
intermediate world. *See* world
interpretation of dreams, 30, 43, 126, 132n6
intuition, 60, 160, 166, 178
  intellectual, 75-8, 81, 159, 169, 175n33, 187, 265
  mystical, 197
al-iqnā', 81, 91n39
irādah, 49, 103-4
'irfān, 28, 161, 232, 234, 241. *See also* gnosis
ishrāqī (illuminationist), 2, 161, 232, 240-1, 245, 247n56, 248, 250
ishrāqī school of philosophy, 2, 248, 269
ishrāqī theory of vision, 240
iṣmah, 193
istinbāṭ (logical conclusion), 81
ittiḥād (union), 202

J

al-jadal (Aristotle's *Topics*), 22
jafr, 132n6
jalāl, 261n29
jism, 50. *See also* bodies
jūdat al-fahm (execellence of understanding), 111
jūdat al-ra'y (execellence of idea), 111
jurisprudence, 12, 43, 86, 123-4, 145, 157, 198, 201n77, 207, 211-2, 220, 223n10, 224, 259, 262n38
  Islamic, 183
jurists (s), 146, 155, 162, 171, 173, 181, 185, 207, 215, 220, 268

K

kalām, 2, 14, 33, 43, 81, 83, 86, 123, 125-6, 133n10, 146-7, 150n67, 156-62, 174, 182-5, 188-9, 196, 198-9, 206, 211-2, 214, 222, 224, 232, 264-7
  Ash'arite, 157, 184
  methodology of, 183-5
  Sunni, 198n9
kamāl, 96, 262
kashf, 160, 175n34, 185, 197

khafīyah (occult), 126. *See also* hidden or occult sciences
khalṭ (process of combination), 53
khayāl (image), 73
al-khiṭābah (Aristotle's *Rhetoric*), 22
khuluqīyah (ethical), 108-9
khuṭābiyah (rhetorical), 84
knowledge,
attained, 195, 203-4, 218, 268
blameworthy, 207, 215-6
discursive, 170, 204
immedidate, 60
inferential, 233n7
intellectual, 205-6, 213, 218-20, 268. *See also* sciences, intellectual
intuitive, 185
metaphysical, 82, 110. *See also* metaphysical sciences
non-philosophical, 236
non-religious, 236
permissible, 207
philosophical, 74, 89, 236, 251. *See also* philosophical sciences
poles of, 44
practical, 204, 217. *See also* sciences, practical
praiseworthy, 207, 215, 223
presential, 195-6, 202n89, 203-4, 218-9, 268-9
prophetic, 168, 193, 218
rational, 91n35, 168
realized, 163, 194-5
religious, 205-6, 213, 217-20, 236, 243, 268. *See also* religious sciences
sapiential, 194-6
scientific, 146, 185
sensual, 195, 204
spiritual, 176n69, 187, 195
theoretical, 110-1, 144, 163, 204, 217
transmitted, 205. *See also al-'ulūm al-naqlīyah*
knowledge of God, 115, 125-6, 150n50, 181, 197, 210, 213, 218, 258, 270

L
language
Arabic, 18, 38, 40, 131, 134n26, 223n11, 260
Greek, 82
laṭīfah, 202n87
Law
Islamic, 214
Maliki, 243
law-giver, 29
learning
Greek, 186
Sunni, 157
legal philosophy. *See* philosophy
legal theory and methodology, 157, 165
legal tradition, Sunni. *See* tradition
lexicography, 128, 254
lexicology, 128
light, 102, 169, 185, 196, 221, 248
geometrical properties of, 137
incorporeal nature of, 102
Muhammadan, 193-4
light rays, 138, 147n9
līmiyā', 132n6
linguistic sciences. *See* sciences
literary criticism, 259
literary history, 128
literature, 12-3, 270
logic, 4, 10, 13-9, 22-3, 33n36, 34, 38, 48, 121, 124-6, 128-32, 133n10, 134n37, 135, 139-40, 142, 157, 175n37, 186-8, 194, 199n33, 200, 209, 213, 220, 237, 241, 248-50, 255-6, 263-5
Aristotelian, 14-5, 17-8, 22-3, 34n41, 38, 161, 184
jurisprudential-theological, 14
Stoic, 14
logical sciences. *See* sciences
logical system, Aristotelian, 135n40
logical theory, 84
logical tradition, 22, 125
logicians, 14
lunar model, 238

M
ma'anā, 131
mabādi' (principles), 96
mabadi' al-barāhin, 142
macrocosm, 50
māddah, 96, 99. *See also* matter
māddah ūlā, 97
madrasah, 27, 157, 164-5, 173-4

magic, 132n6, 253
natural, 254
magnitudes, 100-1, 103, 114, 116, 118n30, 138-9, 148n10
māhiyah, 27, 260. *See also* quiddity
maḥsūsāt, 67n46, 74, 85
makhraqah (trickery), 88
makrūh, 65n20
malakāt, 104
mandūb, 65n20
manṭiq, 14, 34n41, 124, 131. *See also* logic
al-manṭiqiyūn, 14, 34n41, 128. *See also* logicians
maqbūlāt, 67n46, 85
al-ma'qūlāt al-irādiyah. *See* intelligibles, voluntary
al-ma'qūlāt al-ṭabi'iyah. *See* intelligibles, natural
al-ma'qūlāt 55. *See also* intelligibles
al-ma'qūlāt ūlā. *See* intelligibles, first
al-maqūlāt (Aristotle's *Categories*), 22
marātib al-mawjūdāt, 95
ma'rifah, 71, 75, 219, 267. *See also* gnosis
martabah 96
maṣdars 134n21
mashhūrāt, 67n46, 85
mashshā'i, 2, 10, 80, 232. *See also* Peripatetic
mathematical astronomy. *See* astronomy
mathematical entities, 101, 147n2
mathematical forms, 101, 137
mathematical objects, 90n9
mathematical properties, 101, 118n41, 147n8
mathematical sciences. *See* sciences
mathematical shape, 118n33
mathematics, 2, 61-2, 101-3, 112, 114, 116, 120n88, 132n6, 137, 139, 142-4, 208, 217-8, 222, 230-1, 237-8, 240-1, 248-50, 253-5, 261, 264, 266
applied, 112, 116, 137-8
pure, 112, 114, 116
matter, 50-3, 55-7, 61-3, 66n28, 76, 85, 96, 98-9, 101, 105, 117, 118n33, 251-2
celestial, 99, 101, 117n27, 118
prime, 97, 99, 102, 117n27, 118
sublunar, 102
mawādd, 85
mawjūd, 96
al-mawjūd al-awwal, 97
al-mawjudāt, 96, 142, 263
mawjūdāt irādiyah, 103
al-mawjūdāt al-naẓariyah, 110
al-mawjūdāt al-thawānī, 98
mechanics, 62, 101, 103, 112, 237
medical ethics. *See* ethics
medicine, 21, 25, 36n70, 55, 110, 126, 133n7, 141, 148n20, 150, 205, 209, 213-4, 217-8, 220, 223n12, 229-30, 232, 236, 241, 244, 246, 254, 267
Islamic, 242
medieval philosophy. *See* philosophy
mental states, 129
metallurgy, 141
metaphysical sciences. *See* sciences
metaphysical speculation, 114
metaphysical truths. *See* truths
metaphysics, 14, 27-8, 33n20, 38, 46, 61-2, 82-3, 89, 93n54, 97, 100, 102, 104, 106, 111-2, 114-5, 120n88, 123, 125, 139, 142-4, 149n40, 150n50, 171, 189, 209, 234, 237, 240-1, 244, 248-50, 252-3, 261, 263-4, 266-7, 270
metaphysics of light, 232, 239
meteorology, 24, 209, 237, 254-5
method
demonstrative, 187, 189, 265-7
dialectical, 265
esoteric, 191
geometrical, 266
philosophic, 187, 267
method of analysis, 147n7
method of conversion, 93n58
method of induction. *See* induction
method of synthesis, 147n7
methodological basis, 47, 181-2, 263, 266
methodological claim of the mutakallimūn, 158
methodological claim of the philosophers, 158
methodology of knowledge, 43
methods of proof, 84, 89
metre (s), 128, 134n26
microcosm, 50, 75, 91n26
middle term, 57, 60, 84, 93n57, 97

mihnīyah, 55. *See also* faculty of skill
millah, 76, 80-1, 146
mineralogy, 132n6, 141, 209, 254
minerals, 123, 149, 252
mithāl, 73
mithāl awwal, 134n21
moral character, 143
moral good. *See* the good
moral life, 113, 115
moral order, 109
morphology, 128, 134n21
motion, 123
al-mu'allim al-thānī, 3, 30. *See also* The Second Teacher
mubāḥ, 65n20
al-mufakkirah, 53, 61
al-mughāliṭah (Aristotle's *Sophistics*), 22
muḥākāh (similitudes), 73-4
muhlikāt (destructive qualities of the soul), 211
mujaddids, 171
mujtahid, 194
mukashafah, 194, 196
mukhāṭabah, 77
mundus imaginalis, 51
munjiyāt (salvation), 211
muqaddamāh, 77
muqaddimāt, 205, 208, 223n10
muqārib li'l - yaqīn (approximate certainty), 58
mushāhadah, 194-6
music, 16, 19, 25, 38, 62, 123, 137, 139, 147n4, 209, 214, 224n44, 238-9, 253
  practical, 123
  theoretical, 123
musical theory, 25
mutakallimūn, 4, 34n38, 65n22, 146, 158, 174n23, 175n34, 179n116, 181-5, 189, 198-9, 206, 267. *See also* theologians
mutakhayyalah (images), 89
mutakhayyilah (imagination), 53-4, 67
mutammimāt, 223n10
mystic vision, 194
mystical experience, 185
mysticism, 241

N

al-nafs al-insānīyah. *See* human soul
al-nafs al-muṭma'innah, 202n87
al-naḥwīyūn, 14
nairanjiyāt, 209
natural magic. *See* magic
natural objects, 101
natural order of learning. *See* order of learning
natural philosophy. *See* philosophy
natural principles, 80, 256-7
natural sciences. *See* sciences
natural substances. *See* substances
natural world. *See* world
nature (s) of things, 44, 129, 181, 185, 251, 255, 269
navigation, 55, 110, 126
neote hyle, 118n27
nobility, 96
nufūs, 251
nubuwwah, 252. *See also* prophecy
numbers, 100-1, 103, 114, 116, 137, 139, 147, 189, 222, 252, 261
  pure, 100, 103
  practical science of, 122
  theoretical science of, 122
numerical symbolism. *See* symbolism
nūr al-baṣīrah, 221
nuṭq (speech), 49, 66n24, 131
al-nuṭq al-dākhil (interior speech), 131
al-nuṭq *al-khārij* (exterior speech), 131
nutritive faculty, 118n28
nuṭuqīyah, 108

O

occasionalism, 198n20
oneiromancy, 253-4
ontological, 101, 201n77
ontological basis, 47, 66n28, 95, 97, 263, 266
ontological principle, 98-9, 117n25
ontology, 4, 27, 47, 142, 209, 241
opinions
  generally accepted, 67n46, 85-7
  received, 67n46, 85, 87
optics, 62, 101-2, 118n41, 122, 137-9, 147, 148n10, 236-7, 239, 247n53, 253
  theoretical foundation of, 138
order of learning, 130, 266
orthodoxy, 76, 167, 183, 185, 188, 197, 267

exoteric, 166
Islamic, 12, 166
Sunni, 192
orthography, 128

P
people of the ignorant states, 116
perfection, 63, 96, 99, 106, 109, 111-5, 119n53, 132n6, 251, 256, 262
intellectual, 56, 114-5
man's first, 107, 119-n53
man's final, 107, 119n53
spiritual, 114
theoretical, 113-4
ultimate, 109, 144
perfection of the human soul, 109, 115
perfection of the theoretical intellect, 106, 111, 114
philology, 133n12
Arabic, 133n12
philosopher (s)
Greek, 119n53, 187, 200
Islamic, 28, 50
Ismā'ilī, 186
Muslim, 126, 132n6, 151n70, 160, 186, 205, 220
perfect, 77, 114
Peripatetic, 141, 160
theistic, 200
philosopher-scientist, 1-2, 10, 31n6, 32n7, 80, 130, 220, 225, 265
philosophic art. *See* art
philosophical knowledge. *See* knowledge
philosophical school of Ibn Sīnā, 182
philosophical sciences. *See* sciences
philosophical system, 128
philosophical tradition. *See* tradition
philosophization of Ash'arite *kalām*, 157, 174n22
philosophy, 4, 10, 13, 15, 17-8, 22, 32n7, 33, 34n40, 35n56, 38, 40, 64, 69, 76, 79-83, 92, 109, 117n6, 119n66, 126, 130, 132n6, 133, 135n43, 146, 151, 157-61, 175n37, 187, 189, 204, 219, 230-2, 236, 241, 244-5, 250, 256, 263-5, 267, 269
Aristotelian, 15
Christian, 91n26
esoteric, 81
ethical, 108
external, 79, 81
Greco-Roman religious, 80
Greek, 187
Ishrāqī, 240, 248
Islamic, 2, 31n6, 40n110, 91n26, 245
Ismā'ilī, 186
Jewish, 91n26
legal, 147
mathematical, 254
medieval, 4
Occidental, 95
Peripatetic, 182, 231-2, 241, 245
practical or human, 133n10, 143-4, 205, 237, 242, 251, 256-7
rationalistic, 160
theistic Greek, 186
theoretical, 69, 133n10, 143-4, 237, 242, 251-2, 255
philosophy of law, 145
philosophy of mathematics, al-Ghazzālī's, 189
philosophy of mathematics, Islamic, 147n2
philosophy of science, 5, 78, 83
phonetics, 128
physical sciences. *See* sciences
physics, 23-4, 38, 82, 120n88, 123, 137, 209
physiognomy, 132n6, 254
planetary model, 238
planetary motion, 231, 237, 247
poetic verses, 128
poetical, 85, 87-8
poetry, 259, 262n43
rules of, 121
poets, 130
poles of knowledge. *See* knowledge
political education, 145
political movements, 155
political philosophy. *See* philosophy
political science, 29, 100, 103, 106, 112-6, 119n49, 123-5, 133n10, 143-7, 150, 205, 222, 257, 264-7
practical, 145, 150n58, 200n52
political scientist, 113
political theory, 124
Sunni, 162

politicians, 130
prophetic vision, 74
politics, 82-3, 93n54, 114, 143-5, 150n63,·189, 22, 237, 249-50, 256-7
potentiality, 130
practical arts. *See* arts
practical sciences. *See* sciences
practical virtues. *See* virtues
premises, 57, 68n46, 77, 84-9, 91, 93, 111, 140, 183-5, 198-9
  demonstrative, 85-6, 89
  indemonstrable, 57, 85, 87, 91n38
  necessary primary, 89
  rhetorical, 87
  syllogistic, 57
  universal, 60
prime matter, 97
principles
  animate or psychical, 141, 149n38
  incorporeal (second), 79, 96
  metaphysical, 61, 79
  natural, 80, 256-7
  primary, 58-60
  universal moral, 65n20
principles of arithemetic, 62
principles of artificial bodies, 139
principles of being, 62, 67n42, 72, 79-80, 93n52, 117n13, 120n70, 140
principles of the heavenly bodies, 72
principles of household association, 257
principles of jurisprudence, 14, 147, 159, 208
principles of mathematics, 62
principles of natural bodies, 123, 139
principles of political association, 257
principles of the sciences, 61, 150n50
proof (s), 84, 88, 159, 184, 191, 193, 199, 263, 265
  demonstrative, 85, 88, 97, 110, 114, 122
  dialectial, 86-8, 122
  rhetorical, 84, 87
  scienctific, 251
  syllogistic, 89
propaedeutic sciences. *See* sciences
prophecy, 26, 69, 76, 115, 133n10, 150n61, 151n70, 187, 196, 257, 261n26
  theory of, 29
the prophet
  intellectual function of, 71
  law-giving and legislative function of, 72, 90n11
prophetic functions, 71-2, 90
prophetic intellect. *See* intellect
prophetic light, 194, 196
prophetic mind, 74-5
prophetic spirit, 73, 169-79
prophetic vision, 74
proportions, 101
proposition, universal, 60, 140
prosody, 128, 259
prototype, 134n21
psychology, 4, 24, 40, 46, 50, 69, 100, 103, 106, 141, 145, 254-5, 263
  spiritual, 187
  traditional Islamic, 54
public administration, 145

Q

qāḍī, 13-4, 234
qalb, 195
qawānīn (rules), 121
qirā'ah, 121
qiyās, 22. 77, 84, 183. *See also* analogy
qualities of the soul. *See* soul
quality, 148n27, 189
quiddity, 27, 142, 251, 261
quantities
  abstract, 100
  concrete, 100
  continuous, 100, 118n30
  discreet, 100, 118n31
quantity, 18, 148n27
al-quwwat al-fikrīyah, 71, 108
al-quwwat al-ghādhiyah, 49. *See also* faculty, vegetative
al-quwwat al-ḥāfiẓah, 51, 53
al-quwwat al-ḥāssah, 49. *See also* faculty, sensitive
al-quwwat al-muṣawwirah, 51
al-quwwat al-nāṭiqah, 108
al-quwwat al-nuzū'īyah, 109
al-quwwat al-wahm, 51
qaws qazaḥ (rainbow), 239

R

rabābah, 239
radhā'il, 104

raml (geomancy), 132n6
ratiocination, 170, 185
rational animal, 48, 99
rational-data, 158
rational faculty. *See* faculty
rational powers, 115
rational soul. *See* soul
rational system, 81, 265
rational virtues. *See* virtues
rationalism, Muslim, 161
rawiyah, 54, 111
Reason, Universal, 168
reason, 48, 64, 69, 75, 77-9, 91n35, 129, 158, 160, 166-7, 169, 175n34, 179, 183-4, 186, 188, 192, 194, 196-7, 205-6, 257-8, 263-7, 269
  descending or communicating function of, 77
  principle of, 77, 91n35
reasoning
  demonstrative, 130
  discursive, 77, 269
  sophistical, 88
religion, 4, 40, 64, 69, 79-83, 89n8, 92, 93n54, 117n6, 146, 150n66, 201, 241, 258-9, 263-5, 269-70
religious sciences. *See* sciences
religious symbolism. *See* symbolism
remembrance of God, 195
reproductive faculty, 118n28
revealed doctrines, 146, 265, 267
revelation, 44-5, 48, 64, 69-79, 81, 90n25, 91n26, 114, 146, 166-9, 178-9, 184, 188, 206, 218, 259, 263-7, 269
  inner meaning of, 194
  theoretical dimension of, 83
rhetorical, 85, 87-8, 91n39, 146
rhetorical proof, *See* proof (s)
rhetorical statements, 122
ri'āsah, 143
royal crafts, 143
royal functions, 143
rulership, 145
rules, 121-2, 127-9, 131, 140, 144
  grammatical, 129
  logical, 129
rules of correct reading, 121, 128
rules of correct writing, 121, 128

S

al-sa'ādat al-dunyā, 107
al-sa'ādat al-quswā, 106
al-sabab al-awwal, 96
al-sabab al-qarīb, 98
safsaṭah, 17, 22, 35n57
samā', 26
al-ṣanā'at al-khuluqīyah, 150n63
sapientia, 33n20
scholarship, Islamic, 1, 133n10, 166, 234
school of Ibn 'Arabī, 161, 234
school of law, Shāfi'ite, 11, 163
school of philosophy, Spanish, 41n118
schools of Greek philosophy, 186
schools of Muslim philosophy, 186
science (s),
  'aqlī, 257–8
  blameworthy, 210, 213-4, 268
  engineering, 138
  esoteric, 30, 193
  exact, 188
  excellent, 144
  exoteric, 83
  farḍ 'ayn, 203, 210, 224, 268
  farḍ kifāyah, 203, 210-2, 224, 268
  Greek, 126
  hidden or occult, 126, 132n6, 138, 148n12, 221, 255
  intellectual, 189, 197, 203, 205-8, 212-4, 216-22, 223n12, 225, 249, 264, 267-70
  Islamic, 101, 220, 225n68, 236-7, 254
  linguistic, 14, 28, 121, 128-9, 131, 133n10, 134n26, 205, 208, 220, 223, 260, 270
  logical, 25, 189, 200n52, 267
  mathematical, 82, 89, 100-1, 120n88, 122, 138-9, 142, 188, 200n52, 221, 253, 267
  metaphysical, 26, 142-3, 187, 200, 213, 220-1, 241
  modern, 137
  mother of, 82
  naqlī, 257
  natural, 2, 38, 99-104, 106, 115-6, 123, 132n6, 138-43, 148-9, 189, 200,

209, 250, 252-5, 261, 264, 266-7
non-syllogistic, 126
other-worldly, 217-8
philosophical, 2-3, 5, 13-4, 18-20, 33n21, 57, 61-2, 83, 85, 88-9, 97, 99, 110, 112-3, 115, 124-6, 129-31, 157, 188, 200, 204-5, 220, 225n69, 230, 237, 240, 243, 247, 249-51, 255, 264-9
physical, 24-6, 214, 221
practical, 110, 137, 150n55, 203, 217, 249
praiseworthy, 205, 207, 210, 213, 268
pre-Islamic, 3
propaedeutic, 22. *See* sciences, mathematical
quantitative, 188
Qur'ānic, 208, 224, 259
religious, 3, 5, 12-4, 43, 46, 83, 86, 88, 125, 146, 157, 160, 165, 189, 193, 197, 203, 205-7, 210-2, 216, 221-3, 224n39, 225, 229, 233, 236, 241-3, 248-9, 252, 257-8, 261n26, 264-70
science of, 82
syllogistic, 125
theological, 28
theoretical, 120n7, 123, 138, 142, 203, 217, 249
transmitted, 205, 223n10, 249, 260, 269-70
worldly, 217-8
science of astronomical tables and calendars, 253-4
science of the balance, 253
science of calculation, 253-4
science of contractual obligations, 208
science of devotional practice, 210-1
science of divine mysteries, 216
science of dream (interpretation), 132n6, 221
science of the heavens, 122
science of the hereafter, 207, 218, 223n10
science of history, 259, 262
science of ingenious devices, 123, 137-8, 147n1
science of interpretation, 208, 211-2, 223n10, 224, 242
science of irrigation, 253
science (s) of language, 121, 125-7, 131, 236, 243
science of literature, 259-60
science of magic and talismans, 214, 216, 254
science of man and society, 147
science of moral qualities, 208
science of optical devices, 138
science of prophecy, 209, 221
science of prophethood, 207, 252
science of the rainbow, 239
science of reality, 81
science of recitation, 259
science of religion, 204-5
science of religious authority, 252,
science of religious rites, 208
science of revelation, 178n101
science of rhymes, 262n43
science of society, 28, 82
science of the sources of religious knowledge, 208, 211
science of Traditions, 12, 165, 211-2, 233, 242
science of transactions, 208
science of transmission of ḥadīths, 208
science of unveiling, 195, 210, 219
science of weights, 123, 137-8, 253
science of wisdom, 204
scientific institution, 230
scientific knowledge. *See* knowledge
semantics, 259-60
sensation, 49, 51-2, 66n25, 74
senses
external, 50-2, 73
internal, 51-2, 54, 73-4
sense-data, 158
sense-knowledge, 67n46, 85, 89, 93n60
shar'īyah, 21-3, 264
al-shi'r, 22, 121
sīmiyā', 132n6
ṣinā'īyah, 108
ṣinā'at al-manṭiq (the art of logic), 22
siyāsah, 143-4, 257
social change, 145
society, perfect, 29, 145
sociology, 145
socio-political institutions, 80
socio-political order, 29, 145

sophia, 28
sophistical, 85, 87-8, 146
sophists, 130
soul, 49, 66, 70-2, 75, 78, 87, 89n1, 92n45, 96-7, 100, 105, 107, 113, 115, 119n51, 132n6, 141, 188, 193, 195, 201n77, 210-1, 218, 223, 248, 251, 261
  animal, 100, 141, 188
  habitual states of the, 104
  human, 49, 50, 64, 90n18, 100, 102, 104-6, 108-9, 129, 250
  non-rational animal, 118n28
  plant, 118n28, 141
  rational, 55-6, 74, 187
  states of the, 104-5, 107, 109, 257
specialization, 124, 212
species, 140-1, 251
  animal, 141
  plant, 141
speech
  exterior, 66n24
  interior, 66n24, 31
sphere of the fixed stars, 117n20
sphere of the moon, 99
spheres, 117n20
  movement of the, 116n6
spirit, 49, 202n87, 218
  imaginative, 226
  rational, 226
  sensory, 226
spiritual realities, 72, 75, 266
spritual states, 195
substance (s), 142, 148n27, 185, 187, 202n87, 261
  angelic, 209
  animate, 117n14
  celestial, 114
  first, 149n47
  second, 142, 149n47
  separate, 142
  simple, 209
substratum, 75-6, 102
al-ṣufiyah, 181. *See also* Sufis
Sunnah, 28, 194, 208, 223
supra-rational, 4, 77, 158, 170, 175, 187, 193-4, 196, 202n87, 204, 269
ṣūrah (form), 96, 99, 118n33
surveying, 253, 261
syllogism (s), 57, 60, 68n46, 77-8, 84-5, 87-8, 91, 93, 122, 125, 133n7, 135n37, 184-5
  demonstrative, 57, 85, 87, 122, 129-30
  dialectical, 86-7, 122
  figures of, 16, 84, 87
  imperfect, 93n58
  perfect, 84-5
  poetical, 122
  rhetorical, 122
  sophistical, 122
syllogistic, 174n23
syllogistic arts. *See* arts
symbolic interpretation, 192
symbolism
  alphabetical, 192
  numerical, 132n6, 192
  religious, 83
  scriptural, 202n85
symbols, 73-5, 77, 80, 83, 189, 201n77
symmetry, 101

T

ta'qqul, 75, 111, 265
al-ṭabi'ah, 260
al-ṭabi'iyūn, 200
ṭabi'iyat, 23
tadbir-i manāzil, 257
tafṣil (process of separation), 53,
tafsir, 12, 262
tahdhib-i akhlāq, 257
taḥlil (analysis), 147n7,
al-ta'lim, 158, 162, 175n34, 190-4
tanzih (transcendence), 170, 261n29
Ṭariqah, 32n20, 44
tarkib (composition), 53, 147n7
tartib al-ta'lim, 130, 266
taṣaffuḥ, 84
taṣawwur, 58, 251
taṣdiq, 58, 251. *See also* assent
tashbih (immanence), 171,
tashiḥ, 88,
Tawḥid, tawḥid, 43, 170, 224n39
ta'wil, 191-3, 201
terrestrial forces, 209
theodicy, 24, 248
theologians, 1, 130, 146, 155, 157, 182, 185, 206, 215, 230
theological sciences. *See* sciences
theology, 143, 157, 160, 173n13,

176n47, 232, 236
dialectical, 14, 33n35, 43, 123-4, 146. *See also kalām*
Mu'tazilite, 201n77
theoretical beings. *See* beings
theoretical faculty. *See* faculty
theoretical intellect. *See* intellect
theoretical knowledge. *See* knowledge
theoretical sciences. *See* sciences
theoretical virtues. *See* virtues
theories of language, 133n12
theory of atoms, 198n18
theory of causality, 189
theory of dreams, 41n123
Theory of Emanation, 116n6
theory of happiness and human virtue, 143
theory of human society, 150n58
theory of the intellect, 48, 69
theory of knowledge, 27, 38
theory of light, incorporeal, 118n41
theory of methodology, 48, 59, 78, 84, 91
theory of the primary rainbow, 239
Theory of the Ten Intelligences, 116n6
theory of virtue, 107
theurgy, 209, 221, 253-4
tradition
Islamic, 190
philosophical, 76
Prophetic, 12
religious, 75, 146
revealed, 76, 81-3, 265
spiritual, 76
Sunni legal, 162
traditional philosophy. *See* philosophy
truth claims, 84, 88
truth of certainty, 195
truths
divine, 78
eternal, 81-2
intellectual, 72, 78, 83
metaphysical, 78, 92n48, 187-8
philosophical, 80
rational, 186
revealed, 74, 80, 170, 179, 265
self-evident, 159, 184
spiritual, 72-4, 77-8, 83, 189, 195, 204
supra-rational, 77
transcendent, 78, 80, 188, 265, 267

U

al-ukhrawīyah, 217
'ulamā', 86, 194
al-'ulūm al-'aqlīyah, 205. *See also* sciences, intellectual
'ulūm dīnīyah, 225n69
'ulūm ghayr dīnīyah, 225n69
'ulūm ḥikmīy, 249, 264
al-'ulūm al-mashhūrāt, 132
'ulūm al-mukāshafat, 193
al-'ulūm al-naqlīyah, 205, 223n10. *See also* sciences, transmitted
al-'ulūm al-shar'īyah, 205, 225 n69
'ulūm al-ta'ālīm, 100, 122
unity of religious, transcendent, 92n51
universal grammar. *See* grammar
universals, 51, 55-7, 60-1, 142, 251, 255, 261
accidental, 61
essential, 61
single, 61
universe, 104, 119n66, 142, 209, 251, 263, 269
'uqalā', 86
'uqūl, 251
'uqūl al-mufāriqah, 56, 98
usṭuqussāt (elements), 99,
uṣūl al-fiqh, 14, 34n36, 147, 208, 224, 265

V

verification, 88
Verse of the Throne, 44
vices, 104, 107
virtues, 82, 104, 107-10, 113-5, 119n59, 143, 150n55, 257,
artistic, 108, 112
deliberative, 55, 108, 111-4
ethical, 108-9, 112-4
moral, 55
practical, 108, 111, 120n76
rational, 108-9, 111, 120n76
theoretical, 55, 91n29, 108, 110-3
virtuous leadership, 150n59
virtuous state 114, 119n47, 143
voluntary actions, 103, 151n72
voluntary beings, *See* beings

W
wahm, 51-4, 61, 67
waḥy, 48, 69-70, 73, 167, 218. *See also* revelation
wājib al-wujūd (Necessary Being), 261n29,
waṣī, 191
al-wāṣilūn 170,
ways of life, 143-5, 151n72
  ignorant, 145
  virtuous, 145
wilāyah (esotric function), 191
wisdom, 71, 77, 81, 91n26, 110-1, 119n66, 120n88, 132n6, 241, 250, 258, 267
  ancient, 82
  eternal, 82
  the highest, 82
  human, 186, 267
  practical, 71, 76, 110-2, 145
  theoretical, 76
  traditional, 81
  unqualified, 82
wisdom of wisdoms, 82
world
  absolutely incorporeal, 222
  angelic, 226
  celestial, 99
  corporeal, 50
  imaginal, 51, 57, 78-9
  intermediate, 45
  Islamic, 155-6, 160, 233-5, 240, 246n29, 269
  material, 51, 57, 221-2, 226
  natural, 80, 141
  physical, 78-9, 204, 222
  psychic, 50
  spiritual, 50, 195, 204, 219, 222
  sublunary, 63, 99
  subtle, 209, 221
  terrestrial, 99, 116
world of arts and crafts, 80
world of change, 120n88
world of dominance, 221
world of generation and corruption, 99
world of the unseen, 221-2
world-view,.Qur'anic, 269
wujūd, 27
wuqūf, 259
wuṣūl, 202

Z
zamān, 96
zoology, 24, 141, 254-5